TREATISE ON
SOLID STATE CHEMISTRY

Volume 6B
Surfaces II

TREATISE ON SOLID STATE CHEMISTRY

TREATISE ON SOLID STATE CHEMISTRY

Volume 6B
Surfaces II

Edited by

N. B. Hannay

Vice President
Research and Patents
Bell Laboratories
Murray Hill, New Jersey

PLENUM PRESS • NEW YORK-LONDON

Library of Congress Cataloging in Publication Data

Hannay, Norman Bruce, 1921-
    Treatise on solid state chemistry.

    Includes bibliographical references.
    CONTENTS: v. 1. The chemical structure of solids. −v. 2. Defects in solids. −v. 3.
Crystalline and noncrystalline solids. −v. 4. Reactivity of solids. −v. 5. Changes of
state. −v. 6A. Surfaces I. −v. 6B. Surfaces II.
    1. Solid state chemistry. I. Title.
QD478.H35   541'.042'1   73-13799
ISBN-13:978-1-4613-4318-9   e-ISBN-13:978-1-4613-4316-5
DOI: 10.1007/978-1-4613-4316-5

Seven-volume set: ISBN-13:978-1-4613-4318-9

Published by Plenum Press, New York
A Division of Plenum Publishing Corporation
227 West 17th Street, New York, N.Y. 10011

Foreword

The last quarter-century has been marked by the extremely rapid growth of the solid-state sciences. They include what is now the largest subfield of physics, and the materials engineering sciences have likewise flourished. And, playing an active role throughout this vast area of science and engineering have been very large numbers of chemists. Yet, even though the role of chemistry in the solid-state sciences has been a vital one and the solid-state sciences have, in turn, made enormous contributions to chemical thought, solid-state chemistry has not been recognized by the general body of chemists as a major subfield of chemistry. Solid-state chemistry is not even well defined as to content. Some, for example, would have it include only the quantum chemistry of solids and would reject thermodynamics and phase equilibria; this is nonsense. Solid-state chemistry has many facets, and one of the purposes of this *Treatise* is to help define the field.

Perhaps the most general characteristic of solid-state chemistry, and one which helps differentiate it from solid-state physics, is its focus on the chemical composition and atomic configuration of real solids and on the relationship of composition and structure to the chemical and physical properties of the solid. Real solids are usually extremely complex and exhibit almost infinite variety in their compositional and structural features.

Chemistry has never hesitated about the role of applied science, and solid-state chemistry is no exception. Hence, we have chosen to include in the field not only basic science but also the more fundamental aspects of the materials engineering sciences.

The central theme of the *Treatise* is the exposition of unifying principles in the chemistry, physical chemistry, and chemical physics of solids. Examples are provided only to illustrate these principles. It has, throughout, a chemical viewpoint; there is, perforce, substantial overlap with some areas of solid-

v

state physics and metallurgy but a uniquely chemical perspective underlies the whole. Each chapter seeks to be as definitive as possible in its particular segment of the field.

The *Treatise* is intended for advanced workers in the field. The scope of the work is such that all solid-state chemists, as well as solid-state scientists and engineers in allied disciplines, should find in it much that is new to them in areas outside their own specializations; they should also find that the treatment of their own particular areas of interest offers enlightening perspectives.

Certain standard subjects, such as crystal structures, have been omitted because they are so well covered in many readily available standard references and are a part of the background of all solid-state scientists. Certain limited redundancies are intended, partly because they occur in different volumes of the series, but mainly because some subjects need to be examined from different viewpoints and in different contexts. The first three volumes deal with the structure of solids and its relation to properties. Volumes 4 and 5 cover broad areas of chemical dynamics in bulk solids. Volume 6 treats both structure and chemical dynamics of surfaces.

N.B.H.

Preface to Volume 6B

No area of solid-state science offers greater promise of new fundamental understanding, as well as of the development of a new and firm base for the applied science, than does the study of surfaces. An array of powerful new experimental and theoretical tools has been developing in the last several years. They can add immensely to our insights over a range of surface phenomena, ranging from the desired (catalysis, epitaxy, electrochemical reactions, etc.) to the unwanted (corrosion, oxidation). In Volumes 6A and 6B are covered not only the major surface phenomena, but also the new experimental and theoretical techniques for the study of surfaces.

Contents of Volume 6B

Chapter 2
Ion Implantation and Channeling   **125**
Fred H. Eisen and James W. Mayer

Chapter 3
Semiconductor Surfaces   **203**
   S. Roy Morrison

Contents of Volume 6B

Catalysis by Solid Surfaces

T. E. Madey, J. T. Yates, Jr., and D. R. Sandstrom*
Surface Processes and Catalysis Section
National Bureau of Standards
Washington, D.C.

and

R. J. H. Voorhoeve
Bell Laboratories
Murray Hill, New Jersey

This chapter gives an overview of current concepts in catalysis of predominantly simple reactions. Catalysts covered are metals, semiconducting oxides, and sulfides. The emphasis is on the connections between solid-state chemistry/physics, spectroscopy, and surface physics in ultrahigh vacuum on the one hand and catalysis on the other hand. For physicists and materials scientists of various plumage, the chapter is expected to serve as a concise (but not oversimplified) primer.

1. Introduction

We are on the threshold of new insight into the fundamental nature of heterogeneous catalytic processes. This is partly due to recent advances in our ability to accurately and reproducibly measure phenomena at the atomic level on surfaces using modern tools.[1],†

---

* NBS Guest Worker; on leave from Washington State University, Pullman, Washington.
† See also the April 1975 issue of *Physics Today*, which contains several articles on modern surface measurement techniques.

In addition, however, the entry of solid-state and molecular orbital theorists into the catalytic field is just beginning to have a major effect in defining the fundamental features of catalytic processes in terms which may eventually enhance our ability to predict catalytic behavior and to design optimal catalysts on the basis of well-founded principles.

"Heterogeneous catalysis" comprises the phenomena of the enhancement of the rates of chemical conversions in fluids by the presence of solid surfaces, which are called "catalysts." It is evident that heterogeneous catalysis is essentially a problem in chemical kinetics and that theoretical efforts must come to grips with factors affecting the *rates* of reactions—not just with the states of adsorbed species that exist in stable forms under nonreactive conditions. It is the focus on reaction rates which is missing from most theoretical work at present, and it is here that major conceptual advances can now be expected to occur.

This chapter will emphasize certain aspects of experimental catalytic research which have supplied reliable data of use in formulating both general and specific concepts of major importance in heterogeneous catalysis. In addition, theoretical concepts are discussed where it seems clear that the particular concept is verified by sound experimental observations.

The field of catalysis is vast and in this short review it was necessary to pick certain topics while excluding others. We have endeavored to communicate the essence of a variety of topics relevant to the understanding of the field of heterogeneous catalysis, but the reader is referred to the original literature for detailed treatments, which are beyond the scope of this chapter. The reader is particularly directed to the excellent series of reviews appearing each year in *Advances in Catalysis* and in *Catalysis Reviews*. It is hoped that in this review the combination of ideas derived from classical experiments with ideas based on more recent work will help to form a fresh impression of the frontiers of this exciting field.

2. Historical Development of the Understanding of Catalysis

2.1. A Chronology

The development of modern concepts in heterogeneous catalysis began in 1913 with Sabatier, who was the first to suggest that catalysis proceeds through the formation of an intermediate compound on the catalyst surface.[2] This idea was not extensively developed at the time, largely due to the fact that the molecular organometallic compounds which were candidates as catalytic intermediates were unknown to chemists of that era. As will be discussed, the concept of the surface intermediate in catalysis plays a very important role in present-day thinking. This is partially a result of the success of inorganic chemists (since 1950) in the synthesis of analogous molecular compounds of organometallic species.

The concept of "active centers" in catalysis was introduced in 1925 by Taylor,[3] who reasoned that catalytic activity on a solid was restricted to a distribution of specific sites on the surface rather than to all available sites. Since that time, a number of attempts to identify lattice imperfections (such as dislocations and point defects) with catalytically active centers have been made. Although some illuminating facts suggesting a relationship between lattice imperfections and catalysis are available, it is still not possible to formulate a generally satisfactory theory to account for such relationships.[4,5] Many apparent experimental correlations between catalytic activity and lattice imperfections are suspect due to the possible influence of impurities and other interferences. This is particularly so in metals. In compounds such as oxides, sulfides, and halides, point defects are more easily identified as the catalytic sites.[4] One instance of the participation of point defects which appears to be well documented is in Ziegler–Natta catalysis (polymerization of olefin monomers), as will be discussed in Section 6.3.

Historically, catalytic chemists have distinguished between the geometric and the electronic factors in catalysis. This somewhat artificial separation of solid-state properties has resulted in correlations between catalytic activity and lattice spacing on the one hand, and between catalytic activity and electronic character on the other hand.[6] It is now widely recognized that geometric properties of a solid are intimately related to the electronic properties, and that attempts to separate the two can be misleading.

In the 1930's, Balandin developed his multiplet theory of catalysis, which stressed the importance of surface geometry.[7] He reasoned that the activity of a catalyst depends on the presence in the surface lattice of a correctly spaced array of atoms ("multiplet") to interact with the reactive group of atoms in the reactant molecules ("the index"). In addition, the importance of energetic factors was recognized in the later developments of the theory. A priori, the theory needs a rather detailed mechanistic model of the reaction. Since the theory is generally tested by comparing various catalysts with different lattice spacings or different crystal structure, a requirement for such tests is that the mechanism of the reaction is the same on all catalysts compared. For example, according to this theory, the dehydrogenation of cyclohexane and the hydrogenation of benzene by the sextet mechanism should proceed if a suitably spaced hexagonal array of atoms is present in the surface. Therefore, metals and alloys having hcp and fcc structures should be more active than those having bcc structures. Although there is considerable evidence to support this tenet,[7] many counterexamples cast serious doubt on the general validity of Balandin's ideas. For instance, evaporated films of bcc W and Fe are highly active for hydrogenation of benzene, whereas fcc Cu and Ag are inactive.[6] Balandin,[7] recognizing the difficulty, has allowed of the possibility that benzene hydrogenation would also occur through intermediate steps involving only two metal atoms (doublet mechanism). Indeed, Boudart[8] has pointed out that the interconversion of benzene and cyclohexane can be classified as "structure-insensitive" reactions on Pt and Ni catalysts, and such reactions can of course not be used as a test for surface geometry! Here, as elsewhere in the theory of catalysis, concepts which appear to have basic validity can only be tested and applied within a framework of sufficient knowledge regarding the mechanisms of the reactions and the energetics of the intermediate surface complexes. Modern work in catalysis is much more apprehensive about these factors and hence tends to study the examples of template effects in much more detail. Along these lines, Somorjai[9] has provided recent evidence that the surface structure may have a specific influence on certain reactions. He has reported that high-index planes of Pt containing steps and terraces play a special role in the dehydrocyclization of *n*-heptane to toluene; the activity for this reaction is higher on certain stepped surfaces than on the smooth Pt(111) plane. Conversion of neopentane has been shown to require a triplet of Pt sites.[8] In another example where geometry is important, template effects appear to be well established in the catalytic reactions of enzymes[10] but these have to remain outside the scope of this chapter.

The electronic approach to the understanding of catalysis on metals, developed in the late 1940's and 1950's, seeks a relationship between the *bulk* electronic properties of a solid and its catalytic activity. Boudart[11] and Beeck[12] recognized in 1949 that the catalytic activity of a number of transition metals for the hydrogenation of ethylene ($H_2 + C_2H_4 \rightarrow C_2H_6$) could be expressed as plots of the logarithm of the rate constant for the reaction as a function of the percentage *d* character of the catalyst. (In Pauling's theory of bonding in solids,[13] the cohesive energy of transition metals is attributed to the formation of *dsp* orbitals; percentage *d* character is a measure of the extent to which *d* electrons participate in the bonding *dsp* orbitals). As will be discussed below, Sinfelt[14] has recently attempted to correlate activities for ethane hydrogenolysis (to produce methane) with percentage *d* character. As he points out, "while there is a degree of correlation between hydrogenolysis activity and percentage *d*-character, this parameter alone is not adequate for characterizing the catalytic activity of transition metals for hydrogenolysis."

There are several basic limitations to the electronic approach to the catalysis by metals as outlined above. First, there is no simple correlation between the bonding between metal atoms in the bulk and the electronic properties of the surface. In particular, the atomic orbitals available for bonding at the surface will depend on the crystal face. The catalytic activity of a particular solid is, in many cases, dependent on the exposed crystal faces. For example, Gwathmey and

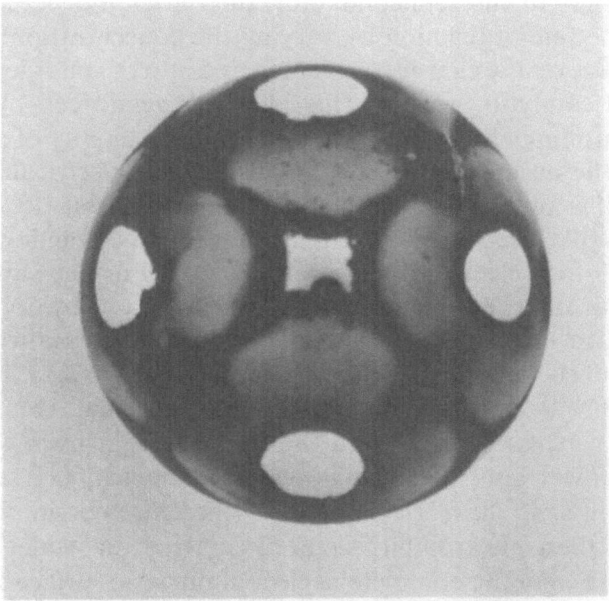

Fig. 1. Crystallographic decoration of a nickel single-crystal sphere by the deposition of carbon in the catalytic decomposition of ethylene. (From the work of Cunningham and Gwathmey,[15] courtesy K. Lawless.)

Cunningham[15] demonstrated the specificity of different Ni crystal planes for decomposition of C_2H_4 and CO (Figure 1), Van Hardeveld and Hartog[16] showed that chemisorption of N_2 on Ni, Pd, Pt, and Ir and H–D exchange into benzene over Ni occurred on sites where the adsorbate would have 5-coordination, and Boudart[8] has discussed several structure-sensitive reactions. Another basic objection to the oversimplified electronic approach given above is that the catalytic process often constitutes a major perturbation to the metal, to the point where metal salts are formed as intermediates. Such was the case in the decomposition of formic acid on a series of metals studied by Fahrenfort *et al.*[17] The electronic properties of the metal might be expected to be of limited relevance in such cases which involve "corrosive chemisorption."[18]

The successes of the theory of semiconductors led in the 1940's and 1950's to the development of an electronic theory of catalysis on semiconductors by Wolkenstein,[19,20] Hauffe,[21,22] Weisz[23] and others. It had been known for some time that chemisorption on semiconductors changed their conductivity.[24] The new theories were based on the concept that if chemisorption changed the number of charge carriers in the semiconductor, then it should be possible to influence the rate of reaction by varying the concentration of charge carriers whenever the charge transfer entered into the rate-determining step of the catalytic process. This idea was reinforced by initially successful attempts by Stone[25] to correlate the activity of a series of oxides for the decomposition of N_2O with their electrical transport properties (*n*- or *p*-type conductors and insulators). It had been established that the rate-determining step for this reaction involves the uptake of an electron by the lattice.[26,27] Further support was derived from the experiments by Parravano[28] and by Schwab and Block,[29] who showed that doping of NiO with Li or Ga would have dramatic effects on the activation energies for the CO oxidation. Enthusiasm was hardly tempered by the fact that the effects in Parravano's experiments[30] had the opposite sign from the effects found by Schwab and Block. However, discussions by Garrett[31] and Garcia-Moliner[32] have clearly shown the extreme simplifications involved in these electronic theories of catalysis on semiconductors. Among these were the complete neglect of intrinsic surface states, the frequent assumption of complete ionization of the surface states induced by adsorption, and the assumption that these states are always to be found in the forbidden gap between valence and conduction bands. As more of these simplifications are removed, more detailed knowledge of the semiconductor is necessary for a quantitative test of the theory. Even the simplest quantitative test, measuring the reaction rate as a function of doping level, requires the knowledge of a multitude of semiconductor parameters unknown for any of the catalysts concerned.[31] In the absence of such tests, qualitative confirmation of the relation between reaction rate and conductivity has been extensively pursued. As will be discussed below, Chon and Prater[33] have related measurements of the rate of CO oxidation with measurements of the carrier concentration for ZnO. Quantitative confirmations may become possible with the use of single crystals for catalytic rate measurements coupled with measurements of the electronic properties of the surface. The ambiguities introduced by the use of polycrystalline samples, such as surface segregation, spurious electric fields, intergranular resistance, and shape effects, would then be removed.

Recent developments in catalysis seem to be returning to the concept originally introduced by Sabatier: Rather than concentrating on the general electronic and geometric properties of the solid, theoretical and experimental work is concerned with the properties of atoms and complexes at surfaces. Among the catalytic chemists, Dowden[34,35] had stressed this viewpoint as early as 1956. However, since then much more comprehensive treatments of the surface complex have become possible. The concepts of coordination chemistry and the local site approach are dominating present thinking concerning catalytic mechanisms.

2.2. The Steady-State Surface of the Catalyst

At this point, it is important to focus attention on the differences between the state of the catalyst per se and the steady state of the catalyst in situ. The catalyst per se can be studied in some experimental atmosphere (e.g., air for microscopy, N_2 for surface area determination, ultrahigh vacuum for surface physics measurements), but the state of its surface is then in general not representative for the catalytically active state. For instance, in air or N_2, it is likely to be covered with oxygen, CO, hydrocarbons, sulfur compounds, and the like. Catalysts prepared in ultrahigh vacuum (UHV) are probably highly unsaturated and in the process of being covered with carbon, oxygen, or hydrogen from residual gases. In contrast, in the catalytic reactor, under the influence of the reaction partners, bonds with the catalyst are continuously made and broken, resulting in a dynamic steady state of the surface. The nature of this state is dependent on the reaction conditions. From the selectivity of the reduction of NO to NH_3 or N_2, as many as five reproducible states have been deduced for Ru supported on Al_2O_3.[36] Few steady-state surfaces have been studied by UHV surface techniques and perhaps not many are amenable to such study. In one notable exception, Joyner *et al.*[37] have found that the Pt surface active for hydrogenation was an ordered carbonaceous overlayer.

For steady-state surfaces, a further distinction is appropriate between catalytic processes in which the intermediates are short-lived and those in which fairly stable surface compounds are formed or, more generally, in which atoms cease to belong to the catalyst and transfer to the surface compound or even to the product. An example is the participation of lattice oxygen in oxidation reactions catalyzed by oxides. The former type is characterized by sufficiently weak interaction between the catalyst and the reacting species that the atoms forming the catalyst surface remain fixed in their lattice positions. At all times, an imaginary plane separates the catalyst and the reacting species. This type of catalytic process has therefore been called *suprafacial catalysis* by Voorhoeve *et al.*[38] It includes catalytic reaction involving extralattice oxygen[30] for the low-temperature oxidation of CO. Another example is the olefin disproportionation reaction,[39] which was discussed by Mango and Schachtschneider[40,41] in terms of a cyclobutane intermediate. The catalyst in the suprafacial catalytic processes primarily provides the electronic bridge for electron flow between the reaction species. It is in these cases that (cooperative) electronic properties of the catalyst are expected to have the most clear-cut influence on the catalytic rate parameters.

There are a large class of catalytic processes in which atoms belonging to the catalyst surface leave their lattice positions and become part of the surface compound or of the product molecules, their place in the lattice being taken by reactant atoms. These types of catalytic systems, in which the separation between catalyst surface and reacting layer becomes diffuse, have therefore been called *intrafacial catalysis*.[38] Corrosive chemisorption is often a forerunner of this type. An example of intrafacial catalysis is the decomposition of formic acid on metal catalysts. Sachtler and Fahrenfort[17] have identified metal formates as the intermediates in this process. Many selective oxidations over metal oxides belong to this class also, since lattice oxygen becomes part of the product. This so-called Mars–Van

Krevelen mechanism, originally proposed for the oxidation of xylene to phthalic anhydride,[42] has been shown to apply for oxidation of propylene and CO over bismuth molybdate at 425°C by Keulks[43] and for many selective oxidation processes[44] (see Section 5.4). In these intrafacial catalytic processes, the kinetic parameters are frequently linked more significantly to the chemical stability of the solid, i.e., to the heat of formation of the metal oxide[45,46] or the heat of formation of the metal formate,[17] rather than to "electronic" properties.

2.3. The Surface Complex: Relation of Heterogeneous Catalysis to Coordination Chemistry

The categories of elementary processes customarily used to analyze heterogeneous catalysis have been listed as follows:[47,48]

(1) Diffusion of reactants to the surface.
(2) Chemisorption of at least one of the reactants on the surface.
(3) Surface diffusion of chemisorbed reactants and subsequent surface reaction.
(4) Desorption of products from the surface.
(5) Diffusion of products from the surface.

Steps 1 and 5 involve transport of material through a gaseous or liquid medium, and need not concern us for the moment. Adsorption and desorption processes are discussed at some length in other chapters of this volume, and will only be discussed here as is necessary to understand catalytic mechanisms. The key step for many catalytic processes is step 3, which is generally believed to involve the formation of surface intermediates which decompose to yield the desired product(s). Surface intermediates or complexes are frequently short-lived compounds whose existence is often inferred from more or less indirect evidence.

A basis for thought about the nature of surface complexes is provided by organic and inorganic chemists, who have identified a wide variety of molecular coordination complexes formed from transition metal atoms, ranging from metal carbonyls $[M_x(CO)_y]$ to complex organometallic compounds. There is abundant evidence from recent experimental studies of chemisorption and catalysis that similar compounds can exist on the surfaces of transition metals. Infrared spectroscopy has been used to demonstrate that a formate (HCOO) is formed when formic acid (HCOOH) is adsorbed on a number of different metallic catalysts; a further conclusion is that decomposition proceeds via the formate intermediate on all of the metals studied.[17] Ultraviolet photoemission spectroscopy has been applied by Demuth and Eastman[49] in studies of the decomposition of C_2H_4 and C_2H_2 on Ni. This work demonstrates convincingly that dehydrogenation of C_2H_4 leads to formation of a chemisorbed acetylenic intermediate, which undergoes further decomposition at elevated temperatures. Burwell[2] has cited extensive work based largely on kinetics and the use of isotope tracers which supports the existence of a veritable "organometallic zoo" of surface organometallic intermediates in reactions involving hydrogen and hydrocarbons over transition metal catalysts.

The ability to directly detect a catalytic intermediate in high concentration on a surface implies that its decomposition to yield the catalytic products is a rate-controlling process. Unfortunately, the direct detection of the elusive surface catalytic intermediate is frequently hampered by its low concentration under reaction conditions. If the rate-controlling step involves formation of the intermediate via adsorption and/or diffusion, then rapid decomposition of the intermediate may result in a steady-state concentration so low as to be undetectable using presently available techniques.

In any event, it appears that an understanding of the fundamental basis of catalytic activity will involve an understanding of the physics underlying the coordination chemistry of organic and inorganic transition metal complexes. This philosophy is the basis for the recent theoretical work of Johnson, Slater, and Messmer,[50,51] who have applied the self-consistent field, X-alpha scattered wave approach to the investigation of catalytic problems. Their calculation of energy level schemes and orbital "maps" for small metallic clusters and organometallic compounds are providing new insights into the electronic structure of potential catalytic intermediates.

2.4. The Role of *d* Electrons in Catalysis

It has long been recognized that the most active catalysts for many reactions are transition metals. (There are notable exceptions. For example, Ag is highly efficient in catalyzing the oxidation of ethylene). Also, as indicated in Table 1, transition metals are more active in chemisorption than the non-transition metals. The basic questions concerning the preeminence of transition metals are: What, if anything, is unique about unfilled *d* orbitals and catalysis? Can we identify the catalytic activity of transition metals with *d* orbitals per se,

TABLE 1
The Activities of Metal Films in Chemisorption*

Metals	Gases					
	N$_2$	H$_2$	CO	C$_2$H$_4$	C$_2$H$_2$	O$_2$
W, Ta, Mo, Ti, Zr, Fe, Ca, Ba	+	+	+	+	+	+
Ni, Pt, Rh, Pd	−	+	+	+	+	+
Cu, Al	−	−	+	+	+	+
K	−	−	−	−	+	+
Zn, Cd, In, Sn, Pb, Ag	−	−	−	−	−	+
Au	−	−	+	+	+	−

* Table taken from Ref. 194, p. 231; data based on adsorption studies on metal films. (+) Gas chemisorbed. Chemisorption takes place over a large part of the surface with great rapidity. (−) Gas not chemisorbed. Little or no adsorption observed between 0°C and temperatures at which physical adsorption occurs.

or are there simply coincidental geometric and electronic factors in "good" catalysts which influence those kinetic and thermodynamic properties of species that are involved in catalytic reactions (stability of intermediates, heats of adsorption of reactants and products, etc.)?

It is helpful to consider the differences between s orbitals and d orbitals in metals. Clearly, s orbitals in free atoms are spherically symmetric; an s band in a solid has a broad, free-electron-like density of states without a high density of localized electronic states anywhere within a few volts above or below the Fermi level E_F. In an isolated atom, completely filled d orbitals also result in a spherically symmetric charge shell. Partially filled d orbitals exhibit a localization in space and energy which promotes electronic overlap with ligands of a reacting molecule. In contrast to the broad, sparsely populated density of states of s bands in solids, the density of states of d electrons in the valence band of transition metals is usually narrow and dense. This localization in energy and space can result in a strong "mixing" between metal surface orbitals and the orbitals of an adsorbing atom or molecule. Whether or not such mixing results in strong bond formation and/or catalytic reaction depends on the details of the positions of the energy levels of molecular orbitals of adsorbed intermediates with respect to the Fermi level.

It is interesting to note that transition metals exhibit a rich coordination chemistry with molecules such as CO and the other π-electron molecules. As discussed in Mango and Schachtschneider's[40,41] treatments of organometallic coordination complex chemistry, *d* orbitals are spatially localized to overlap effectively with ligand orbitals. In addition, the correspondence between *d*-orbital energies and ligand molecular orbital energies makes possible a strong interaction in which electron transfer between ligand and metal and also between various *d* levels in the metal atom can occur. This property of the transition metal atom can promote catalytic processes in the ligands which would be strongly symmetry-forbidden in the absence of the transition metal atom, i.e., an activation barrier is lowered by the action of the transition metal atom.

One may conclude that the spatial and energy localization of *d* electrons in transition metals is related to the unique catalytic activity of these metals, although other geometric and electronic factors have a major influence on reaction kinetics also.

Many of the topics which have been briefly treated in this historical survey will be discussed in more depth in the following sections.

3. The Nature of Heterogeneous Catalysts

3.1. Character and Use of Catalysts

It is appropriate to indicate in general terms the character of practical heterogeneous catalysts so that the reader will have some feeling of the technological aspects of catalytic chemistry. Catalysts are typically prepared in forms dictated by the type of reactor in which they will be used, among other factors.[52] Fundamental to any catalytic process is provision for reactant and product transport to and away from the active area of the catalyst. Since high surface area is desirable in order to increase the activity per unit quantity of catalyst, some of the active area will usually be inside the catalyst particle, and the problem of transport through narrow pores becomes a factor of kinetic importance, as discussed in detail by Weisz and Prater.[53]

A wide variety of processes use metals as catalysts, including hydrogenation–dehydrogenation, reforming of hydrocarbon molecules, ammonia synthesis, etc.[54] Platinum is a particularly important catalyst in petroleum processing, nickel in hydrogenation of fatty oils, and iron in the ammonia synthesis. A large-volume use of catalysts has recently developed in the catalytic treatment of automotive exhaust to remove carbon monoxide and unburned hydrocarbons. The extremes of temperature and contamination met in this environment have resulted in an intense effort to develop metallic catalysts and new ceramic supports.[55] Among the more unusual applications of metal catalysts in large-scale chemical manufacture is the use of copper catalysts to accelerate a gas–solid reaction, i.e., the reaction of organic halides with silicon to form the precursors for silicones.[56,57]

To obtain high specific area, metal catalysts are usually present in highly dispersed form, either as a powder, or more commonly as particles supported in a nonmetallic matrix. Supported metal catalyst particles are typically prepared on high-area silica or alumina powders by reduction of metal salts deposited by drying and calcining (i.e., high-temperature heating in a controlled atmosphere) the salt-impregnated support material. Metal content may range up to 50% by weight, but values in the 1% range or lower are common for noble metal catalysts.

Catalyst manufacture is best described as an art and few standard procedures are available, according to Ciapetta and Plank,[58] who have listed a number of useful preparative methods for various technically important catalysts.

Other forms of metal catalysts include high-area metals and alloys bonded to screens, tubes, or other structures for support and heat transfer. An example is Raney nickel catalyst, which is prepared by NaOH treatment of nickel–aluminum alloys; this alkali leaching process increases the active surface area of the Raney nickel by dissolution of aluminum from the alloy.[59]

Often, metals are supported on active insulator supports in order to achieve a dual function catalyst, in which different classes of reactions are performed successively on different catalyst sites, as discussed in an excellent general review by Haensel and Burwell.[60]

Oxide catalysts play an exceedingly important role in the refining of petroleum products.[52] An example is the catalytic cracking of petroleum distillate fractions, for which natural aluminosilicates, amorphous synthetic silica–alumina combinations, and crystalline, synthetic silica–alumina compounds (synthetic zeolites) exhibit catalytic activity attributed to acid sites.[61] Some chemicals manufactured on a very large scale, such as butadiene, acrylonitrile, and ethylene oxide, are produced by selective oxidation over metal oxide catalysts, one of the most popular being bismuth molybdenum oxide. Similar selective oxidation catalysts play an important role in the manufacture of pharmaceuticals.

Sulfides such as cobalt molybdenum sulfide and nickel tungsten sulfide are extensively used for partial hydrogenation of sulfur-containing feedstocks in the petroleum industry to remove sulfur as H_2S and in hydrocracking of heavy distillates.[62]

Chlorides are extensively used as polymerization catalysts, e.g., $TiCl_3$ for polypropylene.[63] Another important use is that of a CuCl catalyst in the recovery of Cl_2 from HCl.[64]

Catalytic reactors take many forms, dictated by the nature of the catalytic process as well as the manufacturing process desired. Commonly employed for gaseous reactants with solid catalysts are "fixed-bed" reactors, in which the reactant gases flow through a packed bed of pellets, spheres, or coarse grains. However, these are impractical where continuous movement of the solid is required for cooling of an exothermic reaction or for transport of the solid to a regenerator in order to restore the catalyst to useful activity. In such cases fluid bed reactors are common,[65] in which the gas or liquid flows upward through the particulate solid at a sufficient velocity to suspend the catalyst particles in the fluid phase. For hydrogenation reactions of heavy distillates, trickle reactors are used to effect two-phase flow of gas and liquid through a fixed, packed catalyst bed. For some processes, such as polymerizations, fine catalyst particles are suspended in a stirred liquid and the reactions occur at the liquid–solid interface. Laboratory as well as commercial versions of these and other reactor designs are widely used.[66,67]

The design of a commercial catalyst has as its central factors high rate of conversion for reactants (activity); preferential or exclusive formation of desired products rather than side or waste products (selectivity); stability of both activity and selectivity under the conditions of the process, including impurities in the reagents; and mechanical strength. The latter is crucial for fixed-bed reactors to withstand the hydrostatic pressure of the self-supporting catalyst bed and also for fluid beds, where attrition of the suspended particles would otherwise lead to excessive losses of catalyst blown out of the reactors.

In the following sections, more details concerning the nature of finely divided catalysts will be discussed.

3.2. Catalyst Characterization: Surface Area and Dispersion Measurements

For the comparison of the catalytic activity of different high-area catalysts, it is desirable to separate the two factors involved, namely the available surface area of the active phase and its activity normalized per unit of surface area. The determination of the surface area of the catalysts may be done indirectly using a number of physical techniques, such as X-ray or electron diffraction line broadening, low-angle scattering of X-rays, and electron microscopy.[68] The average particle size as determined in this manner may then be related to catalyst surface area. However, to do this with acceptable accuracy, the distribution of particle sizes is required.

A more direct method involves measuring the adsorptive properties of the catalyst, where the quantity of gas adsorbed per unit mass of active catalyst is measured. This may be done in several ways, as follows.

(a) For unsupported catalysts, such as high-area Pt black, the surface area may be determined by the physical adsorption of a gas that is not chemisorbed by the catalyst. The adsorption isotherm is fitted to some model (such as the BET or Brunauer–Emmett–Telle. model),[69] and the adsorbate surface area is determined directly, using well-known areas for the physically adsorbed molecule.[70,71]

(b) For supported catalysts, it is often possible to selectively chemisorb molecules such as O_2, CO, H_2, or NO on the catalytic material. These measurements, carried out to saturation of the monolayer, are often directly reported (as μmoles/g of catalytic component). Assumptions about the monolayer capacity (molecules adsorbate/ surface atom) can be made so that the number of surface atoms in the catalyst may be estimated.[72]

(c) For supported catalysts, where interference in chemisorption measurements occurs due to support adsorption, or to absorption in the bulk of the catalyst, it is often possible to titrate a chemisorbed layer on the catalytic surface with a second gas. This has been done for O(ads) on supported Pd[73] and Pt[70] using $H_2(g)$.

Methods (a), (b), and (c) have been intercompared with each other.[74] An example of this may be found in the work of Sinfelt, who compared H_2 and CO chemisorption results for a number of different supported metals at different degrees of metal dispersion.[72] The ratio of hydrogen chemisorption per metal atom was found to be equal to the ratio of CO chemisorption per metal atom over a wide range of dispersion of the metal. In the limit of highest dispersion every metal atom was able to adsorb either one H or one CO.

The observation that for a given catalyst the monolayer capacity for H and CO chemisorption agrees with fair accuracy may imply a simple stoichiometric relation between the number of atoms or molecules in the monolayer and the number of surface metal atoms.

15

This has, in fact, been checked for Pt by comparing the BET surface area with oxygen chemisorptive uptake.[70] There is, of course, some uncertainty about the correct average Pt site density to assume for the catalyst; choosing the average of the site densities for Pt(100), Pt(110), and Pt(111) as 8.4 Å^2/site, an O/Pt ratio of 1.1 was obtained for Pt surface atoms. It was found by titration that 2.16 H reacted with 1 O(ads) in this experiment, in agreement with the stoichiometry of H_2O product as well as with calorimetric measurements made on Pt black.[75]

For the characterization of oxides, volumetric measurements of the chemisorption of NO coupled with infrared absorption measurements are good techniques for the determination of the number of oxygen vacancies and their coordination. For example,[76] on chromia, NO can be adsorbed on a single oxygen vacancy as an NO_2 chelate and on a double vacancy as a *cis*-dimer:

$$
\begin{array}{ccccc}
 & \overset{N}{\diagup \diagdown} & & & N{-}N \\
 & O \quad O & & & \diagup \qquad \diagdown \\
\diagdown \diagup \quad \diagdown \diagup & & & O \qquad O \\
Cr + NO \rightarrow Cr & & \text{or} & & Cr + 2NO \rightarrow Cr
\end{array}
$$

The identification of such infrared-active complexes can be materially reinforced by the use of isotopically labeled ^{15}NO (Figure 2).

The *dispersion D* of a metal catalyst is defined as the fraction of the total number of metal atoms that are at the surface of the metal particles. Figure 3 illustrates this property for model cubooctahedra of Pt.[77] The chemisorptive uptake may be used to calculate a value of the catalyst dispersion by assuming or measuring surface stoichiometries (atoms of adsorbate per surface atom) for the system considered. To obtain the average particle size from chemisorption data,[73] one must make further assumptions regarding particle shape and morphology, and the accessibility of the surface to the chemisorption process. For Pd, the relation $d = 0.9/D$, where d is the average particle diameter (nm), is found to give fair agreement ($\sim 30\%$) with x-ray line broadening values for d.

The absolute measurement of surface stoichiometries of monolayer chemisorption on single crystals would contribute to a better understanding of the chemisorption measurement of surface area and dispersion of high-area catalysts. For instance, as pointed out by Benson *et al.*,[73] there is uncertainty about the stoichiometry of the Pd–CO system since both linear (Pd–CO) and bridged (Pd$_2$CO) species are detected by infrared spectroscopy on supported Pd

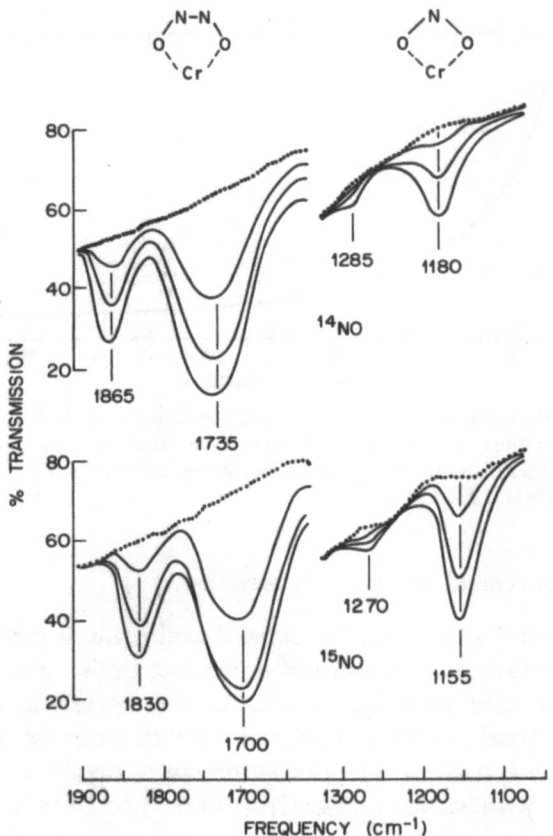

Fig. 2. Adsorption of ^{14}NO and ^{15}NO on self-supported Cr_2O_3. Infrared absorption spectra. Assignments of the two sets of bands as indicated at the top. Curves with decreasing percentage transmission correspond to increasing NO coverage. (After Kugler et al.[76])

catalysts.[78] Furthermore, recent low-energy electron diffraction (LEED) studies of the CO + Pd(100) system indicate that at monolayer coverage CO adsorbs out of registry with the underlying Pd(100) lattice.[79]

Measurements of the absolute surface coverage of hydrogen and oxygen monolayers on the (100) face of tungsten have recently been made using calibrated molecular beam techniques.[80] Such measurements are difficult at present and are accurate to about $\pm 15\%$.

17

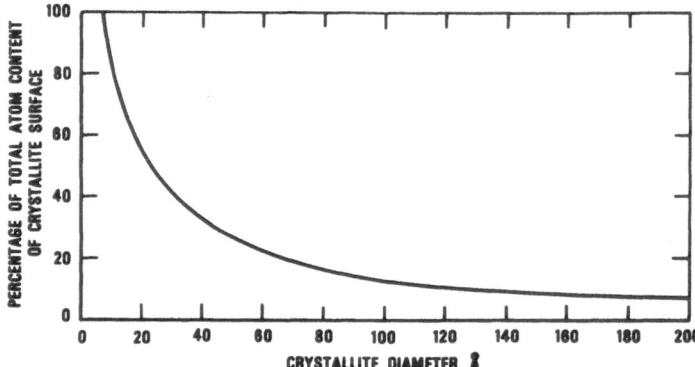

Fig. 3. Variation of the ratio of surface atoms/total atoms for small Pt crystallites as a function of crystallite diameter; the model crystallites are assumed to have face-centered-cubic packing. (From Stonehart and Zucks.[77])

3.3. The Measurement of Catalytic Activity

The "activity" of a catalyst is defined as the rate of conversion of a reactant or the rate of formation of a product under specified conditions of temperature, partial pressures or concentrations of reactants and products, total pressure, and flow rate of the reacting medium through the catalyst mass. The measurement of catalytic activity can be carried out with various objectives. One is to obtain parameters for the design of a catalytic reactor, such as an automotive exhaust reactor with Pt catalyst for the oxidation of CO and hydrocarbons.[81] Another purpose may be to compare the usefulness of catalysts for a given process, e.g., the disproportionation of olefins.[82] Frequently, it is desired to characterize the properties of a surface in terms of the activity for a chosen "test reaction," as in the comparison of *d* metals for ethane hydrogenolysis.[14]

In order to compare the activity of different catalytic surfaces for the same reaction, it is important to make kinetic measurements which can be intercompared, and in the limit, to determine the absolute rates of elementary reaction steps leading to a catalytic product. Absolute rate is defined as the rate per surface site. Since catalytic activity is a surface property, it might appear that reproducible measurements are a hopeless task, since it is well known that the properties of "clean" surfaces are only reproduced by painstaking efforts involving UHV techniques. However, this viewpoint ignores the fact that catalytic surfaces are used and measured under conditions of a kinetic steady state. Impurities are controlled by the occurrence of both deposition and removal reactions. An example is the use of high hydrogen pressures in hydrogenation reactions to prevent the formation of carbonaceous deposits. The very widespread use of catalysts in the chemical industry and the near-unanimity on the preferred catalyst for a given process attest to the general reproducibility of catalytic surfaces. In detail, rate measurements by different workers on catalysts of the same provenance often agree remarkably well.[83–86]

To determine absolute reaction rates, it is necessary to know the number of active sites and, in general, for multiphase mixtures, the catalytically active phase has to be identified. This identification may be difficult, as illustrated by the several years of effort expended to establish that the active phases in supported and unsupported CO–Mo–S and Ni–W–S catalysts are the intercalated trigonal prismatic layer sulfides MoS_2 and WS_2.[62]

Most catalytic reactions are carried out at high pressures, where conditions of molecular flow are not present, and where the transport properties of the gas phase are involved in determining the rate of impingement of reactant molecules on the catalyst surface. In addition, for dispersed catalysts, one is dealing with nonideal geometric surfaces, where it is important to consider diffusion processes within the pores of the catalyst.[87] In order to make meaningful *comparisons* of catalytic activity of different catalysts under those conditions of high pressure and nonideal substrates, it is common to operate under identical conditions of temperature T and partial pressure of reactants P_i. Often a measurement of the initial rate or the rate at low conversion ratios is made to avoid complexities due to changes in the catalyst or to back reactions. The rate of the catalytic reaction R is expressed per unit area of catalyst surface as

$$R = \left[\frac{\text{moles product}}{\text{catalyst area} \times \text{time}} \right]_{T = \text{const}, P_i = \text{const}}$$

A slight variation on this is to express rates as the turnover number N, defined as

$$N = \left[\frac{\text{molecules product}}{\text{number catalyst sites} \times \text{time}} \right]_{T = \text{const}, P_i = \text{const}}$$

19

In both cases, chemisorption measurements are often used to estimate surface area or the number of catalytic sites involved, as discussed previously.

By means of detailed studies, it is often possible to separate the rate of chemical conversion per unit area into a concentration- (or partial pressure-) dependent term and a concentration-independent factor,[87]

$$dn/dt = k_s f(c_i)$$

The factor k_s is characteristic of the catalyst and the catalytic reaction and in some cases may be factored into a preexponential factor A and a temperature-dependent term involving the activation energy for the reaction E_{act}:

$$k_s = A e^{-E_{act}/RT}$$

Thus, assuming that a single activation energy characterizes the reaction, it is possible to intercompare catalytic activities measured at somewhat different temperatures.

Determination of the identity or the rates of the *detailed steps* of a heterogeneous catalytic reaction by the aforementioned kinetic flow technique has rarely been possible. The presence of the surface hampers the use of many spectroscopic techniques successfully used in the elucidation of detailed mechanisms in homogeneous catalysis. In addition, the precise nature of the catalytic surface could rarely be established. Fortunately, the scope of catalytic rate measurements has been extended recently by the introduction by Madix *et al.*[88,89] and Bernasek and Somorjai[90,91] of reactive scattering of molecular beams from surfaces. Some results obtained by these techniques will be discussed in more detail in Section 8. Suffice it to say at this point that rate constants for individual steps can be measured by these techniques.

The coupling of UHV characterization techniques such as low-energy electron diffraction and Auger electron spectroscopy with catalytic rate measurements at conventional pressures has been made by characterizing the single-crystal Pt catalyst in UHV, isolating it in a small part of the UHV apparatus by a bellows mechanism, and introducing the reactants, which after conversion were measured by gas chromatography.[83] These measurements, on the hydrogenation of cyclopropane, were compared with earlier data on supported Pt catalysts of 7–8 orders of magnitude smaller particle size. The turnover numbers for the two systems were equal within a factor of two. These kinds of comparisons are of great value in defining the meaning of catalytic studies on single crystals, where factors such as crystal orientation and surface composition may be well controlled.

The adsorption and desorption steps in a reaction can also be measured in UHV. Adsorption rates have been measured and expressed as a sticking coefficient S:

$$S = \frac{\text{rate of adsorption}}{\text{rate of impingement}}$$

measurements of S can be made by measuring either the capture rate or the reflection rate at known impingement rate. For active chemisorption, often $S = 1$ at zero coverage[80] and S decreases as coverage increases. These measurements have been made for evaporated films[92] and for single crystals.[80,93] This technique has increased in importance with the introduction of molecular beam kinetics for the study of reactive gas–surface interactions.

3.4. Active Site Character and Measurement

The concept that a catalytic surface possesses a distribution of surface sites having different intrinsic catalytic activities was first enunciated by Taylor[3] in 1925. The opposite concept that a chemisorbed monolayer saturates free and *equivalent* surface valences had been proposed by Langmuir in 1922.[94] Taylor's active site hypothesis was a refinement of Langmuir's concept of a uniform "checkerboard" surface. Taylor conceived the surface as being nonhomogeneous by virtue of the presence of a distribution of substrate atoms having different degrees of coordination with other substrate atoms, i.e., a catalytic surface was not a perfect crystalline surface. In Taylor's model, the sites with the highest catalytic activity were assigned to surface atoms possessing the lowest degree of coordination prior to adsorption, and these were called "active sites." The early experimental evidence for the existence of active sites was obtained from annealing and poisoning experiments, which Taylor summed up as follows: "It has been shown that active catalysts manifest an extraordinary sensitivity to heat treatment and to the action of poisons. Both alter adsorptive capacity and reactivity to different degrees, the reduction in catalytic activity being much more pronounced than the reduction in adsorptive power." The concept of a distribution of active sites was consistent with the "varying capacity of the surface to effect catalysis as revealed by progressive poisoning (experiments). These indicated that the amount of surface which is catalytically active is determined by the reaction catalyzed."

The basic concept of active sites on catalysts persists today. The challenge to modern surface chemistry is to induce and characterize these sites and to determine their surface density, structure, etc. Many experiments attempted to increase the active site density by means of irradiation,[95] ion bombardment[96] and cold working,[97] stressing,[98] etc. Following such treatment, enhanced catalytic activity has often been measured, and in some cases subsequent treatment to remove the active sites has caused a reduction in catalytic activity. One difficulty with many of the early experiments is that the results may be influenced by impurities introduced or removed during treatment of the catalyst. For example, Farnsworth and Woodcock,[96] working under ultrahigh vacuum conditions, found that the activity of Ni for the hydrogenation of C_2H_4 could be increased 100-fold using Ar^+ bombardment. A tenfold increase in Pt activity was also produced upon Ar^+ bombardment. Annealing caused the activity to return to lower values in both cases. We now know, from Auger electron spectroscopy (AES) studies, that Ni often contains S impurities which diffuse to the surface from the bulk upon annealing, and which may be removed by ion bombardment. Thus, one cannot be certain about the interpretation of these ion bombardment experiments. Later, ion bombardment work was done by Sosnovsky[99] where single crystals were bombarded and then transferred in air to a reaction chamber, where the activity for HCOOH decomposition was found to vary in a complex way with Ar^+ bombardment energy. Again, one would have to question whether the effects measured are related only to the induction of dislocations by ion bombardment, or whether surface impurities introduced by exposure to the atmosphere may somehow be involved.

The identity and properties of active sites are often inferred from the catalytic and chemisorption properties, e.g., in series of oxides with cation or anion vacancies.[38,100,101] Similarly, the so-called B-5 sites on Ni were inferred from N_2 chemisorption and particle size determinations.[102] It is preferable to have independent means to study the structure of the sites and to measure their number.[103] Recently, a number of well-controlled experiments involving spectroscopic methods have been used to help characterize the electronic and structural nature of active sites. In the first examples cited below, correlation of the catalytic activity with the concentration of active sites has been directly observed. More often, the predominant sites for chemisorption (which are not necessarily the sites for catalytic activity) have been characterized spectroscopically and one example of this is given also.

3.4.1. Active Sites on WS$_2$- and MoS$_2$-Based Catalysts

Many electron spin resonance (ESR) studies of catalysts have served to suggest active sites for catalysis.[104] The first observation of a *quantitative* relation between the number of active surface sites identified and characterized by ESR and the rate constant of a heterogeneous catalytic reaction appears to have been made by Voorhoeve[105] for the hydrogenation of benzene on WS$_2$ and Ni–WS$_2$ catalysts. These catalysts were prepared in seven different ways, with different nickel and sulfur contents. In all catalysts, an ESR signal with an apparent g value of 2.015 was found, which was assigned to W^{3+} ions at the surface in a sulfur coordination of low symmetry.[106] The linear correlation between the intensity of the ESR signal and the rate constant is given in Figure 4, covering more than three orders of magnitude in activity. The identity of the W^{3+} ion as a surface center was established by the proportionality between surface area and ESR intensity for a series of Ni$_{0.53}$WS$_2$ catalysts ground to various surface areas and by the response of the signal to exposure of the catalysts to

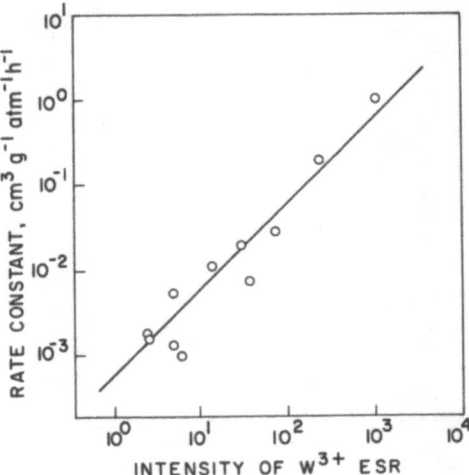

Fig. 4. Identification of W^{3+} as the active site in benzene hydrogenation over WS$_2$-based catalysts by the linear relation between the ESR intensity and the hydrogenation rate constant. (After Voorhoeve.[105])

benzene and cyclohexene. The dependence on equilibration of the catalyst with H_2S/H_2 mixtures showed that the W^{3+} center was associated with an S^{2-} anion vacancy and the temperature dependence of the signal intensity indicated that tungsten was in a low-spin configuration with spin $S = \frac{1}{2}$. On the basis of this information and stereochemical considerations, the structures of Figure 5 have been suggested for this active center.[107,108] In the WS_2 platelike layer structure, the W^{3+} sites must on energetic grounds occur at the edge of the plates. Benzene is postulated to adsorb as a π-bonded complex.

The active site for hydrogenation of butylmercaptan over MoS_2 has been proposed to be a Mo^{3+} ion on the basis of electrical conductivity measurements[109] and Schuit *et al.*,[62,110,111] on the basis of the analogy with Ni–WS_2 and extensive chemical evidence, have concluded that the active sites in Co–MoS_2 hydrodesulfurization catalysts are Mo^{3+} ions in similar coordination as the W^{3+} sites in Figure 5.

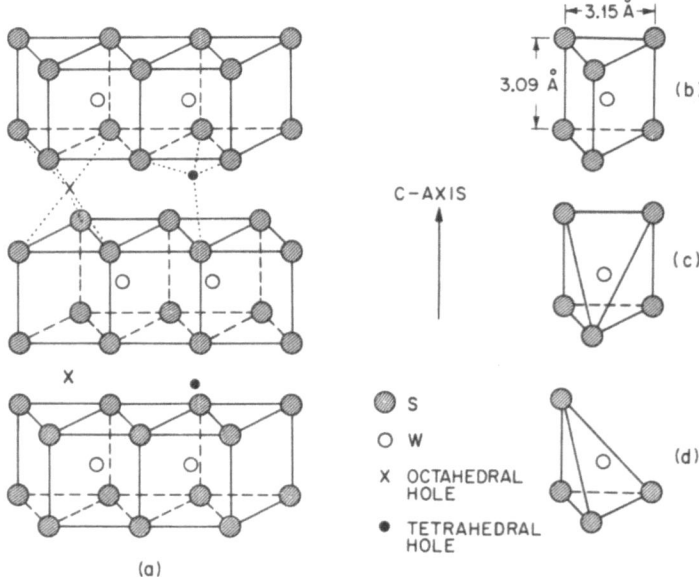

Fig. 5. The structure of tungsten disulfide: (a) The stacking of layers illustrating the position of octahedral holes, which may be partly occupied by nickel; (b) the site symmetry of tungsten ions in the bulk; (c) same for a side face; (d) same for an edge parallel to the *c* axis. (After Voorhoeve and Stuiver.[107])

3.4.2. Active Sites for H_2–D_2 Exchange on MgO

Isotopic exchange between H_2 and D_2 molecules can occur at low temperatures over some transition metal catalysts and over some transition metal oxide catalysts. This process is thought to proceed by dissociative hydrogen chemisorption on the transition metal or its ion.

MgO is also capable of catalyzing H_2–D_2 exchange at 70°K[112] and Harkins *et al.*[113] and Lunsford[114] had suggested that an ESR-active V_I center (a trapped hole) was the active site. Boudart *et al.*[115] have related the rate of exchange to ESR measurements of the concentration of the active site (EPR signal at $g = 2.0030$ and designated V_I) responsible for the catalytic activity. It was found that various MgO powders could be prepared by heating in such a way that their specific activity for exchange varied over a wide range. Heating removes protons, deactivates the catalyst, and causes the ESR signal to decrease. The catalytic activity and the ESR signal

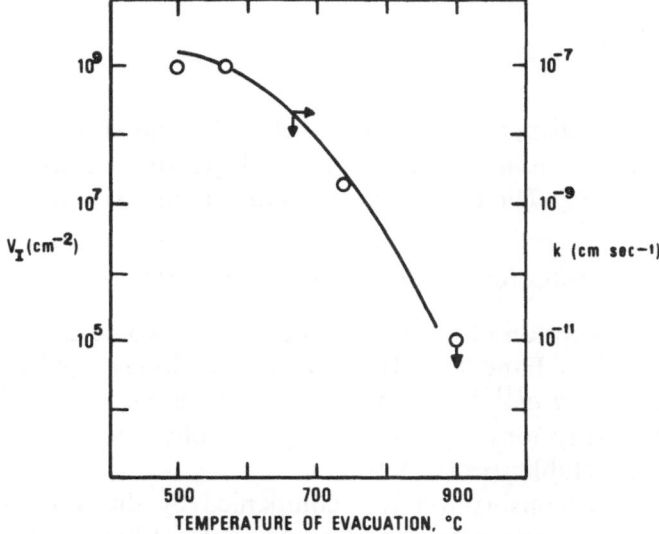

Fig. 6. Correlation of catalytic activity with active site density. The rate constant k for H_2–D_2 exchange at 78°K on MgO following heating in vacuum at various temperatures is shown. The solid line represents the rate constant data and the points represent the measured surface concentration of paramagnetic centers responsible for catalytic exchange. The 900°C point is plotted at the lower limit of sensitivity of the ESR spectrometer. (From Boudart *et al.*[115])

25

may be restored by treatment of the heated MgO with H_2O. Figure 6 shows that the H_2-D_2 catalytic exchange activity (solid line) varied in parallel to the concentration (circles) of V_I sites over a wide range. No ESR signal was obtained for the catalyst with the lowest activity and the corresponding point at 900°C is plotted at the lower limit of sensitivity of the ESR spectrometer.

The paramagnetic V_I centers are thought to be at the MgO surface because of the rapidity of response of characteristic ESR signals upon exposure to $O_2(g)$ and to $H_2O(g)$, i.e., the sites may be quantitatively titrated. Boudart *et al.*[115] argue that the active site consists of three O^- ions arranged in a triangle on the MgO (111) surface. Adjacent to this site is an OH^- group, which participates in exchange with a D_2 molecule adsorbed between the two closest O^- ions, in an orientation that allows a triangular transition complex to form,

$$
\begin{array}{c}
O^- \\
D \\
OH \cdots \quad O^- \\
D \\
O^-
\end{array}
$$

permitting exchange. Such a triangular O^- site should coordinate effectively with lone-pair electrons in H_2O or O_2, with resultant line broadening. The observed spectrum is in agreement with this.

3.4.3. Active Sites for H_2 Chemisorption by ZnO

The adsorption of H_2 by ZnO occurs via two parallel processes, designated Type I and Type II. Type I adsorption is rapid and reversible. Eischens *et al.*[116] found that it yields intense infrared spectra. Type II adsorption is slow and irreversible, and does not yield infrared-detectable species.[116]

Type I chemisorption is accompanied by the development of two infrared bands corresponding to ZnH (1709 cm^{-1}) and OH (3489 cm^{-1}) vibrations.[116] Recently, Kokes *et al.*[117] have probed the nature of the active sites involved in Type I H_2 chemisorption by chemisorption of pure HD. If we envision the active site to be schematically

Zn–O

then HD can chemisorb either as

$$
\begin{array}{cc}
\text{H} & \text{D} \\
| & | \\
\text{Zn} & \text{O}
\end{array}
\quad \text{or as} \quad
\begin{array}{cc}
\text{D} & \text{H} \\
| & | \\
\text{Zn} & \text{O}
\end{array}
$$

These two structures can be easily distinguished spectroscopically. Upon adsorption at 78°K, the kinetic isotope effect favours the "Zn D" species significantly, while at room temperature the "Zn H" species is favored thermodynamically; at 78°K the Zn D is spectroscopically observed to be favored over the Zn H. Warming to ~250°K causes a rapid and thermally *irreversible* switch as the thermodynamically favored "Zn H" species is formed, as shown in Figure 7. These simple measurements thus indicate that the active site in this case is in fact a Zn–O *pair* site since intensity ratios cannot easily be explained by models involving large separation of the H and D atoms following HD chemisorption at 78°K.

Thus we see from these three examples that in favorable cases it is now possible to correlate spectroscopic measurements with catalytic activity although an underlying physical interpretation may be missing.

3.4.4. Active Sites on Metal Catalysts

If one asks about the status of active site characterization on metal catalysts, it must be said at present that our understanding is not as good as in the specific cases for the oxides described above. As will be discussed later, Boudart[8,103] has divided catalytic reactions

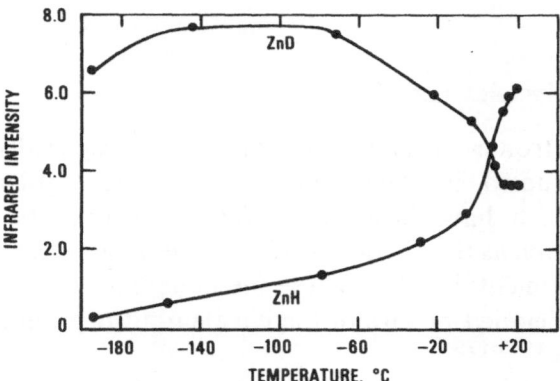

Fig. 7. Infrared measurement of the thermal equilibration of species produced by adsorption of HD on ZnO at −195°C. (From Kokes *et al.*[117])

on metals into two general groups: the *facile* reactions and the *demanding* reactions. Facile reactions are classed as those where specific catalytic activity seems to be kinetically independent of particle size. Demanding reactions occur on surfaces having specific catalytic activity that depends upon catalyst particle size and method of preparation. It is inferred that facile reactions occur generally on most surface sites, whereas demanding reactions may proceed only on specific active sites whose nature can be inferred by indirect means. Several examples of crystallographic effects on surface reactivity of metals are discussed in Section 6.1.

4. The Modern Theory and Spectroscopy of Chemisorption

In recent years, there has been a surge of interest from theorists and experimental physicists in problems associated with surfaces, chemisorption, and catalysis. The theoretical efforts include studies of vibrational surface modes,[118-120] which are important in the understanding of phase transitions at the surface and in phonon-assisted surface chemical processes. Other studies stressing the collective nature of the solid deal with the magnetic properties of surfaces,[121] a subject that is of particular relevance for catalysis on ferromagnetic catalysts. On these catalysts, spin density waves are thought to contribute directly to the kinetics of desorption,[122] and a mean field theory of surface spin fluctuations is being developed.[123,124] The dynamics of dielectric surfaces, a subject that is especially interesting for the effect of ferroelectric phase transitions on the catalytic rate, has been reviewed recently by Kliewer and Fuchs.[125]

4.1. First-Principles Theories

The electronics of semiconductor surfaces, especially Si and Ge, has been studied in detail with self-consistent pseudopotential methods, which have been successful in treating the bulk band structure as well as the intrinsic surface states in good agreement with optical experiments.[126,127] These methods have also been applied to obtain a detailed picture of the covalent adsorption of atomic H on Si and Ge.[128,129]

Chemisorption on transition metals is in a much less clear-cut position. Various theories have been developed, which vary in their approach from an entirely collective picture taking into account the itinerant electrons of the metal as a first determinant of the chemisorption system, to a free molecule picture in which only one or two metal atoms react with the chemisorbing molecule to form a free compound molecule. Most of these theories are still rather remote from experimental tests and we will not be able to devote much space to them within the scope of this chapter.

The most widespread models for the chemisorption on metals are:

(a) The Anderson model, which was originally given for impurities in dilute magnetic alloys[130] and which uses a virtual level and restricted Coulomb and exchange interactions. It was applied by Newns[131] to chemisorption on refractory metals.

(b) The density functional formalism as used by Smith *et al.*,[132] which treats the substrate as an electron gas.

(c) The induced covalent-bond mechanism introduced by Schrieffer and Gomer,[133] which is based on the surface molecule approach also used by Grimley.[134]

The last approach has been reviewed recently.[135,136] The application of the Anderson model to the adsorption of hydrogen and oxygen on Pd, Ni, etc., leads to difficulties due to the neglect of correlation, while the induced-covalent-bond mechanism overemphasizes correlation. A phenomenological model has been proposed which seeks a middle ground between these models.[137]

Schrieffer *et al.*[133,136,138] have further developed the induced-covalent-bond theory. In analogy with the Heitler–London treatment of H_2, the overlap between the orbital of the adsorbed hydrogen atom and the orbitals of the metal is of crucial importance. Calculation of the local density of states for the atomic d orbitals on the surface atom(s) forming the adsorption site provides a chemically appealing way of judging the mode of chemisorption.[136] The methods used are rather approximate, in that matrix elements of the Hamiltonian are taken to be zero on any but nearest-neighbor atoms, while the remaining matrix elements are fitted to the known bulk energy bands.

The concept of a "surface molecule" as an approximation to the description of a chemisorbed species has been considered by Grimley[134] and Thorpe.[139] A surface compound could be constructed by using in chemisorption the atomic orbitals of metal surface atoms that are not involved in bonding to the bulk. But the energies of the orbitals formed must lie outside the energy bands of the metal with which they interact, or they will be so highly delocalized as to destroy the concept of a surface compound with discrete orbital energies (although a particular geometric configuration may be retained with delocalized bonding electrons). Grimley points out that if the coupling of metal atomic orbitals to the adsorbate is stronger than the metal–metal coupling, then it is still useful to start with the concept of a surface molecule, calculate the orbital energies and wave functions, and then allow for interaction with the itinerant electron states of the rest of the metal. If this latter interaction is weak, the levels will be shifted and broadened even in the vicinity of the conduction band. Grimley has applied these general ideas to the chemisorption of CO on Ni and other metals and he shows that reasonable but arbitrary choices of chemisorption parameters give metal–CO bond strengths which are in qualitative agreement with experimental measurements.

Less approximate calculations are feasible if the size of the adatom–substrate system is limited to a few atoms. Einstein[140] has proceeded this way, taking into account a three-atom chain with adatom Coulomb repulsion, adatom–substrate hopping, and substrate bandwidth as variables. These calculations show that for a weakly bonded adatom, second-order perturbation theory yields a very good approximation of the binding energy. For strong binding, a surface complex (diatomic molecule) perturbatively rebonded to the indented chain gives a solution for the intermediate case which is more accurate than Hartree–Fock approximations. The rebonded surface complex is formed as a single orbital detaches from the chain to form a diatomic covalent molecule with the adatom, in much the same way as the "surface molecule" of Grimley.

A more drastic simplification of the system is provided by the "reaction" of two metal atoms in a dimer with the hydrogen molecule.[141,142] The advantage of this scheme is that it becomes possible to calculate the *energy surface* of the four-center system and to follow the importance of the contribution of various orbitals in the basis set in a generalized valence bond calculation.[141] This was done for the $Ni_2 + H_2 \rightarrow 2NiH$ and $Cu_2 + H_2 \rightarrow 2CuH$ reactions, with the interesting result that the Ni $3d$ orbitals had their main contribution in the lowering of the activation barrier associated with orbital symmetry effects.[141]

The use of molecular orbital theory to describe chemisorption leads naturally to the calculation of the energy levels of a molecular aggregate containing a limited number of substrate atoms in addition to the chemisorbed molecule. Therefore, a computational scheme developed for large molecules and clusters appears to be of particular use in chemisorption problems. This is the SCF-Xα-SW cluster method developed by Slater and Johnson[50,51] from Johnson's multiple-scattered-wave formalism combined with Slater's self-consistent field method using a statistical α parameter to include exchange correlation. Fundamental to the method is the partition of the cluster into Wigner–Seitz spheres surrounding the atoms. The regions between the spheres are represented by a volume-averaged potential and the whole cluster is enclosed in a sphere, outside of which again a spherically averaged potential is used. Schrödinger's equation is solved in each region and the wave functions and their first derivatives are joined continuously across the boundaries separating the adjacent regions; this matching is achieved using a multiple-scattered-wave approach. The charge density throughout the cluster is then computed, and used (together with Poisson's equation and the Xα exchange-correlation expression) to generate a new molecular potential for which Schrödinger's equation can again be solved. The whole procedure is iterated until self-consistency is reached. The method has been shown[143] to give the correct orbital orderings consistent with Hartree–Fock calculations for the SF_6 molecule and to give orbital energies in very good agreement with ESCA results. It has subsequently been used to calculate energy levels of clusters M_i, where M = Li, Cu, or Ni and $2 \leq i \leq 13$, as well as the levels for NiO_6^{10-}, MnO_4^-, and SiO_2.[51] In chemisorption problems, it has been applied to the adsorption of chalcogens on the (001) "surface" of a Ni_5 cluster[144] and to the chemisorption of CO on the same type of Ni_5[145,146] cluster. A somewhat similar technique, in which no spherical averaging of the potential is used but which is not self-consistent, has been applied to the same adsorption system, CO on (001) Ni_4.[147] The preliminary results by Batra and Bagus are shown in Figure 8 with the early experiments of Eastman and Cashion.[148] The correspondence is reasonably good, but is rather sensitive to the parameters used (cf. Ref. 145 and 146). Interestingly, interaction of CO with the Ni surface perturbed the CO levels to the point where the ordering of the 5σ (lone pair) and 1π levels of adsorbed CO are reversed with respect to their ordering in gas-phase CO. This result is consistent with more extensive calculations using the SCF-Xα-SW procedure on a series of metal carbonyls.[149] The calculations show general support for the idea of "back bonding", i.e., charge transfer from *d* orbitals of the metal into the lowest π* orbital of CO, but they also show that *s*, *p*, and *d* orbitals on the metal are all involved in the

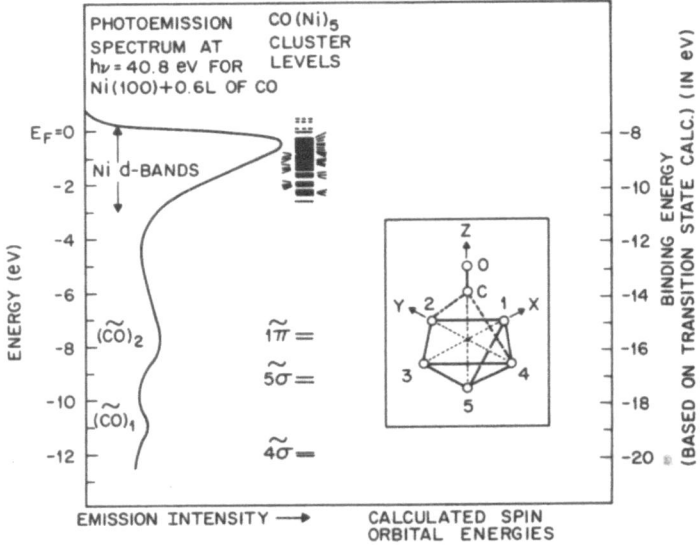

Fig. 8. Comparison of theoretical spin-unrestricted energy levels for the CO(Ni)₅ cluster and experimental photoemission spectrum of CO chemisorbed on Ni(100). The insert shows the model cluster. (Experimental data from Eastman and Cashion,[148] calculations and figure from Batra and Bagus.[146])

bonding. The experimental identification of the levels for CO on Ni is discussed in Section 4.3.

4.2. Empirical and Semiempirical Methods

The foregoing methods were quite restricted in the size of the system that could be treated, or in the approximation necessary to obtain a tractable model. Empirical methods, on the other hand, have shown a wide range of application to catalytic problems. The price paid is the danger of "overinterpretation" of the results. We will discuss the CFSO–BEBO model and extended Hückel calculations as applied to chemisorption and catalysis.

4.2.1. The Crystal Field Surface Orbital–Bond Energy Bond Order Model (CFSO–BEBO)

This highly empirical model devised by Weinberg and Merrill[150,151] allows kinetic and thermodynamic predictions regarding adsorption and reaction of gases at surfaces. The model combines features of the bond energy–bond order relationships found in molecular spectroscopy with the concepts of crystal field theory. The basic assumptions of the CFSO–BEBO approach are as follows:

(1) The surface bond is a localized covalent bond involving primarily d electrons of the solid; the electronic structure of the solid is taken from the Engel–Brewer rules.

(2) Adsorption positions on the surface are those sites where strong directional bonding is expected based on predictions of crystal field theory. "Dangling bonds" at the surface are oriented in directions consistent with bonding in the bulk.

(3) The bond energy between two atoms involved in the surface bond is assumed to vary with bond distance and bond order as predicted by spectroscopic measurements of model compounds.

(4) The energy of a single order bond is derived from data on analogous bulk compounds.

To illustrate the model, the CFSO d orbitals associated with the Pt(111) surface are indicated in Figure 9.[151] There are no d orbitals that emerge from the surface atoms orthogonal to the surface plane; they emerge so that each interstitial position on the surface is a possible center for strong bonding. Half of these positions are out-of-the-plane intersections of three orbitals with e_g symmetry, and half

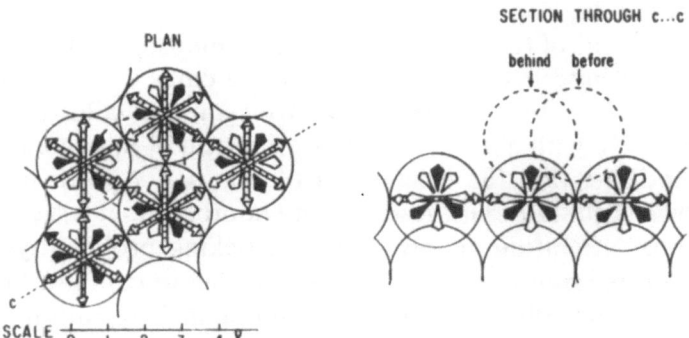

Fig. 9. Crystal field surface orbitals (CFSO) emerging from the Pt(111) surface. Filled arrows: e_g orbitals emerging at an angle of 35°16′ with the (111) plane and intersecting at an angle of 90° in the plane of intersection. Open arrows: t_{2g} orbitals emerging at an angle of 54°44′ with the (111) plane and intersecting at an angle of 60° in the plane of intersection. Cross-hatched arrows: bonding orbitals in the plane of the surface. (From Weinberg *et al.*[151])

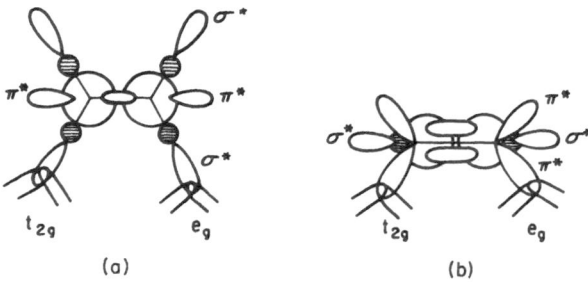

Fig. 10. Orbital models illustrating symmetry-allowed interactions of ethylene on a Pt(111) surface which lead to associative adsorption: (a) formation of Pt–H surface bonds; (b) formation of Pt–C surface bonds. CFSO–BEBO calculations indicate that adsorption into configuration (a) is an endothermic process, whereas configuration (b) is an exothermic interaction. (From Weinberg *et al.*[151])

with t_{2g} symmetry. Based on the model assumptions, predictions of several relevant quantities in gas–surface interactions have been made, namely: (1) binding energies for both atomic and molecular adsorbed species and (2) activation energies for both dissociative and nondissociative chemisorption. The model has been applied to such systems as H_2, CO, H_2O, CO_2, and C_2H_4 on Pt and Ni surfaces with a surprising degree of success.

Predictions of the adsorption and decomposition of ethylene on Pt(111) have agreed well with experimental data.[151] CFSO–BEBO calculations have been made for the two symmetry-allowed interactions of C_2H_4 with Pt(111), which are illustrated in Figure 10. The results indicate that the orientation of Figure 10a, with the initial formation of Pt–H bonds, leads to an endothermic interaction with appreciable activation energy and low sticking probability, clearly an unfavorable situation. In contrast, the formation of Pt–C bonds, as shown in Figure 10b, is exothermic, resulting in dissociation of C_2H_4 to form an acetylenic complex and two adsorbed H atoms. In addition, the calculations predict that the acetylenic surface complex in the first monolayer can adsorb a second ethylene by a type of hydrogen bonding, and that heating the saturated surface leads to further decomposition, resulting in desorption of H_2 and nucleation of a graphite layer. All of these predictions are in good agreement with experiment.

In summary, the CFSO–BEBO treatment is a highly empirical model calculation which provides a useful insight into the energetics and mechanisms of chemisorption and catalysis.

4.2.2. Extended Hückel Methods and Orbital Symmetry

Extended Hückel molecular orbital methods have been extensively applied to organic molecules by Hoffmann[152] and to transition metal complexes by Ballhausen and Gray[153] and Fenske *et al.*[154] Mango and Schachtschneider[40,156] have used a self-consistent charge and configuration (SCCC–EHT) variation of this method in an intercomparison of the effects of various metals on ligand reactions. They point out: "In our opinion the iterative extended Hückel method in any of its variations gives at best a crude description of the ground state of the molecule and hopefully the correct ordering of the highest few filled and lowest few empty molecular orbitals. It should be used only for comparative calculations in a series of related molecules or configurations or to obtain a crude picture of the ground state molecular obitals."[40] The ordering of the highest few filled and lowest few empty orbitals is of importance to apply the concept of "symmetry-allowed" and "symmetry-forbidden" reactions to the catalysis by transition metal surfaces.[40] In view of the prominence of these ideas, based on the Woodward–Hoffmann orbital symmetry conservation rules, in present-day catalytic thought,[157] it is appropriate to give a short outline here, following Mango and Schachtschneider. However, at the outset it should be stressed that the application of the Woodward–Hoffmann rules to catalysis is based on the assumption of *concerted* reactions with high-symmetry transition states, whereas alternative mechanistic schemes are sometimes more likely.[158]

The formation of cyclobutane 1,2 and 3,4 σ bonds by the concerted fusion of two olefin π bonds is a symmetry-forbidden reaction (Figure 11a). This can be understood from the symmetry of the molecular orbitals of cyclobutane and the symmetry of the molecular orbitals formed by the interacting olefinic π systems. The symmetry elements are the ZX and YZ planes in Figure 11a, with Z being the fourfold symmetry axis. The orbitals of the olefinic systems ($\pi_{1,4} \pm \pi_{2,3}$ and $\pi_{1,4}^* \pm \pi_{2,3}^*$) and of the cyclobutane ring (σ and σ^*) are classified as symmetric (S) or antisymmetric (A) about the ZX and YZ planes (Figure 11b). This correlation diagram shows that the highest occupied MO of the reactant system has the same symmetry

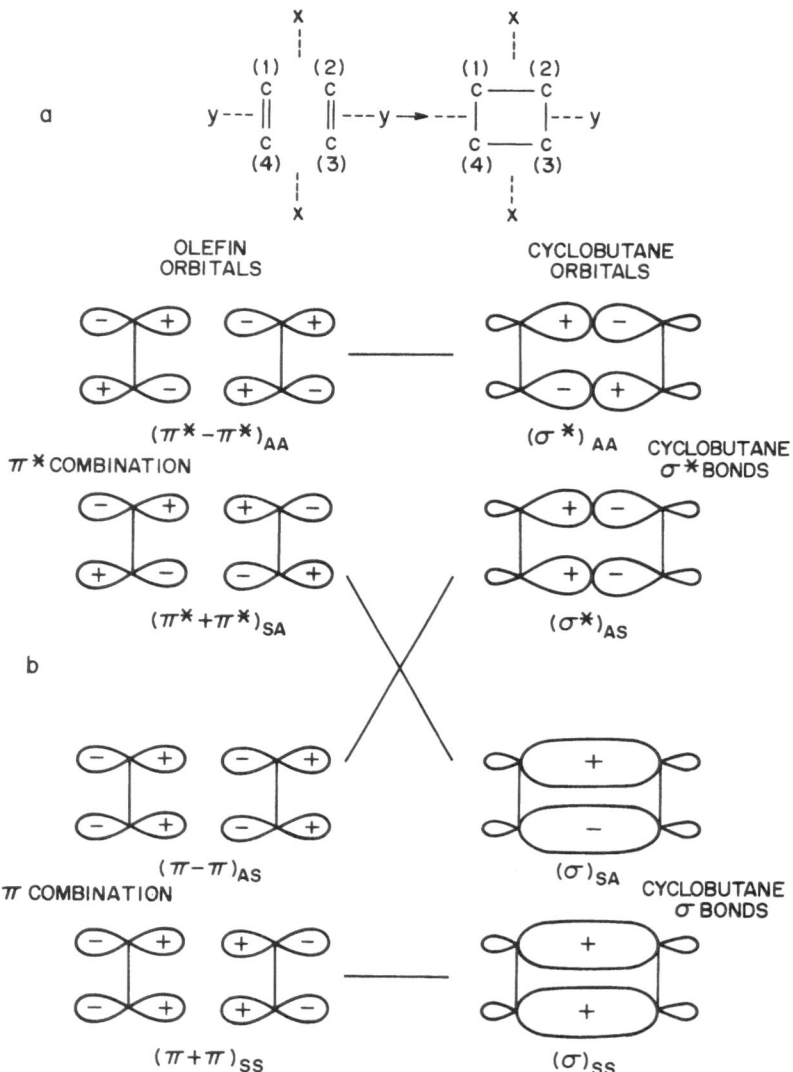

Fig. 11. (a) Transition state symmetry and (b) orbital correlation diagram for the homogeneous cyclobutanation of ethylene. In (a), the hydrogen atoms have been omitted for clarity. (After Mango and Schachtschneider.[40])

as the lowest unoccupied MO of the product, which leads to a high-energy excited state of the cyclobutane. The postulate of conservation of orbital symmetry during concerted reactions[159] makes it impossible to change symmetry during the reaction and hence prevents transition from the ground state $[(\pi + \pi)^2_{SS}, (\pi - \pi)^2_{AS}]$ to the ground state $[(\sigma)^2_{SS}, (\sigma)^2_{SA}]$. This postulate has been shown by Pearson to stem from the requirements of nonzero overlap between high-energy, occupied MO's of the initial state and low-energy, empty MO's of the final state.[160] It has very general validity.[157,160]

The symmetry restriction can be removed if the olefin molecules are coordinated to an appropriate transition metal atom. This is the basis of the widespread occurrence of catalysis of symmetry-forbidden

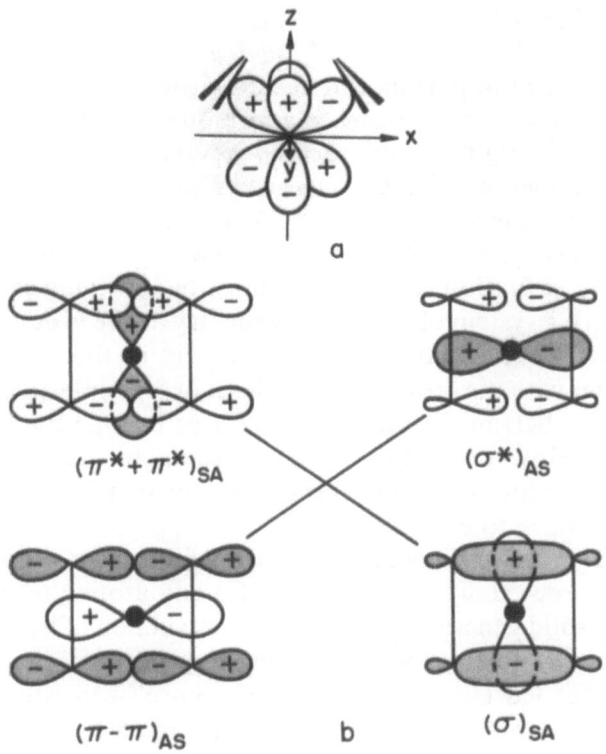

Fig. 12. Catalysis of a symmetry-forbidden reaction by a *d* metal. (a) Two olefin ligands occupying adjacent positions in the metal complex, with d_{yz} and d_{zx} orbitals shown. These have SA and AS symmetry, respectively. (b) The exchange of electron pairs between ligands and metal proceeding through orbitals of SA and AS symmetries. Shaded orbitals are occupied. (After Mango and Schachtschneider.[40])

reactions.[40,156,161] As an illustration, we return to the [2 + 2]-cycloaddition of olefins. In Figure 12a, the olefins are coordinated to the d orbitals of a metal atom on the Z axis under the XY plane. The d orbitals have the proper symmetry to accept electrons from the orbitals that are being raised in energy during the reaction [i.e., $(\pi - \pi)_{AS} \leftrightarrow (\sigma^*)_{AS}$] and to donate electrons to the orbitals that are being lowered in energy during the reaction [i.e., $(\pi^* + \pi^*)_{SA} \leftrightarrow (\sigma)_{SA}$]. This is shown graphically in Figure 12b, where the shading indicates occupied orbitals. To perform this function effectively, the metal can use nearly equivalent nonbonding orbitals of the proper symmetry and a pair of nonbonding electrons. The energy of the d levels used has an optimum position with respect to the π-orbital combinations.[40] The ordering of the d levels of the transition metal atom is of crucial importance in deciding the effect of a given metal complex or metal surface atom on a forbidden reaction. The calculation of the ordering has been done by Mango and Schachtschneider for metal complexes, using the extended Hückel method.[40] The real question is not the positions of the lowest unoccupied and highest occupied levels in a set of MO's. For a chemical reaction to occur with a reasonable activation energy, there must be low-lying excited states for the reacting system of the same symmetry as the ground state.[160] In the exact wave function, we have to mix in the infinite sum of excited states with the ground state, at least in principle, to arrive at the energy of the barrier. In practice, it suffices to take the few lowest-lying states. However, if the density of states in this region is high (many excited states in a small energy increment), then an appreciable number will have to be included.

Orbital symmetry rules define a class of reactions which require a catalyst to proceed. The availability of many atomic orbitals at the surface of a solid generally permits the construction of hybrid orbitals of the proper symmetry for the reaction to proceed.[162]

The extended Hückel method in a non-self-consistent form has been applied to the dissociative chemisorption of diatomic molecules and ethylene on transition metal surfaces, e.g., Ni and W.[163] The lack of self-consistency leads to unrealistic charge distributions for some of the adsorbates. It appears to be too early to decide whether this application of the extended Hückel method gives useful results. The effects of changing the parameters used in the calculation on the results obtained have not yet been examined. In this respect we recall Mango and Schachtschneider's cautionary remarks for the iterative SCCC–EHT quoted above, which must apply a fortiori to the simple EHT.

4.3. Spectroscopy of Surface Molecules and Localized Orbitals at Surfaces

The concept that a chemisorbed species involves localized covalent bonding to atoms in a surface was present in the earliest models of chemisorption. The surface bond(s) formed upon chemisorption have long been thought to be analogous to bonds in chemical compounds.

It is only within the past 20 years or so that spectroscopic methods for probing surfaces have been generally available for detailed characterization of bonding with the chemisorbed layer. The first of these techniques was infrared spectroscopy, followed recently by electron–diffraction, ion scattering, and various electron spectroscopy methods.[164,165] Within the past ten years, theoretical models involving the concept of a local surface molecule have begun to prove useful in calculations relating to chemisorptive bonding, as has been discussed previously.

Infrared spectroscopy was first used to characterize chemisorbed CO species in 1954 by Eischens *et al.*,[166] although the method was used by Terenin[167] in the early 1940's to observe hydroxyl groups present on the surface of oxides. Eischens[166] applied transmission infrared spectroscopy to supported catalysts containing metallic Cu, Ni, Pt, and Pd. The C–O stretching vibration for chemisorbed CO was observed by passing infrared radiation through the supported catalyst samples. Following this pioneering work, the transmission infrared technique has been widely applied to surfaces, and many interesting results have been reviewed in two books.[168,169] Infrared spectroscopy is currently used widely for studies of catalytic materials.

In general, it has been found that the principles useful in interpreting the infrared spectra of chemical compounds are applicable to species on surfaces also. Thus, the concept of group vibrational frequencies, where particular structural groups of atoms are associated with characteristic vibrational frequencies, applies well on surfaces just as it does in molecules. Minor variations in the group vibrational frequency associated with a particular group of atoms may often be attributed to interactions either within the molecule, or between adsorbed molecules, or between the molecule and the surface. Because of the limited spectral range and the low absolute intensities often available in infrared studies on supported surfaces, it is usually not possible to observe a complete infrared spectrum of the chemisorbed species, so that the determination of force constants and molecular

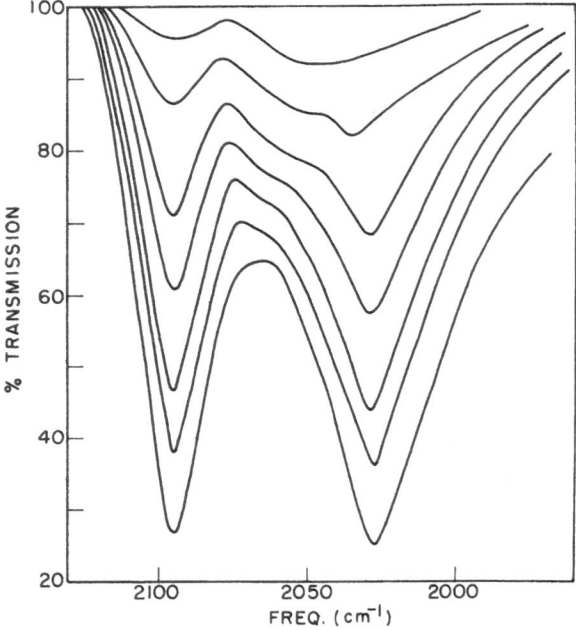

Fig. 13. Infrared spectrum of chemisorbed CO on an unsintered Rh catalyst (2%) supported on Al_2O_3. The development of the carbonyl absorption bands corresponds to increasing the equilibration pressure and surface coverage of CO. Curves are for 1.3×10^{-3} Torr (top curve) to 25.6×10^{-3} Torr (bottom curve) at 300°K. (From Yang and Garland.[170])

geometry of the adsorbed species cannot usually be done as with the more complete spectra of molecules. Thus, arguments about the structure of surface species are often based upon spectral analogies with compounds of known structure. An excellent example of this method, illustrated below, is the comparison of the infrared spectrum of CO chemisorbed on Rh with the spectrum of a Rh carbonyl complex.[170]

In Figure 13, a series of infrared spectra corresponding to the C–O vibrations for increasing CO coverage on highly dispersed Rh is shown. It is seen that two major absorption bands develop together at 2095 and 2027 cm^{-1}. The two major bands are thought to be due to symmetric and antisymmetric C–O stretching vibrations, which could arise through coupling in an adsorbed species, such as

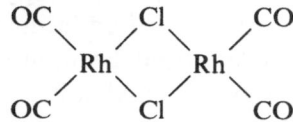

A *single* C–O stretching vibration would be expected to occur at an intermediate frequency for the adsorption of a *single* CO by Rh, and such a species may be responsible for the smaller 2045 cm^{-1} band observed. A comparison with the infrared spectrum (Figure 14) of the Rh$_2$(CO)$_4$Cl$_2$ molecule of known structure,[171] namely

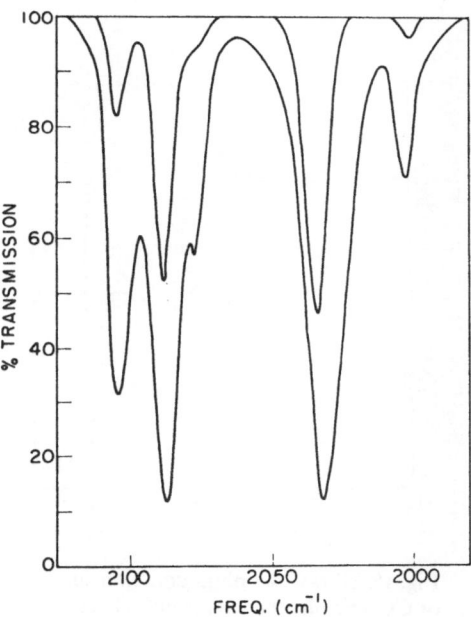

Fig. 14. Infrared spectrum of Rh$_2$(CO)$_4$Cl$_2$ in the carbonyl stretch region. Upper curve: CCl$_4$ solvent. Lower curve: Nujol mull. The strong similarity between these spectra and those of chemisorbed CO in Rh (Figure 13) suggests that two CO molecules chemisorb per Rh atom. (From Yang and Garland.[170])

41

demonstrates the strong correspondence between the spectra and gives confirmation to the assignment of a double-CO surface species. In the carbonyl spectrum, symmetric and antisymmetric coupling between CO ligands gives two strong bands at 2033 and 2088 cm^{-1}, accompanied by weaker overtones at 2022 and 2104 cm^{-1}. This example shows the usefulness of structural analogies between surface species and chemical compounds. The analogy between chemisorbed CO and metal carbonyls has been extended[172] to include intensity comparisons as well as small spectral shifts due to local electron donation effects in the vicinity of the adsorbed CO species. Recently, infrared spectra of CO adsorbed on atomically clean metal surfaces have been obtained by reflection spectroscopy[173] and compared to spectra on supported metals.[174] In addition to observations of the C–O stretching vibration, it would be useful to observe the lower frequency M–C vibration also, and this has been done by Blyholder,[175] who suspended the metal adsorbent particles in a transparent oil. Measurements in an oil suspension are likely to be influenced by

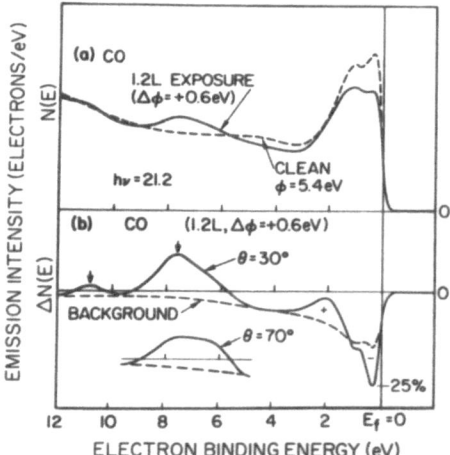

Fig. 15. Ultraviolet photoemission spectra of CO chemisorbed on a Ni(111) crystal. (a) Spectrum of clean surface (dashed curve) compared to spectrum of surface exposed to 1.2 Langmuirs of CO at 300°K. (b) Difference spectra at two angles of incidence of photon beam showing extra features assigned to various molecular orbitals. (From Eastman and Demuth.[176])

contamination effects, although the adsorption of CO may displace some contaminant species.

The character of a surface molecule may be further defined by studies of the distribution of electronic states associated with the chemisorbed species. These measurements are now being made by means of photoelectron spectroscopy using monochromatic ultraviolet excitation. Photoelectrons emitted from the surface containing a chemisorbed layer are energy-analyzed and compared with the photoelectron spectrum from the clean surface. Typical data[176] for the adsorption of CO by Ni(111) are shown in Figure 15, where 21.2-eV ultraviolet radiation has been used for excitation. It is seen that CO adsorption leads to attenuation of emission from the Ni d band, 0 to ~ 3.5 eV below the Fermi edge E_F, and to new photoemission features at about -7.5 and -10.7 eV. The new features increase in intensity as CO coverage is increased, and are characteristic of the local surface molecule. There is some controversy about the assignment of the spectral features to particular molecular orbitals.[145,146,164,176,177] The assignments based on the SCF-Xα-SW calculations of Batra and Bagus[146] discussed in Section 4.1 were given in Figure 8. Calculations have also been done by Blyholder,[177] using the CNDO (complete neglect of differential overlap) method. This calculation also suggests that the -7.5-eV feature is caused by the overlap of the 5σ and 1π levels. The -10.7-eV feature is assigned to the 4σ level. A similar assignment has been made by Eastman *et al.*[176,178] and by Fuggle *et al.*[179] In view of the different assignments by the theoreticians and the difficulty in making assignments by analogy with gas-phase spectra, the question is still open as to the order of the 1π and 5σ levels. The experimental data[176,178] and Figure 15 further indicate a lowering of the intensity of the emission from the Ni $3d$ states near the Fermi level and an increase near -1.5 eV (or, equivalently, a shift of the d states to higher binding energy). This is indicative of the role of the d orbitals in the binding and is a generally observed phenomenon on d metals.[164,176,180]

More recently, Demuth and Eastman[49,176] have applied ultraviolet photoelectron spectroscopy to the study of hydrocarbon adsorption and subsequent catalytic decomposition. In their studies of benzene chemisorption by Ni(111) the π level of benzene is preserved by interaction with the surface, but is shifted to higher binding energy. In the case of the chemisorption of ethylene by Ni(111), it was found that at low temperatures ($\sim 100°$K) the chemisorbed ethylene yields a spectrum similar to that of gaseous ethylene

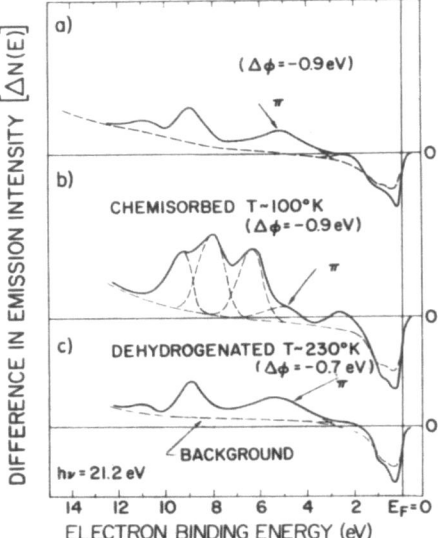

Fig. 16. Ultraviolet photoemission spectra for hydrocarbons adsorbed on Ni(111). (a) Acetylene chemisorption at ~100 or 300°K. (b) Ethylene adsorption at ~100°K. (c) Conversion of chemisorbed ethylene to chemisorbed acetylene achieved by heating (b) to ~230°K. Note the similarity between the spectra of chemisorbed acetylene and ethylene following dehydrogenation by heating. (From Demuth and Eastman.[49])

except for a shift of the π level to higher binding energy, as was seen for chemisorbed benzene. However, upon heating chemisorbed C_2H_4 to about 230°K, dehydrogenation is observed, leaving behind an acetylenic species whose photoemission spectrum is virtually identical to the spectrum found for chemisorbed C_2H_2. This result is shown in Figure 16 and demonstrates the usefulness of photoelectron spectroscopy in understanding catalytic reactions.

4.4. Long-Range Electron Interactions in the Surface Region

The models involving localized surface complexes suggest methods of treating chemical reactions at surfaces in terms of molecular aggregates and therefore do not logically include the possibility of longer range interactions between adsorbed species. One of the major experimental discoveries about adsorbed layers in the last 15 years is the existence of long-range adsorbate ordering effects on single crystals containing less than a monolayer of adsorbed species.[181] The periodicity of the ordered overlayer is often two, three, or four substrate lattice spacings in a particular direction. These large unit cells may also be due in some cases to short-range interactions, e.g., size mismatch may give rise to coincidence lattices.[181] Coulomb repulsion of charged adsorbates also gives rise to large spacings. For instance, alkali metal adsorbates on Ni or W form, at low coverages, hexagonal arrays with a spacing which varies continuously with increasing coverage.[182] The large unit cells found on clean surfaces of Si and Ge have been explained by a variety of short-range interactions, such as the pairing of neighboring "dangling bonds" or the presence of surface vacancies combined with the formation of buckled "aromatic-like" rings of surface atoms.[183,184]

Long-range effects (such as a charge density wave caused by an instability in the Fermi surface) have been suggested as explanation for the large unit cells of Si and Ge surfaces.[185,186] Prior to the experimental observations of long-range adsorbate ordering, Koutecky[187] and Grimley[188–190] suggested that adsorbate–adsorbate interaction through the substrate might be important in chemisorption. As schematically indicated in Figure 17, the ground-state wave functions for electrons associated with two separated atoms in vacuum decrease quickly toward zero in the space between the nuclei;

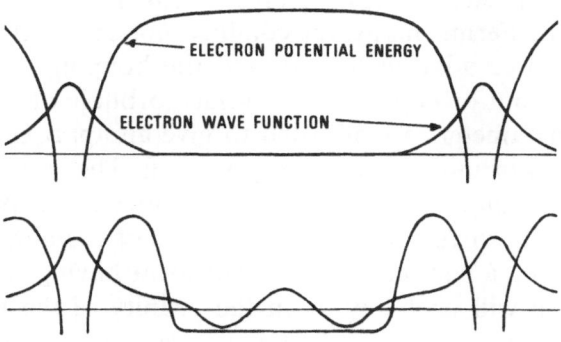

Fig. 17. Schematic illustration of the interaction between atoms. Upper figure: isolated atoms in vacuum, which undergo no interaction. Lower figure: atoms that interact indirectly when placed apart on a metal surface. (From Grimley.[188])

if, however, the nuclei are placed in contact with a metal surface, it is now possible for the electrons to move through the metal between the nuclei because of the lowered potential in the metal. An oscillatory wave function exists along the metal surface, giving rise to an indirect interaction between adsorbate atoms as shown in Figure 17. Grimley[188] first calculated the magnitude of this effect in the asymptotic limit of large spacing between adsorbate atoms on a structureless surface. Representing an adsorbate molecule by a broadened virtual level, he showed that interaction between two molecules will split these levels. Depending on the shape of the resulting double hump in the density of states, the electron occupancy may be either enhanced or decreased, provided that the Fermi level is within the energy range of the hump. For this to be the case, it is generally important that the virtual level is broad compared to its distance from the Fermi level.

Thus, the adsorption of alkali metals would be expected to undergo such indirect interactions because of their low ionization energies, whereas chemisorbed CO, with its highest occupied molecular orbital far below E_F, would not be expected to undergo indirect interaction in this model. However, H(ads) would be expected to undergo such interactions because of an extreme rise of its $1s$ atomic level to the vicinity of E_F due to the large Coulomb integral for this system.

Einstein and Schrieffer,[191] extending the ideas of Grimley, have calculated the indirect interaction for the (100) surface of a simple s-band solid. They succeed in explaining the frequently occurring (2×2), (2×4), etc., superstructures. Their model involves four parameters, the Fermi energy, the conduction bandwidth, the atomic energy of the free adatom orbital, and the hopping matrix, which mixes the adatom orbital with the surface orbitals of the substrate atoms. The parameters are adjusted to give an approximate chemisorption energy in agreement with experiments. The indirect adatom–adatom interaction is then calculated as a function of spacing. They conclude that the virtual level approximation of Grimley is not valid in general, since a surface complex will form having new bonding orbitals, which will probably lie in the vicinity of the Fermi level. However, in agreement with Grimley's general conclusions, it is found that the indirect interaction varies from repulsive to attractive as the spacing is changed between adatoms. Reasonable parameter choices [to simulate H(ads) + W(100)] suggest that the $c(2 \times 2)$

lattice structure should be stable and that the disordering temperature observed by LEED for this open surface structure is consistent with the interaction energy calculated. For the pair interaction, it is found that the indirect interaction energy is of the order of 5–10% of the chemisorption binding energy and decreases by a factor of 5–10 per lattice spacing. Furthermore, multibody forces are small compared to pair interactions even at high coverage.

These new concepts related to adsorption are of importance to heterogeneous catalysis in a general way. For instance, the displacement of one adsorbed species by another may be understood more easily if such adsorbate–adsorbate interactions are involved. Such displacement processes could alter the course of a catalytic reaction by changing the nature of the adsorbed layer at steady state. One experimental study using thermal desorption methods has shown that adsorbed CO significantly weakens the strength of adsorption of adsorbed H on the W(100) surface, and that the sign of the surface dipole changes during this interaction.[192] Similar work function results have been obtained on Ni films when H(ads) and CO(ads) interact.[193] It is possible that catalytic poisoning or catalytic activity enhancement by promoters may conceivably involve electronic effects similar to those discussed above. However, Schrieffer and Soven[136] stress that more realistic models must be studied before detailed comparison of the theory of indirect interactions with experiment will be meaningful.

5. Electronic Effects in Catalysis

In this section, we explore the influence of the electronic structure of the catalyst on the various steps in the catalytic process: chemisorption, nature of catalytic intermediate, rate of reaction (catalytic activity). In keeping with common usage, we distinguish between the "electronic factor" in catalysis and other electronic effects more specific to the catalyst surface. As discussed previously, the term "electronic factor" is used by most catalytic chemists to describe attempts to correlate catalytic activity with the bulk electronic properties of the catalyst. Such correlations will be discussed; in addition, present thinking regarding catalytic processes and the molecular orbital character of the catalyst surface will also be examined.

5.1. The "Electronic Factor" in Catalysis by Pure Metals

Atoms at the surface of a metal possess bonding capabilities not exhibited by either free atoms or atoms in the bulk of the solid. There exist at the surface of a solid free orbitals which may be available for bonding to an adsorbate; such orbitals are usually saturated in the bulk and may not be present at all in the free atom. Despite the clear uniqueness and complexity of the electronic character of the surface, it has frequently been assumed that there is some correspondence between the properties of bulk and surface atoms. It has been hoped that this correspondence will be reflected in a correlation between chemisorption behavior and bulk electronic properties, and ultimately, to correlation between catalytic activity and bulk electronic properties.[6]

In support of these ideas, there are distinct trends which have long been known from studies of chemisorption. Whereas oxygen is almost universally adsorbed on clean metal surfaces, certain metals are very specific to hydrogen or nitrogen chemisorption[194] (see Table 1). In general, heats of adsorption for molecules such as H_2 and N_2 are low for the noble metals in group VIII (Rh, Pd, Ir, Pt, etc.), and high for group VA and VIA metals (Ta, Nb, W, Mo, etc.). However, when one attempts to establish a *quantitative* correlation between the heats of adsorption and some electronic property of the solid, problems arise. Thus, the first problem to be faced in seeking such a correlation is: Is there an electronic property of a metal catalyst which *quantitatively* relates to its ability to chemisorb and participate in catalytic reactions? The two theories of the metallic state which have been considered are the band theory of solids and Pauling's valence bond theory.[13] It is clear that recent advances in band theory based on relativistic density-of-states calculations provide a fundamental view of the electronic structure of transition metals and alloys. Pauling's valence bond theory is an empirical formalism which ascribes metallic bonding in transition metals to *dsp*-hybridized orbitals, and assigns the metallic and magnetic properties to electrons in atomic *d* orbitals. The valence bond approach has often been the basis for correlations involving the "electronic factor" in catalysis[6]; the primary reason for this is the fact that a quantitative percentage *d* character can be assigned to the metallic bond in transition metals. Band theory has not yielded a simple quantity which by itself characterizes the bonding of the solid, and has not been as extensively applied to the studies of the electronic factor. There is, of course, a conceptual correspondence between percentage d character and the existence of unfilled d bands in the band theory picture.

Attempts at correlations in the past have frequently been confused by inadequate and discordant results as well as the inadequate state of theory.[195] Recent attempts to relate catalytic activity to properties of the bulk solid have been discussed by Sinfelt[14] and by Anderson and Baker.[196] Sinfelt has systematically examined the specific catalytic activity of a series of metals for ethane hydrogenolysis, i.e.,

$$C_2H_6 + H_2 \rightarrow 2CH_4$$

A comparison of supported metals as hydrogenolysis catalysts reveals a dramatic variation in activity, even among similar metals. Figure 18 shows the specific activity of the group VIIA and group VIII metals for this reaction. The figure has three separate parts, representing transition metals of the first, second, and third transition series. The catalysts used in these experiments were silica-supported metals; specific activities were all relative reaction rates per unit surface area of metal at 205°C, and ethane and hydrogen partial pressures were 0.030 and 0.20 atm, respectively. Also shown in Figure 18 are plots of the percentage d character of the metallic bond, based on Pauling's valence bond theory.

Among the interesting observations which have been drawn from these data are the following.

(1) The maximum activity for ethane hydrogenolysis in both the second and third transition series is observed for the group VIII_1 metals, Ru and Os.

(2) The percentage d character correlates moderately well with catalytic activity in the second and third transition series.

(3) There is little correspondence between specific catalytic activity and percentage d character when one compares different series of the periodic table, e.g., the percentage d character of Ni is slightly less than that of Pt, but Ni is about 10^6 times more active than Pt. Sinfelt[14] has concluded that percentage d character is *not* an adequate parameter for characterization of the catalytic activity of transition metals for hydrogenolysis.

Sinfelt's specific observations for this reaction support a conclusion which has long been recognized, namely that a simple unifying theme for catalysis based on the "electronic factor" is probably illusory. Significantly, minute changes in percentage d character would have to be responsible for huge changes in catalytic activity. Still, it is anticipated that the observation of trends such as those

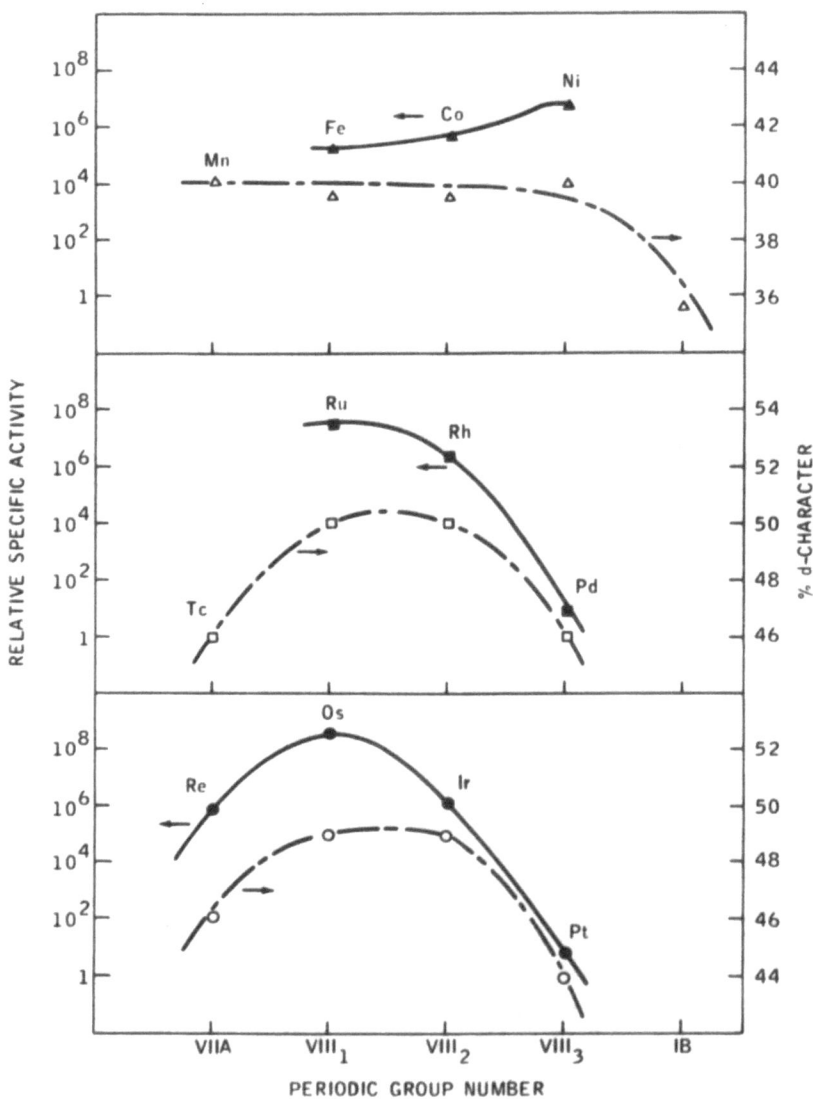

Fig. 18. Relationship between catalytic activity for ethane hydrogenolysis $(H_2 + C_2H_6 \rightarrow 2CH_4)$ and the percentage d character of the metallic bond of the bulk solid catalyst, based on Pauling's valence bond theory. Three separate panels are shown in the figure to distinguish the metals in the different long rows of the periodic table. (From Sinfelt.[14])

shown in Figure 18 will continue to spur investigators to seek a key relating catalysis to solid-state physics and chemistry.

5.2. The "Electronic Factor" in Catalysis by Alloys

5.2.1. Relationship between Bulk and Surface Properties

Metal alloys have been investigated by a number of workers attempting to study the electronic factor in catalysis.[197,198] The existence of qualitative correlations between catalytic activity and partially filled *d* bands in transition metals has led to the expectation that catalytic activity of alloys would change abruptly once these vacancies were filled by alloying with an appropriate metal. These expectations were based on the rigid band theory of alloys,[199] which predicted that the *s* electrons contributed by the addition of IB metals to the metals of group VIII would fill the unoccupied *d* states on the group VIII metal. Examples of catalytic behavior in line with the rigid band model have been reported in several specific cases for alloy systems such as Pd–Au.[200] However, this close correlation between catalytic data and the rigid band theory proved fortuitous. Photoemission data on the electronic structure of Cu–Ni and Pd–Ag alloys have disproved the validity of the rigid band model[201] and in fact, have shown that the Anderson virtual bound state model is appropriate.[201,202] These data are shown in Figure 19. In addition, many more recent experimental observations have revealed that a simple band-theory picture of catalytic activity is inadequate:

(1) The catalytic activity of alloys depends very much on mode of preparation of the alloy (thin films, powders, wires, and ribbons).

(2) In general, there are major differences between bulk and surface composition of alloys. A further complication is the fact that the surface composition may change during the course of a catalytic reaction.

The activity of alloys formed from the catalytically different group VIII and group IB metals (Ni–Cu, Pd–Ag, Pd–Au) have been studied most extensively. The Pd–Au and Pd–Ag systems form a continuous series of fcc solid solutions over the entire composition range, i.e., they appear to have complete miscibility.[197] In contrast, alloying of Ni with Cu is an endothermic process which results in (a) the formation of single-phase solid solutions at high temperatures, and (b) phase separation due to a miscibility gap at low temperatures.[198] The influence of the expected phase separation on the surface composition of Ni–Cu alloys in the temperature range of

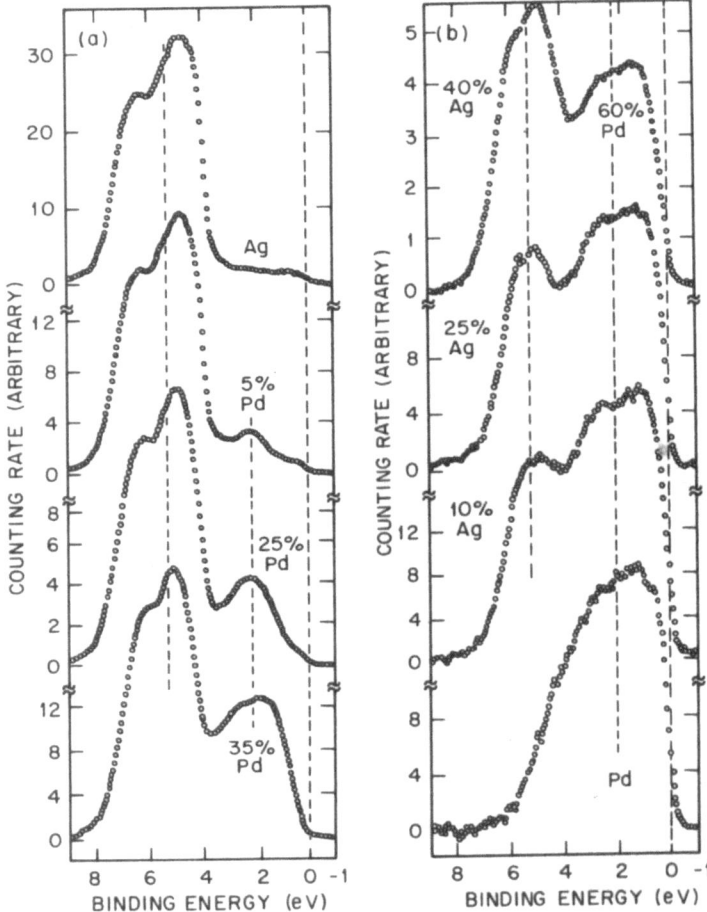

Fig. 19. Valence-band X-ray photoelectron spectra of AgPd alloys obtained with monochromatized X-rays. The alloy compositions refer to atomic fractions. Dashed vertical lines are drawn at the Fermi level and the approximate center of the *d* band for pure Ag and pure Pd. (From Hufner *et al.*[201])

catalytic interest ($\lesssim 400°C$) is an additional complication in the interpretation of catalytic data.

Sachtler and co-workers[203,204] first demontrated the difference between bulk and surface composition of Ni–Cu alloy films prepared by successive evaporation of the component metals under ultrahigh vacuum conditions. They predicted on the basis of a thermodynamic analysis[203] that alloy films sintered at 200°C should consist in two phases at equilibrium; Phase I contains 80% Cu and phase II

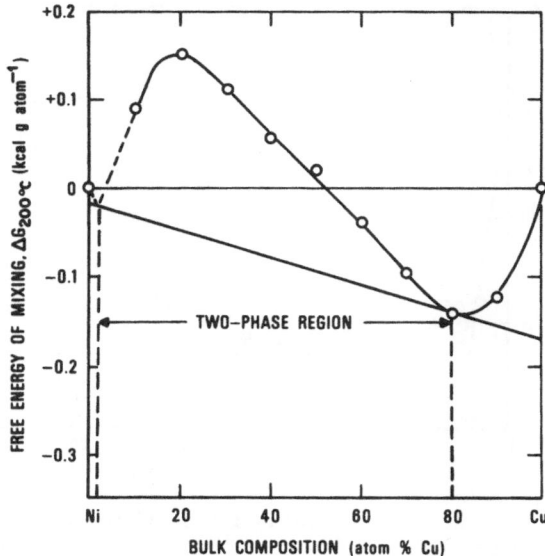

Fig. 20. Calculated values of the free energy of mixing at 200°C as a function of bulk composition for copper–nickel alloys. The minima at bulk Cu concentrations of $\sim 80\%$ and $\sim 2\%$ define the range of a miscibility gap; at equilibrium, all alloys having bulk compositions in this range must consist of two phases. (From Sachtler and Jongepier.[203])

contains 2% Cu. They reasoned that alloys within the miscibility gap (cf. Figure 20) had a constant surface composition (80% Cu–20% Ni), independent of nominal bulk composition. They employed a variety of methods (photoemission work function measurements, chemisorption of hydrogen, catalytic hydrogenation of benzene), which yielded results in support of this point of view. Figure 21 is a plot of the rate of benzene hydrogenation[204] at 150°C over Cu–Ni alloy films as a function of nominal bulk composition. The data indicate a nearly constant hydrogenation rate over a wide range of bulk compositions, as would be expected if the surface composition is invariant over this range. At Cu concentrations of $<2\%$ the Cu-rich phase is absent, the surface becomes Ni-rich, and the rate of benzene hydrogenation increases correspondingly.

The surface composition of alloys have been determined more directly using Auger electron spectroscopy (AES). This technique is particularly useful for studies of surfaces because it is sensitive to

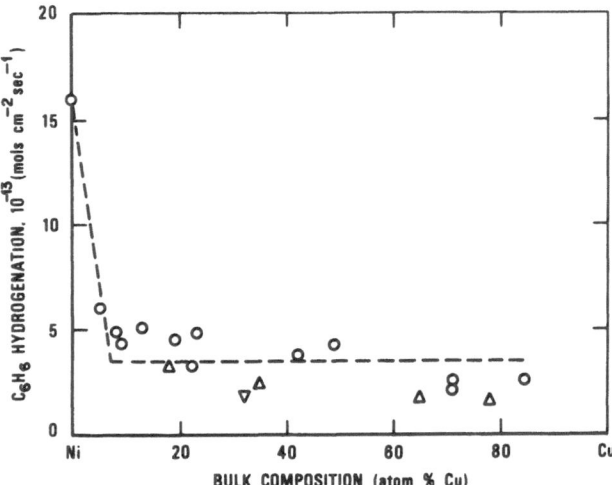

Fig. 21. Rate of benzene hydrogenation to form cyclohexane as a function of the bulk composition of Cu–Ni alloy films. The constancy of reaction rate over a wide range of bulk composition is interpreted as evidence for the near-invariance of the *surface* composition over this range. (From Van der Plank and Sachtler.[204])

the topmost atomic layers of the solid, to a depth of 5–20 Å beneath the surface. Ono and co-workers have applied AES to the study of Cu–Ni alloys,[205,206] and found that the composition of the surface is a sensitive function of the surface treatment. Argon-ion bombardment yielded a Ni-rich surface due to preferential sputtering of Cu; thermal annealing and oxidation–reduction treatments resulted in a Cu-rich surface, in qualitative agreement with the predictions of Sachtler.[203,204] Helms[207] has confirmed these resultes for a 50% Cu–50% Ni alloy by resolving Auger electrons associated with low-energy Ni and Cu transitions (100 and 106 eV, respectively); at these energies, the mean electron escape depth is only ~ 5 Å, so that the near surface contribution is emphasized. His data are summarized in Figure 22, in which the surface enrichment of Cu in the thermally annealed surface is clearly evident.

Yu *et al.*[208] have examined the surfaces of Cu–Ni alloys using ultraviolet photoemission techniques, and found that the surface electronic structure changes significantly from the bulk electronic structure on the annealed alloy surface, and correlates with the surface composition as measured by AES.

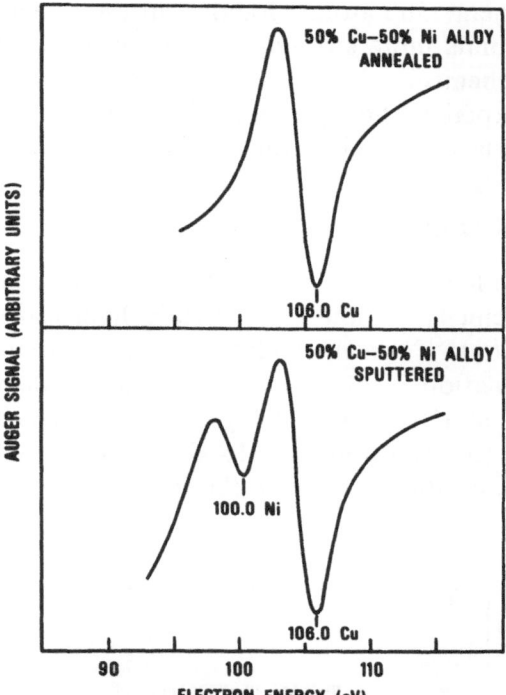

Fig. 22. Auger electron spectra of the surface of a 50% Cu–50% Ni alloy sample. The feature at 100.0 eV indicates that Ni is present along with Cu on the surface after sputtering (ion bombardment) with 300-eV argon ions. The Ni feature disappears after annealing in vacuum at 600°C, demonstrating that copper segregates to the surface under these conditions. (From Helms.[207])

A theory which allows predictions of the surface composition of binary alloys based on bulk thermodynamic data has been developed by Williams and Nason.[209] The model is developed using a pairwise bonding approach, with a broken bond surface. The alloy free energy is minimized by allowing exchange of atoms between surface layers and the bulk. The surface atom fraction is different from the bulk atom fraction for all compositions. In general, it is found that the alloy element with the lower heat of vaporization is enriched with respect to its bulk atom fraction. This conclusion is in agreement with the recent calculations of Van Santen and Sachtler.[210] The model of Williams and Nason[209] also predicts the effect of chemisorption of a foreign component. The atomic fraction in the surface of the alloy element that bonds more strongly to the adsorbing species should increase after chemisorption and surface equilibration. The model qualitatively explains the behavior of clean, thermally annealed Ni–Cu alloy surfaces, as well as the influence of chemisorption.

5.2.2. Catalysis by Alloys

A remarkable aspect of catalysis by alloy surfaces is the strong element of specificity exhibited for certain hydrocarbon reactions. Sinfelt's recent data[14] on the hydrogenolysis of ethane to methane and dehydrogenation of cyclohexane to benzene demonstrate this clearly, and his data are shown in Figure 23. The data are plotted as specific activity (reaction rates at 316°C) as a function of *bulk* alloy composition. The catalysts were in the form of finely divided metal

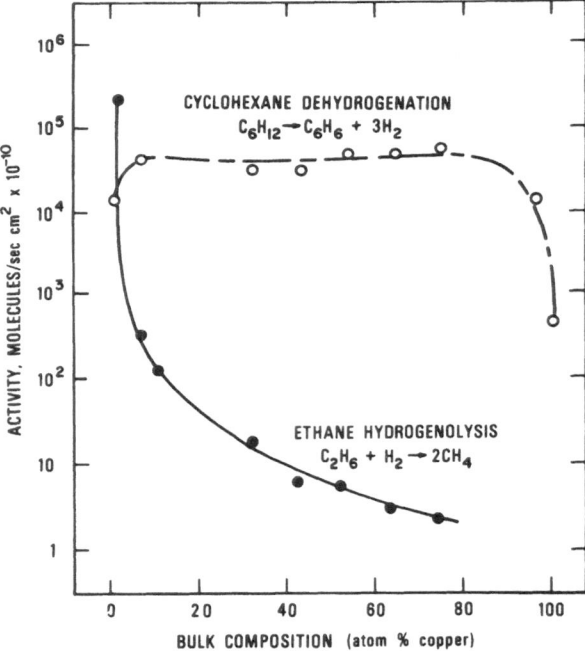

Fig. 23. Variation in the activity of copper–nickel alloy catalysts for different hydrocarbon reactions. Activities for cyclohexane dehydrogenation and ethane hydrogenolysis are plotted as a function of the bulk alloy composition. (From Sinfelt.[14])

granules having surface areas of 1–2 m²/g. The catalytic activity of Ni for ethane *hydrogenolysis decreases by orders of magnitude with the addition of a small fraction of Cu, but the rate of cyclohexane dehydrogenation is constant over a wide range of bulk composition.* Sinfelt observes that intermediates in hydrogenolysis reactions are generally assumed to be unsaturated hydrocarbon residues multiply bonded surface metal atoms, so that one would expect such reactions to be sensitive to surface structure and composition (geometric effect). In addition, the presence of Cu may affect the strength of bonding of the intermediates to the Ni surface atoms (electronic effect). In contrast, cyclohexane dehydrogenation is relatively insensitive to these same factors. It may also be argued that the nature of the reaction itself affects the surface composition (Cu/Ni ratio), as predicted by Williams and Nason[209] for chemisorption.

According to Soma–Noto and Sachtler,[211,212] one of the most spectacular discoveries in heterogeneous catalysis in the last decade is the marked effect of alloying on the catalytic *selectivity* of metal catalysts (i.e., the relative rates of parallel reaction paths for formation of different products from the same reactants). Even for a binary alloy AB where A is inactive for the reaction considered, selectivity can differ from that of the pure metal B. They ascribe this phenomenon to two possible causes:

(a) A geometric "ensemble effect." An adsorbing molecule can form bonds with one or with two or more surface atoms; during the course of further reaction, each of these different chemisorption complexes will result in a different product. Selectivities will change as the surface atoms are diluted by inert atoms simply because the relative concentrations of ensembles having a small number of atoms will increase upon alloying with an inactive metal.

(b) An electronic "ligand effect." The strength of the chemisorption bond between even a single metal atom and an adsorbed atom will depend upon the chemical nature of the neighbors of the metal atom. Changing the chemisorption bond by alloying will result in a change in selectivity.

To examine the relative contributions of the "ensemble effect" and "ligand effect" in alloy catalysis, Soma-Noto and Sachtler[211] investigated infrared spectra of carbon monoxide adsorbed on supported Pd–Ag alloys. CO is strongly chemisorbed only on the Pd atoms. The CO spectra yielded information regarding the relative concentrations of monoadsorbed (linearly bonded) CO and diadsorbed (bridge-bonded) CO; they also indicated the manner in which the infrared stretching frequency depends on the chemical nature of the metal atoms adjacent to the adsorbing Pd atom. The results indicated that the relative concentrations of linear to bridge-bonded CO changed with alloy composition, but the infrared frequencies remained almost constant. They concluded that for this system, the geometric "ensemble effect" is much more pronounced than the electronic "ligand effect." Further evidence for the dominance of the "ensemble effect" in adsorption of CO on Ni–Cu alloys was reported in another infrared study by Soma-Noto and Sachtler.[212] On the other hand, Williams and Boudart[213] have reported that oxygen does not adsorb on a Ni–Au alloy if the surface contains less than 30% Ni, suggesting that an electronic or "ligand effect" may be operative in this case.

5.2.3. "Bimetallic Cluster" Catalysts

Sinfelt[214] has examined the catalytic activity of highly dispersed bimetallic catalysts containing copper and ruthenium or osmium; the total metal concentration on the silica support was of the order of 1–2% by weight. Despite the low metal concentration and high dispersion, and despite the fact that copper is almost totally immiscible with Ru or Os in the bulk, the addition of the Cu drastically affected the specificity of Os and Ru for hydrogenolysis and dehydrogenation of cyclohexane. In view of the strong interaction between the metallic components (as demonstrated by the catalytic data), it was concluded that "bimetallic clusters" were formed on the support. The term "bimetallic clusters" rather than alloys was used because the metallic combinations employed do not form bulk alloys. The nature of these bimetallic clusters[215] has yet to be resolved completely (are the metals miscible in small particle form, or does one metal form a coating on the surface of the second?), but these investigations have opened up a new area of interest in catalysis by bimetallic systems.

5.3. Electronic Effects in Carbides

Levy and Boudart[216] have recently observed that tungsten carbide exhibits catalytic behavior which is typical of platinum but which is not characteristic of W. Specifically, WC catalyzes the formation of water from H_2 and O_2 at 300°K and the isomerization of 2,2-dimethylpropane to 2-methylbutane. They conclude that the surface electronic properties of W are modified by carbon in such a way that they resemble those of Pt. Bennett *et al.*[217] have recently studied the band structure of W, Pt, and WC using ESCA (X-ray photoelectron spectroscopy) and conclude that the valence band density of states at the Fermi level of WC is more like Pt than W. Houston *et al.*[218] have presented evidence to the contrary, so the matter is far from settled. Nonetheless, this situation illustrates nicely the trend of present-day thought in catalysis: Progress in understanding the role of electronic factors in catalytic processes will very likely come from a combination of the careful kinetic measurements of the catalytic chemists and the surface spectroscopic measurements of the surface chemists and physicists.

5.4. Electronic Effects in Catalysis by Semiconductors

As in other areas of catalysis, the views on the electronic factor in catalysis by semiconductors have gradually shifted from an emphasis on collective effects to an emphasis on localized electronic effects and point defects. However, the complexity of catalytic phenomena is such that the hope for a unified theory of catalysis on solid surfaces appears utopian. The application of modern electron spectroscopies to oxide surfaces is expected to provide a wealth of solid-state information with which catalytic properties may be correlated.[219]

5.4.1. The Collective Properties of the Semiconductor as a Factor in Catalysis

On the heels of spectacular developments in the theory and understanding of the nonmetallic solid state, attempts were made during the late 1940's and the 1950's to relate the catalytic activity of these solids to their collective electronic properties, namely ferromagnetism, ferroelectricity, conductivity type, the position of the Fermi level, and the band gap. Suggestive relations between chemisorption and conductivity had already been found by Wagner and Hauffe[24] and for some perovskite-type oxides, catalytic excursions near the ferroelectric[220] and ferromagnetic[221] Curie temperatures were found by Parravano.

The electronic theory of catalysis on semiconductors as developed by Hauffe,[21,22] Wolkenstein,[19,20,222] and others and subsequently elaborated by Garrett,[31] Lee,[223,224] and Krylov[225] starts from the simple idea of a chemisorption event coupled with

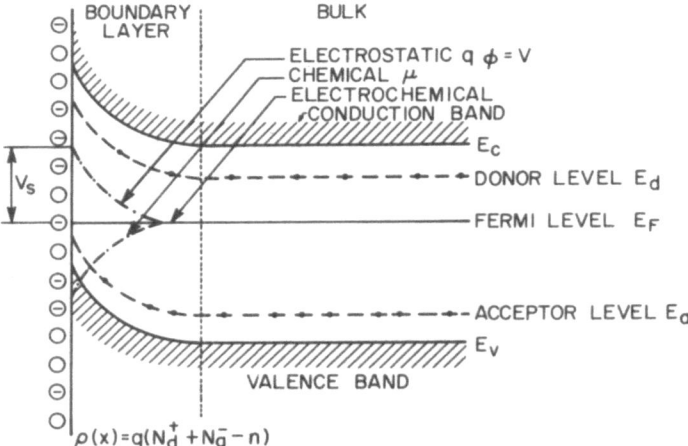

Fig. 24. Simple band picture for anionic chemisorption. ($\ominus$) Charged adparticles; ($\bigcirc$) uncharged adparticles; ($\bullet$) an electron occupying an impurity state. For electrons (negative charge) the electrostatic potential energy has opposite sign to the electrostatic potential. Band bending is due to electrostatic charge on the surface. Note partial ionizaton of adsorbed particles and impurities. (From Garcia-Moliner.[32])

transfer of an electron from an adsorbate to the solid or *vice versa*. If charge transfer leads to a change in the concentration of mobile carriers in the bulk, i.e., the adsorbed particle acts as a donor or acceptor, then the electrochemical potential of the carriers in the bulk will in turn have an effect on the equilibrium concentration of chemisorbed species. The same potential will then also affect the rate of a catalytic reaction in which these chemisorbed species might participate. The concentration of active species at the surface will be determined by the position of the electrochemical potential at the surface (in the absence of an external field, this is the Fermi level E_F) with respect to the positions of the donor or acceptor levels (induced surface states) associated with chemisorbed particles (Figure 24). The calculation of the concentration and degree of ionization of the chemisorbed species may be done in a self-consistent scheme similar to that used for bulk impurities using the Poisson equation.[32,222,226] Wolkenstein and Garcia-Moliner find that a Fermi-statistical treatment of the concentrations of ionized and neutral species adsorbed at the surface leads to a dependence on the Fermi level as shown in

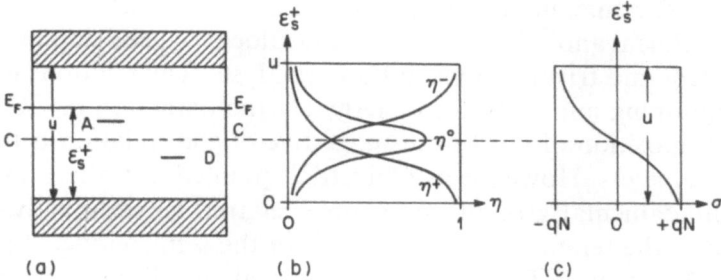

Fig. 25. The charge state of the adsorbate as a function of the position of the Fermi level: (a) The adsorbed molecule is characterized by one donor level D and one acceptor level A. (b) Relative abundance of adanions (η^-), adcations (η^+), and neutral adparticles (η°). (c) Average charge state. In (b) and (c) the vertical axis is the position of E_F. C is band center. (From Vol'kenshtein.[222])

Figure 25. This stage of the theory, which is an equilibrium theory in that it relates the catalytic activity to the equilibrium concentrations of electrons and holes, can be tested by the relation it predicts between changes in the Fermi level and changes in the catalytic activity, *provided the mechanism of the reaction has been established first.* The changes in catalytic activity are then expected to follow the concentration profiles sketched in Figure 25. Due to changes in the character of the rate-limiting step (e.g., from electron transfer to the surface to electron transfer to the adsorbate) as the Fermi level is changed, the correlation between E_F and the rate constant k may not be monotonic.[222]

The tests of this theory have involved the measurement of the rate of some test reaction as a function of the bulk doping of the catalyst, or as a function of the band gap of the catalyst. The latter method, which is described in detail by Krylov,[225] has been criticized, particularly on the basis that the induced surface states for a particular reaction on the various catalysts bear no simple relation to the band gap. Effects of the E_F on the rate may, in principle, be studied by changing E_F in one catalyst by illumination, by an external field, or by straining the catalyst sample. Unfortunately, side effects such as the creation of defects, trap centers, or dislocations and the polarization of adsorbed species in the external field make the interpretation of such experiments hazardous unless several of these techniques are employed concurrently. Apparently, this has not been done. Common are attempts to relate the reaction rate to the doping level of the bulk. This approach was used most extensively for the two best-studied semiconducting-catalyst systems, NiO and ZnO. For example, Parravano[28] and Schwab and Block[29] used doping with monovalent and trivalent ions in these catalysts. The relations found between doping and activation energy for the oxidation of CO and the decomposition of N_2O were suggestive of a correlation between these parameters. However, as Garrett[31] pointed out, there should be, on fundamental grounds, no simple relation between activation energy and the temperature dependence of the semiconductor parameters. Moreover, in Parravano's and Schwab and Block's measurements, the relation between the rate constant (as opposed to the activation energy) and the doping level is not at all clear-cut. In fact, the effects of doping level are most pronounced at high temperatures (200–400°C) where the simple electronic picture of the catalytic process is probably invalid for two reasons: The semiconductor is in the intrinsic range, with the Fermi level close to the middle of the gap, and the mechanism of the reaction is intrafacial, with participation of lattice oxygen. Gravelle *et al.*[227,228] have additionally argued that the oxidation of CO over doped and pure NiO follows two different mechanisms, through a CO_3^-(ads) species on NiO(Li) and through O(ads) on NiO and NiO(Ga). Deren *et al.*,[229] working with single-crystal NiO(Li) catalysts, have shown only minor effects of doping on the rate of CO oxidation at 250°C, and changes of less than a factor ten at 400°C, while, moreover, the changes were not monotonic with increasing Li content. In view of the complexities of the transport properties of NiO discussed by Adler,[230] including the partial compensation of Li doping with oxygen vacancies, it is not surprising that the large amount of work on NiO catalysis has not materially advanced the status of the electronic theory of catalysis.

Changing the scene to the catalytic properties of ZnO, for which catalyst a large body of data is also available, does not provide any more encouragement. The effects of doping are inconclusive for similar reasons as for NiO: variable mechanisms of the test reactions and small effects on the rate constant, while the only clear effects are on the activation energy. Parenthetically, it may be remarked here that two effects are likely to interfere in this type of measurement: the phase segregation of low-melting compounds at the surface and the variation of the oxidation state of the catalyst with temperature. The first effect may yield anomalously low rates,[231,85] while the second effect may yield apparent activation energies which are very dependent on the history of the sample and the experimental details.[85,232]

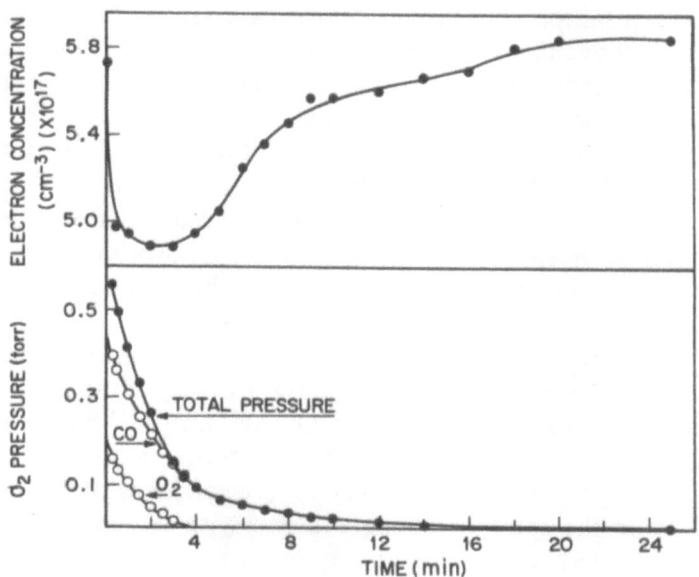

Fig. 26. Electron concentration as measured by the Hall effect and ambient gas composition change observed during carbon monoxide oxidation over an indium-doped ZnO sample at 350°C. (From Chon and Prater.[33])

The foregoing has stressed the difficulties which have arisen in experimental tests of the relation between semiconductor properties of catalysts and catalytic properties. Nevertheless, strong evidence of a semiquantitative nature is provided by the many measurements of the conductivity of catalyst samples during the catalytic process, and particularly by some remarkable experiments by Chon and Prater on ZnO during CO oxidation.[33] These authors measured the Hall voltage on a bulk polycrystalline sample during catalysis and thereby established the concurrent changes in the number of carriers and the changes in the reaction rate (Figure 26). This type of measurement eliminates the difficulties associated with interpretation of conductivity measurements, namely intergranular resistance and mobility changes, and establishes the correlation (if not the causal relation) between carrier density and activity.

Further progress in the electronic approach to the catalysis on semiconductors is likely to arise from a more complete picture of the transport properties of the semiconducting catalyst, preferably using single-crystal samples. The theory of the catalysis on semiconductors as outlined above is grossly oversimplified in the following respects:

(a) It is based on the band model, which may be less than appropriate for the relevant states of catalysts such as NiO.[230,233]

(b) It ignores pinning of the Fermi level by surface states. As pointed out by Wolkenstein,[20] following Bardeen,[234] a high density of surface states resulting from surface defects, intrinsic surface states, or surface states created by adsorption of reactants results in a Fermi level at the surface which is independent of the bulk Fermi level.

(c) It ignores trapping levels, which intervene in the transfer of carriers between the adsorbed species and the bulk bands.

Detailed studies of ZnO and TiO_2 by Gray[235] have shown a very large variety of trapping levels extended over a large fraction of the gap. Modern thinking is very much concerned with the effect of trap states on the catalytic properties and with the complexities introduced by changes in carrier lifetime, changes in ionization of donors, and interaction of surface and bulk trap states with the bulk bands.[235–238] The variation of the photoinduced catalytic activity

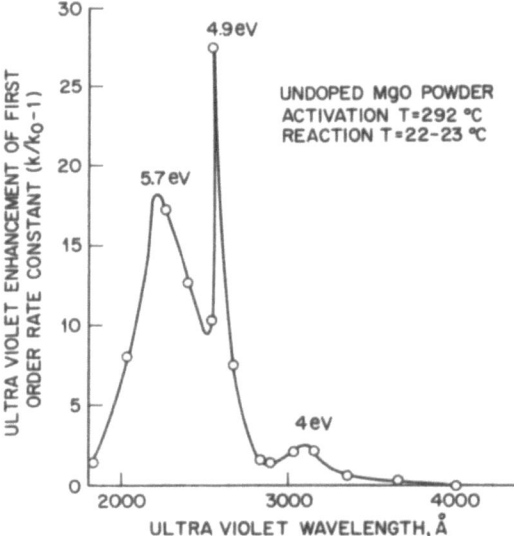

Fig. 27. Reaction rate spectrum for the H_2–D_2 exchange on irradiated MgO powder. Enhancement factor as a function of wavelength, showing the activity of various traps in the bandgap. k_0 is the rate constant in the absence of irradiation. (From Harkins *et al.*[113])

with photon energy is an example of the way these problems may be experimentally elucidated.[113] The results for MgO are shown in Figure 27. The emphasis on defects forms a natural bridge with the work of Kokes *et al.*,[239] which is almost exclusively focused on the defect structure of the surface (of ZnO catalysts) rather than on the semiconducting properties.

In the above discussion of the electronic theory of catalysis on semiconductors, the chemisorption of gas molecules was taken to involve mobile carriers, which were released again (or taken up again) during the further course of the surface reaction or during desorption of the product. In contrast to this equilibrium approach, there have been attempts to correlate catalytic reactivity with the nonequilibrium electronic processes. Wolkenstein[222] has discussed the possibility that excitons diffusing to the surface might provide the energy to surmount the activation barrier for a reaction. In the case of electron–hole pairs (Mott excitons), the measured activation energy would then be diminished by the energy of the band gap. A more specific discussion limited to the annihilation of electron–hole pairs during the surface reaction has been given by Lee.[223] Consider the reactions

$$A + n \leftrightarrows A^-$$

$$D + p \leftrightarrows D^+$$

$$A^- + D^+ \rightarrow AD$$

in which A and D are the reaction partners and n and p are the carriers. The energy required to create a pair of carriers, i.e., $(E_c - E_v)$, will contribute to the measured activation energy if the rate-determining step is the creation of these pairs. For an intrinsic semiconductor, a correlation of linearly increasing activation energy with increasing band gap is then expected. Although a correlation has indeed been observed (between the activity and the band gap) for an important number of catalysts (Figure 28), such relations must have a somewhat more complicated cause, since in general the pre-exponential factor also changes appreciably.[225] Finally, it has to be stressed that in the presence of large numbers of donor and acceptor impurities, no relation with the band gap is expected.

Collective properties are of course not limited to transport properties. There is no room at this point to discuss collective properties arising from spin or dipole ordering,[122,220,221,240,241] but some attention is warranted for the relation between catalytic

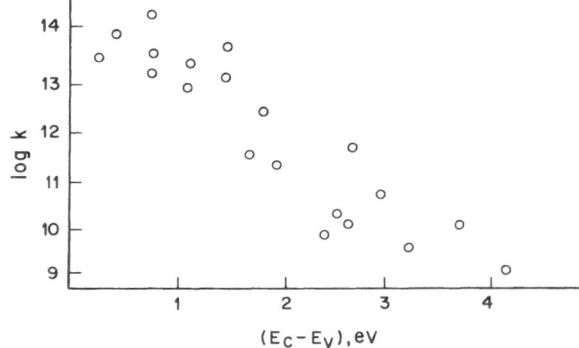

Fig. 28. The dependence of the rate constant k for dehydrogenation of isopropanol at 200°C on the width of the bandgap of III–V, IV–IV, II–VI, and I–VII semiconductors. (After Krylov.[225])

activity and thermodynamic functions of the catalyst. This relation was established for the decomposition of formic acid on metals by the work of Fahrenfort *et al.* discussed in Section 8.3. Following this, correlations have been sought and found between the averaged heat of oxide catalysts and their activity for oxidation reactions. The important parameter is expected to be the metal–oxygen bond strength in the adsorption complex, taken to include the atoms making up the environment of the active site. Instead, however, the average bond energy is used, since this is more easily available from tabulated data. The correlations found for the oxidation of propylene over a series of metal oxides[242,243] are quite satisfactory (see Figure 29) and can be understood on the basis of the intrafacial mechanism of the reaction at high temperatures, with lattice oxygen participating in the process. Refining this concept to include the selectivity in addition to the activity, Sachtler *et al.*[244] have correlated these catalytic parameters with the differential enthalpy and free energy of lattice oxygen in the oxidation of benzaldehyde to benzoic acid over MnO_2, V_2O_5, and V_2O_5–SnO_2 catalysts. However, a localized bond model requires correlation with the metal–oxygen bond in the active complex and such a correlation was recently established for the selective reduction of NO to dinitrogen products over perovskite-like oxide catalysts[245] (Table 2).

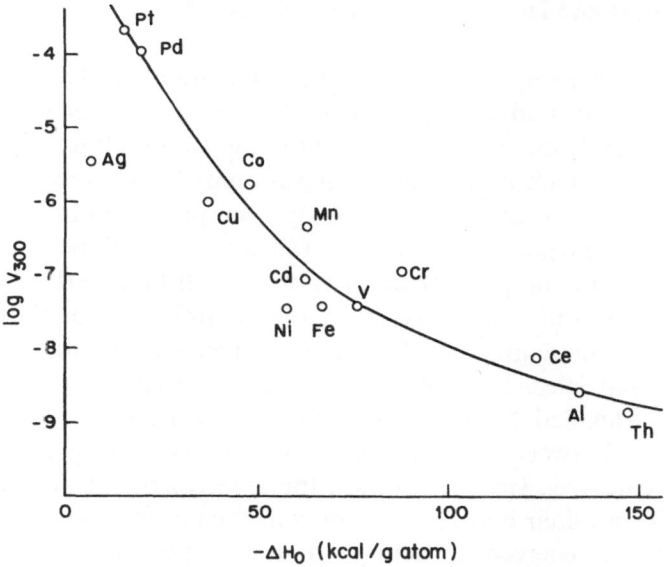

Fig. 29. The dependence of the rate of the oxidation of propylene at 300°C (V_{300}) in an excess of oxygen on the enthalpy of formation per oxygen atom (ΔH_0) of the oxide catalyst. Data for BaO, MgO, and ZnO omitted because of immeasurably low V_{300} values. (From Morooka and Ozaki.[242])

TABLE 2

Correlation between Binding Energy of Surface Oxygen and $N_2 + N_2O$ Yield

Ranking in order of decreasing $N_2 + N_2O$ yield	$\Delta(A-O) + \Delta(A'-O)$,* kcal/mole
Bi, K	-25
La, ☐	-34
La, Rb	-41
La, K	-42
La, Na	-42
La, Pb	-47
La, Sr	-54

* Calculation of $\Delta(A-O)$ from $\Delta(A-O) = (\Delta H_f - m\,\Delta H_s - \frac{1}{2}nD_0)/12m$, where ΔH_f, ΔH_s, and D_0 are, respectively, the enthalpy of formation of one A_mO_n, the enthalpy of sublimation of A, and the dissociation energy of O_2.

5.4.2. Local and Defect Properties of the Solid

It has been emphasized above that the influence of defects on the semiconductive and catalytic properties of solids has received increasing attention. In agreement with this trend, efforts to identify the structure of active sites are of paramount importance in modern catalysis. Some examples have already been presented in Section 3.4. In Section 6, the geometric aspects of surface sites will be discussed in some detail. In the present discussion we will be mostly concerned with the electronic nature of the active complex, when the latter is taken to be the aggregate of the site, its nearest neighbors, and the adsorbed particles. This approach to catalysis has in modern times been spearheaded by Dowden.[34,35] His work provides a logical connection between coordination complex chemistry and heterogeneous catalysis. In recent times, the role which cation and anion vacancies and their coordination play in the mechanism of oxidation reactions has received particular attention, and some examples will be included in this section.

The crystal field theory of catalysis takes as its starting point the change of coordination of the surface ion in the act of adsorption. For instance, on the (001) surface of NiO, a Ni^{2+} ion has a square pyramidal coordination, which upon adsorption of a ligand becomes more symmetric, approaching more nearly the octahedral coordination in the bulk. Calculations[35,246] of the crystal field stabilization energy contribution to the heat of chemisorption found that a minimum resulted for d^0, d^5, and d^{10} (as expected) and pronounced maxima at d^3 and d^8. In correlating these results with patterns of catalytic activity for oxides of the transition series (Ti^{4+} to Zn^{2+}), two caveats have to be heeded.[35] The first is that comparison should properly be made only for oxides of the same stoichiometry and lattice structure, e.g., in the series MgO–MnO–NiO or Al_2O_3–Cr_2O_3–Fe_2O_3. The second is that during chemisorption and catalysis the valence of the ions should not change, since the energy differences accompanying valence change are large compared with the crystal field stabilization effects. The "twin-peaked" pattern suggested by the crystal field stabilization terms was found experimentally for H_2–D_2 exchange at temperatures of -195 to $20°C$.[34] Similar patterns have also been found for the disproportionation of cyclohexene, the dehydrogenation of propane, and the hydrogenation of ethylene, as reviewed by Cimino.[247] These correlations suffer from the fact that the oxides compared are not isostructural. Cimino and co-workers have made

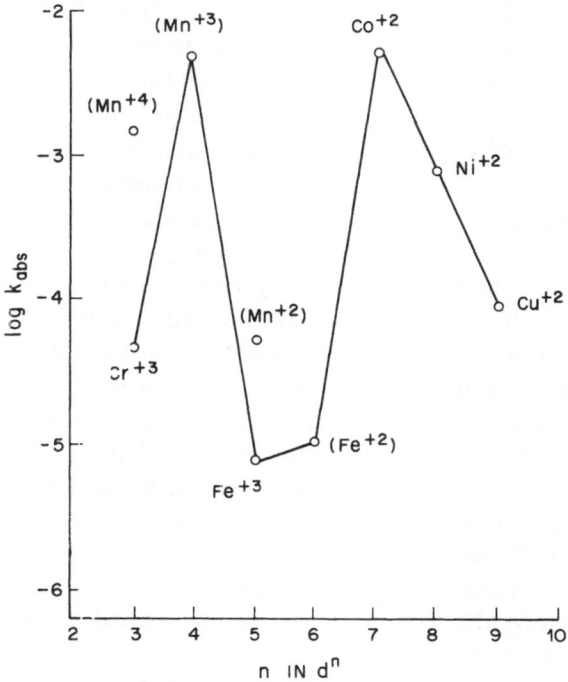

Fig. 30. Correlation of the absolute rate constant per transition metal atom for N_2O decomposition at 315°C with number of d electrons. Solid solutions of 1 at % in MgO. Valence states in parentheses may be slightly impure; i.e., Mn^{4+} containing a slight amount of Mn^{3+}, etc. (From Cimino.[247])

extensive studies of MgO catalysts with transition metal ions atomically dispersed on Mg sites, thus ensuring an isostructural series, and have used these in the decomposition of N_2O. They have again found a marked "twin-peaked" pattern (Figure 30) with maxima at Mn^{3+} and Mn^{4+} (d^3 and d^4) and Co^{2+} (d^7) and a minimum at Fe^{3+} and Fe^{2+} (d^5 and d^6). The fact that N_2O decomposition involves the charged adsorbed species N_2O^- and O^- indicates that the metal ions must change valence during the reaction. It is surprising that this does not destroy the "twin-peaked" pattern, since the crystal field stabilization effects are likely to be small compared to the valence change energies.

A study[38] which heeds both caveats discussed above, and avoids valence changes while working with an isostructural series of semiconducting ternary oxides, employed the cubic perovskite-like oxides

69

$\{La_{1-x}^{+3}M_x^{+3\pm1}\}[TM_{1-x}^{+3}TM_x^{+3\mp1}]O_3$, where $M = Ce^{4+}, Pb^{2+}, Ca^{2+},$ Sr^{2+} and TM = Cr, Mn, Fe, Co. The test reaction was the oxidation of CO. The activity pattern showed a maximum with the Fermi level near the bottom of the d_{z^2} level. This is the position which is optimal for the bonding of CO to the transition metal ion, since the nearly empty d_{z^2} level is capable of accepting the carbon lone pair, while the lower lying and occupied d_{yz} and d_{zx} levels can effectively donate electrons into the π^* CO orbital. This study showed how useful the principles of the crystal field theory are in understanding activity patterns, even though in the neutral adsorption of CO the "crystal field stabilization" invoked by Dowden for ionic adsorbates is not applicable and a twin-peaked pattern is not found.

Isolated ions and similarities with coordination chemistry are particularly important in the catalytic chemistry of the transition metal ion-exchanged zeolites (aluminosilicates, which, due to their cagelike structure, have been called "molecular sieves"). The diamagnetic matrix makes it advantageous to use optical and ESR spectroscopy. The detailed information obtained in these studies is well illustrated by the work of Chao and Lunsford on the adsorption of NO on Ag in Y-type zeolite[248] and that of Klier *et al.*[249] on the π-adsorption complexes of olefins and acetylene on Co(II) ions in zeolite A.

In the oxidation of olefins over bismuth molybdate catalysts and in the redox reaction of NO with CO and H_2 over perovskite-like oxides of manganese, the catalytic activity has been directly linked with defects in the lattice. These studies are of special relevance in the present context since they demonstrated the importance of defects in contradistinction to the transport properties of the catalysts.

The system Bi_2O_3–MoO_3 has been studied in detail by Batist, Schuit, Matsuura, and co-workers[250,251,100] to establish the relation between the solid-state chemistry and catalytic action of this commercially important selective oxidation catalyst. Using the oxidative dehydrogenation of 1-butene to butadiene as their test reaction in a kinetic, crystallographic, and chemisorption study, they concluded that the most active compound in the Bi–Mo–O phase diagram was koechlinite, Bi_2MoO_6. The process was found to be an example of intrafacial catalysis, involving lattice oxygen. Two types of adsorption centers were distinguished by the chemisorption measurements on reduced $Bi_2MoO_{5.5}$ and oxidized Bi_2MoO_6. The first is called the *A* center, which is a single site which slowly but strongly adsorbs butadiene. The second, so-called B center, consisting of several surface oxygen atoms, adsorbs 1-butene, *cis*-2-butene, butadiene, and *trans*-2-butene in a fast process leading to relatively weak binding. The conversion of butene to butadiene involves the A and B centers simultaneously. At temperatures below 400°C, only the A center is involved in the incorporation of O_2 in the lattice, with rapid diffusion of O^{2-} providing the oxygen necessary for the oxidation reaction. These results are explained by a model of the structure of koechlinite. The structure consists of plates of $(Bi_2O_2)^{2+}$ and $(MoO_2)^{2-}$ in alternate stacking, with a layer of bridging O^{2-} between (Figure 31). One-fourth of the oxygen at the edge of the Bi_2O_2 layer is removable and is replenished at low temperature by O^{2-} from the same layer, whereas at $T > 400$°C, O^{2-} can diffuse from one layer to the next. The olefin dissociatively adsorbs on a B site (Figure 31) as an allyl radical, transfers to a bismuth vacancy, and loses a second H atom to a second B site while the Mo^{IV}/Mo^{VI} redox reaction provides balancing of the charge states.[251]

The role of Bi vacancies postulated in the above-discussed work on Bi_2MoO_6 has a dramatic parallel in the work of Sleight, Aykan, and co-workers.[252-254,101] They have approached the study of the selective oxidation catalysts by adopting catalysts based on the

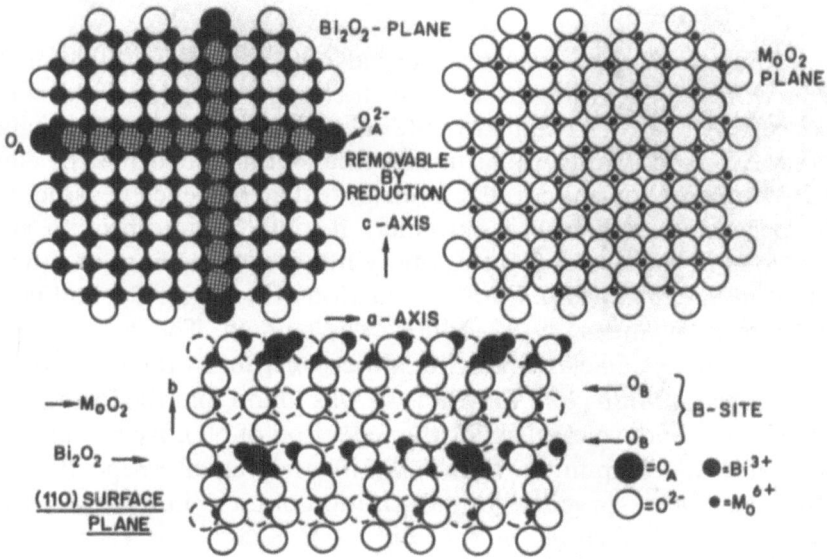

Fig. 31. The surface boundary (110) plane of Bi_2MoO_6 showing the Bi_2O_2 and MoO_2 layers (top) and the alternate stacking with bridging oxygen (bottom). (From Matsuura and Schuit.[100])

scheelite ($CaWO_4$) structure. In this structure, extensive substitution for Ca and W is possible and point defects are easily introduced, notably vacancies on the Ca site. For the AMO_4 structure, A is coordinated to eight different MO_4 tetrahedra. The structure type is very flexible, allowing A to be from monovalent to tetravalent and M to be from monovalent to octavalent. A-site vacancies may be introduced by proper choice of A, M, and the stoichiometry to obtain $A_{1-x}\phi_x MO_4$, where ϕ represents the vacant A site. Scheelites with M = V, Mo, and W have been made and studied as catalysts.[253,254] In a very large series of these catalysts, it was observed that the simultaneous presence of A vacancies and of Bi was necessary to obtain high activity and selectivity for the ammoxidation of propylene:

$$H_2C{=}CH{-}CH_3 + NH_3 + \tfrac{3}{2}O_2 \;\rightarrow\; H_2C{=}CH{-}C{\equiv}N + 3H_2O$$

or the oxidative dehydrogenation of butene to butadiene. The role of bismuth apparently is to rapidly replenish the active site with oxygen, similar to the Batist–Schuit–Matsuura mechanism above. The role of the defect is to attract a proton from the olefin by Coulombic forces. The activity of $Pb_{1-3x}Bi_{2x}\phi_x MoO_4$ for the 1-butene oxidation accordingly increases linearly with x in the range of $0 < x < 0.06$.[252]

The role of cation vacancies has also been demonstrated for the reduction of NO to dinitrogen products ($N_2 + N_2O$) over perovskite-like catalysts.[38,255] The perovskite structure, of the formal stoichiometry ABO_3, contains A ions in a dodecahedral hole formed by eight BO_6 octahedra sharing corners. Even more than the scheelite structure, the perovskite structure is extremely flexible with respect to substitutions in the A and B sites. Many hundreds of compounds with this structure are known. For the reduction of NO with CO and H_2, a series of manganites $A_{1-x}\phi_x MnO_3$ was studied. The reaction was shown to be intrafacial, involving lattice oxygen. In the comparison of $La_{0.91}\phi_{0.09}MnO_3$ and $LaMnO_3$, it was found that the former was a much more active catalyst for the reduction of NO. By comparison of a series of compounds with approximately equal proportions of Mn^{3+} and Mn^{4+} prepared by substitution of La^{3+} in $LaMnO_3$ with appropriate proportions of K^+, $Bi^{3+} + K^+$, Sr^{2+}, or vacancies, it was shown that the activity for NO reduction was directly related to the contribution of the occupants of the A sites to the binding of surface oxygen (Table 2). This contribution was calculated for the process

$$Mn{-}O{-}Mn \;\rightarrow\; Mn{-}V_O{-}Mn + O_{gas}$$

$$\quad\;\; A \quad\; A' \qquad\qquad A \quad\; A'$$

where Mn–O–Mn stands for the surface of the oxide and V_O for an oxygen vacancy. The latter was proposed to be the active site for the dissociative chemisorption of NO.

The above examples show how in modern catalytic work the solid-state chemistry, and particularly the defect chemistry, of the catalyst plays a major role. Even though the proposed mechanisms may prove inaccurate in detail, they appear to be sufficiently well-founded to prove the essential importance of localized and defect influences on catalytic activity. This is especially significant since the same catalysts, bismuth molybdates and manganite perovskites, have featured prominently in collective solid-state physics approaches to catalytic activity.

6. Geometrical Effects in Catalysis

In the introduction, it was already noted that the separation between geometric and electronic effects is an artificial one. It will have surprised no one that as the discussion in the last section progressed from collective properties to localized and defect properties, the geometry of the active site was explicitly considered together with the electronic effects. In the present section, oxides will therefore not be discussed, except to note that a general rule has been found stating that transition metal ions in tetrahedral coordination are much less active in redox reactions than those in octahedral coordination.[247] Ion scattering studies of Co^{3+} in octahedral and tetrahedral coordination have provided an explanation for this rule.[256] At the surface, tetrahedrally coordinated Co ions were found to be largely shielded from the gas phase, in accordance with observations that CO chemisorbs on CoO with the NaCl structure, but not on CoO with the ZnS structure.

6.1. Correlations between Kinetics and Surface Structure of Metallic Catalysts

High-area metal catalysts are often prepared by dispersion of the metal on the surface of a support. This dispersion process leads to more efficient use of the active metal component, reaching a limit where each metal atom is exposed to reactants (dispersion $= 1$). The widespread use of highly dispersed catalytic metals has led to an important theme in catalytic research which asks whether the catalytic activity per exposed atom of active material varies as the degree of dispersion changes. As pointed out by Poltorak and summarized in an excellent review by Boudart,[8] one should expect major changes in the physical nature of metal crystallites as their sizes decrease below about 50 Å. This is because the fraction of atoms in the surface of a crystallite increases markedly in this range as well as because an appreciable fraction of abnormal surface atom coordination situations begin to occur at ~ 50 Å particle size (see Figure 3). Thus, on this basis, one might expect electronic as well as geometric differences to occur as crystal size is decreased below 50 Å.

Within the last several years, careful kinetics experiments designed to answer the question of the relation of specific catalytic activity to dispersion have been carried out. It has been found in certain catalytic systems that little change in *specific activity* occurs over many orders of magnitude variation in dispersion, and these catalytic systems have been designated "structure-insensitive" or "facile" by Boudart.[257] Other reactions whose specific rate depends on catalyst structure are termed "structure-sensitive" or "demanding" reactions. Boudart[8] points out that, in view of the general historical irreproducibility of kinetic measurements on catalysts, some of these results "may be viewed as opening a new era in our confidence in catalytic data." In addition, since the comparison of catalytic activity for certain facile reactions has been extended all the way to bulk metal foils[257] and to single crystals,[258] the way is now opened for meaningful catalytic experiments to be done on single crystals where modern techniques of surface characterization can be applied[1] and where important parameters can be well controlled. A corollary assertion about "structure-sensitive" catalytic reactions has been advanced by Boudart,[103] namely, that reactions proceed *on all sites available*, although not necessarily with exactly equal rates. This observation is supported by controlled poisoning experiments by Kral,[259] who demonstrated that the fraction of active sites on a catalyst varied over a range from about one to about 10^{-3} for various hydrogenation reactions. This fraction is termed the Taylor ratio. With a ratio near unity, one is dealing with a "structure-insensitive" reaction.[8] It should be pointed out that in some cases[257] where "structure-insensitive" reactions are being studied, when the weight percent of metal is made very low (~ 0.1–0.05%) the specific activity is also found to be low. This may be due to uncontrolled poisoning effects, specific support effects which are only detected at low metal concentrations, or to reduced precision in the kinetic measurements. A number of structure-insensitive reactions are listed in Table 3.

Since many factors may be responsible for changes in catalytic activity for supported catalysts having different degrees of dispersion, it is important to carefully study changes in specific activity before concluding that a reaction is "structure insensitive" or "structure sensitive." One way to do this is to examine parallel reactions starting from the same reactants but proceeding through different intermediates to yield different products. If one of these reaction paths is "structure insensitive," then it may be used as an internal control in experiments involving different catalyst preparations. The ratio of the rate of the second reaction to the control reaction may then be used to judge whether the second reaction is structure insensitive or not. Changes in relative rates for various catalyst dispersions are termed selectivity changes and control of catalyst selectivity is of great importance in practical catalysis. Examples of reactions proceeding in parallel are found in studies of neopentane isomerization (to isopentane) compared with neopentane cracking (to isobutane and methane) on Pt.[8] Another example is the comparison of benzene hydrogenation (to cyclohexane) with benzene isotopic exchange with deuterium on Ni.[16] In both of these cases, the first reaction was found to be "structure insensitive."

When one comes to speculate about the factors which are responsible for a reaction being either sensitive or insensitive to catalyst structure, details of the reaction mechanism must be invoked. Thus, a structure-insensitive reaction might involve the formation of intermediates in the rate-controlling step which do not demand very specific geometric arrangements of the atoms of the catalyst.

A second possibility advanced for the meaning of structural insensitivity has to do with the possibility that these reactions are associated with the structural properties of the intermediates themselves, and therefore are little influenced by the surface structure of the catalyst itself. This explanation has been advanced by Yates *et al.*[260] to explain the similarity in the kinetics of liberation of CH_4 from W(111) and W(100) single crystals covered with a layer produced by adsorption of formaldehyde. In these studies, it appears that CH_4 is produced in two distinct thermal processes on *both* crystal faces with maximum liberation rates at about 270 and 500°K as shown in Figure 32. In addition, work function changes for the adsorbed layer on the

TABLE 3
Structure Insensitive Catalytic Reactions

Reaction	Catalyst	Dispersion range	Specific activity range	Ref.
	Pt/SiO$_2$	70×	1.7×	306
	Ni/SiO$_2$ Ni/SiO$_2$ · MgO Ni/SiO$_2$ · Al$_2$O$_3$	20×	~2×	307
	Pd/SiO$_2$ Pd/Al$_2$O$_3$ Pd black	30×	~1.6×	307
	Pt/SiO$_2$ Pt/Al$_2$O$_3$	6×	1.1×	307
	Ni/MgO	30×	2.7×	308
	Pt/Al$_2$O$_3$ Pt/SiO$_2$ Pt foil Pt single crystal	2 × 10^4×	2.5×	257, 258
	Pt/Al$_2$O$_3$	3×	~1.3×	309

Reaction	Catalyst			Ref.
(alkane → alkane)	Pt/Al$_2$O$_3$	3×	~1.3×	309
(→ benzene) + 4H$_2$	Pt/Al$_2$O$_3$	3×	~1.3×	309
—C—C=C + H$_2$ → (alkane)	Pt/SiO$_2$	4×	~2×	310
H$_2$ + D$_2$ → 2HD	Pt/SiO$_2$	4×	~2×	310
2SO$_2$ + O$_2$ → 2SO$_3$	Pt/SiO$_2$, Pt	—	—	311
2H$_2$ + O$_2$ → 2H$_2$O (in excess O$_2$)	Pt/SiO$_2$, Pt	7×	2.1×	262, 313
—C—C— + H$_2$ → 2CH$_4$	Ni/SiO$_2$	3×	1.5×	312
(cyclohexene) + H$_2$ → (cyclohexane)	Pt/SiO$_2$, Pt/Al$_2$O$_3$	7×	1.09	262
2CO + O$_2$ → 2CO$_2$	Pd(100)	—	—	299
	Pd(110)	—	—	
	Pd(111)	—	—	
	Pd(polycrystalline)	—	—	
2H$_2$CO → CH$_4$ + CO$_2$	W(100)	—	—	261
	W(111)	—	—	

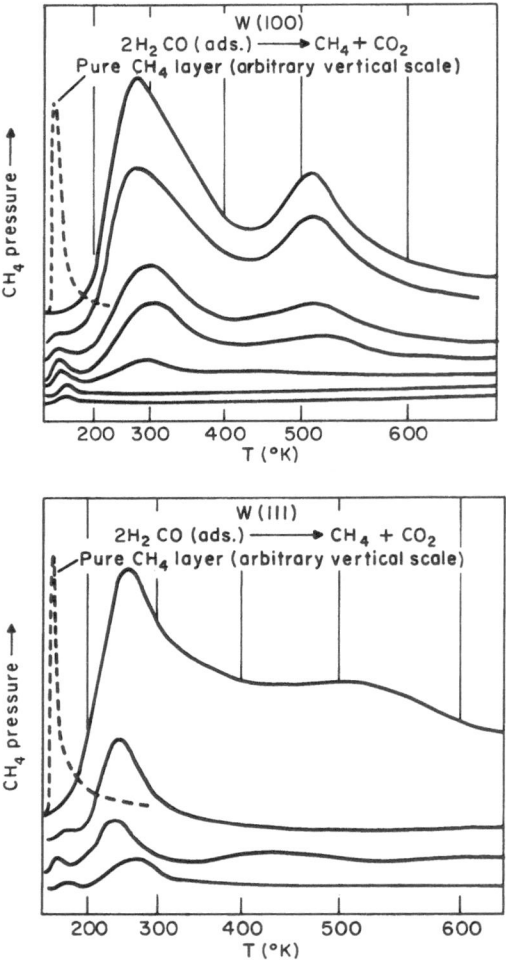

Fig. 32. Structure insensitivity of the catalytic production of CH_4 from adsorbed H_2CO on two single-crystal planes of tungsten. Following adsorption of varying amounts of H_2CO on the crystals at $\sim 100°K$, heating causes liberation of CH_4 in two stages in both cases. The kinetics of CH_4 evolution differ significantly from that observed for CH_4 desorption from pure physically adsorbed CH_4 layers, and is thought to reflect the catalytic decomposition of intermediates at elevated temperatures. Upper figure: Desorption traces from W(100) surface. Lower figure: Desorption traces from W(111) surface. (From Yates *et al.*[260])

two crystal faces are very similar, indicating that the experiments are probably detecting intrinsic properties of surface complexes which are not strongly dependent on underlying crystallography. Recently, Boudart[261] has made a similar suggestion regarding the $H_2 + O_2$ reaction on Pt, which is structure sensitive in excess H_2, but structure insensitive in excess O_2. The oxygen is postulated to chemisorb in such a way as to erase surface anisotropies and reaction occurs on the "platinum oxide molecule" surface layer. This means that in the latter case, we deal with an intrafacial catalytic process.

Recently, Somorjai and his co-workers have carried out several pioneering experiments designed to test whether specific surface geometric factors such as step density can affect the rate of catalytic reaction. In the first work of this type, the isotopic H_2–D_2 exchange reaction was examined using a chopped molecular beam.[262,263] From a planar Pt(111) surface, no HD production could be measured, whereas above a stepped surface containing nine-atom-wide Pt(111) terraces separated by one-atom-high steps, about 5–10% of HD product is detected. In both cases, the Pt surface temperature was 1000°K while the beam temperature was 300°K. As shown in Figure 33, the spatial distribution of unreacted H_2 or D_2 is peaked at the specular angle (45°) while HD product is desorbed with a cosine distribution, indicative of thermal equilibration of the product HD. These results suggest that a stepped Pt surface provides sites capable of catalyzing an efficient exchange mechanism, possibly H(ads, at step) + D_2(ads) → HD(ads) + D(ads).

At $T < 500°K$, Bernasek and Somorjai[90] suggest that the rate-limiting step is the diffusion of D_2 across the (111) terraces to H atoms chemisorbed at the step. The activation energy for this process is 4.5 ± 0.5 kcal/mole. At $T > 600°K$, an Eley–Rideal process involving the collision of an incident D_2(g) molecule with an adsorbed H atom at the step begins to become important.

It should be pointed out that these experiments using a chopped molecular beam plus phase-sensitive detection of product are designed to detect surface events which occur on a time scale of the order of milliseconds for this system; processes which occur more slowly are not measured. Recently, Rye and Lu[264] have used a steady-state method to compare the rate of HD production on various single-crystal faces of Pt. This method samples surface events occurring on a much longer time scale than the molecular beam method. They too find that the Pt(111) surface is less reactive than a stepped Pt(111) surface, but the factor difference in rates is only about two

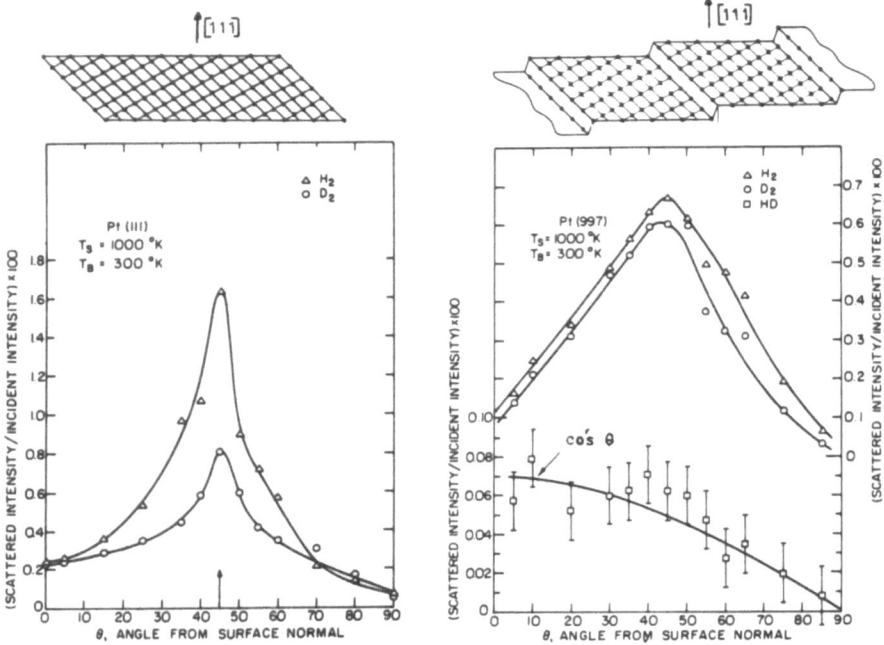

Fig. 33. Angular distribution of isotopic hydrogen molecules from Pt(111) and a stepped Pt surface containing (111) steps. Left figure: H_2 and D_2 scattered from $D + Pt(111)$. No HD production is observed. The H_2 and D_2 reflected beams have maximum intensities at the specular angle (45°). Right figure: H_2, D_2, and HD scattering from stepped surface. The HD scattering appears to be cosine-like, indicating that HD product has come to thermal equilibrium with the surface prior to desorption. (From Bernasek and Somorjai.[90])

at 700°K. This smaller difference in observed ratios of rates is understandable if the Pt(111) surface is more active for exchange when exchange events occurring over a longer time scale are counted.

Somorjai and co-workers have attempted to extend these comparisons of planar and stepped Pt(111) surfaces to surface-catalyzed hydrocarbon reactions of importance in petroleum chemistry. In particular, the reactions of *n*-heptane on platinum have been studied, and various reactions have been observed as shown below:

$$CH_3-(CH_2)_5CH_3 \xrightarrow{H_2, Pt} \begin{cases} \bighexagon CH_3 + 4H_2 \quad \text{(dehydrocyclization)} \\ CH_4, C_2H_6, C_3H_8, \ldots \quad \text{(hydrogenolysis)} \\ \underset{\text{(isomerization)}}{H_3C-\overset{\overset{\displaystyle CH_3}{|}}{CH}-\overset{\overset{\displaystyle CH_3}{|}}{CH}-CH_2-CH_3, \ldots} \end{cases}$$

Both the hydrogenolysis and isomerization processes seem to exhibit little sensitivity to step density; in addition, a disordered carbonaceous layer forms when these reactions take place at low pressures, and this layer does not hinder the reactions in any way. In contrast, the dehydrocyclization reaction to form toluene occurs only when an *ordered carbonaceous layer* is formed on a *stepped* platinum surface, and terraces of six atom width seem to exhibit the highest catalytic activity for this reaction.[9,265] Recently, irreproducible factors in these experiments have been discovered; the unique activity of six-atom stepped surfaces is presently being reinvestigated.[266]

The relation between studies of catalytic processes on dispersed catalysts and on single-crystal substrates is evident from the above examples. Unfortunately, at this time, few examples are available demonstrating the use of single-crystal substrates as catalysts at high pressures. Recently, Somorjai and co-workers have utilized a Pt single crystal as a catalyst for the hydrogenolysis of cyclopropane.[258] Initial specific rates were identical to within a factor of *two* when compared with dispersed Pt catalysts. This work was carried out in an auxiliary high-pressure reaction chamber. Figure 34 shows the

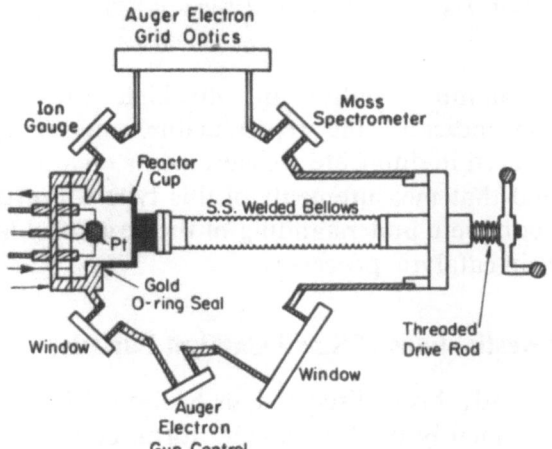

Fig. 34. Schematic view of an ultrahigh vacuum system containing a high-pressure reactor cup for measuring the catalytic activity of a single crystal of Pt. The reactor cup may be isolated from the vacuum chamber and catalytic reaction rates may be measured using gas chromatography in a recirculating flow system. (From Khan *et al.*[83])

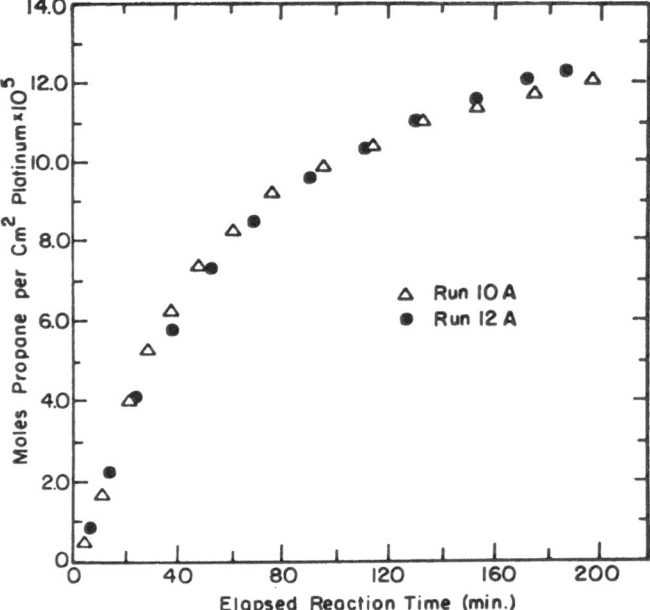

Fig. 35. Reproducibility of the rate of cyclopropane hydro-genolysis $(C_3H_6 + H_2 \rightarrow C_3H_8)$ on a stepped Pt single crystal surface [Pt(s)–6(111) × (100)]. $P_{C_3H_6} = 135$ Torr, $P_{H_2} = 675$ Torr, $T_{crystal} = 74°C$. (From Khan *et al.*[83])

high-pressure chamber within an ultrahigh vacuum apparatus and Figure 35 indicates the reproducible rate of cyclopropane formation observed in duplicate measurements on a Pt single crystal. It is anticipated that measurements of this type will eventually lead us to a more complete understanding of the structural factors which are significant in catalytic processes.

6.2. Recent Investigations of Small Catalyst Particles

As has already been discussed in Section 3.1, practical metal catalysts are frequently used in powder form, or in supported form (finely dispersed metal particles supported on a high-area insulating substrate). An understanding of the physical and chemical properties of such small particles has a clear relevance to an understanding of catalytic processes. The existence of "demanding" reactions, in which catalytic activity and selectivity are known to vary greatly as a function of particle size,[8] has served as a spur to studies of small particles.

In particular, there has been a great deal of theoretical interest demonstrated recently in the geometric and electronic properties of small metallic "clusters" of particles.

6.2.1. Pairwise Interaction Models

For many years, it has generally been assumed that microcrystals of a material have the same crystal structure and symmetry as the bulk macrocrystals. This assumption is the basis for particle size determinations using electron diffraction techniques, for example.[267] More recently, however, it has been recognized that microcrystals frequently exhibit fivefold rotational symmetry—a symmetry which is forbidden in long-range crystalline lattices by the requirement of periodicity. Figure 36 is an electron micrograph showing several pentagonal gold crystals which were observed following deposition of thin gold films on a mica substrate.[268] Bagley[269] has shown that it is possible to produce a space-filling, hard sphere structure which has fivefold symmetry; this structure, shown in Figure 37, is simply not a periodic structure. It can be regarded as a fivefold twin of close-packed material surrounding a small seed of fivefold rotational symmetry.

The origin of the seed of fivefold symmetry has been discussed by Hoare and Pal[270] and by Burton.[271] If one examines the growth sequence of a crystal in which the atoms interact by simple Lennard-Jones type of pairwise interactions, the structures will deviate from the normal fcc structure. A surprising result was the observation that

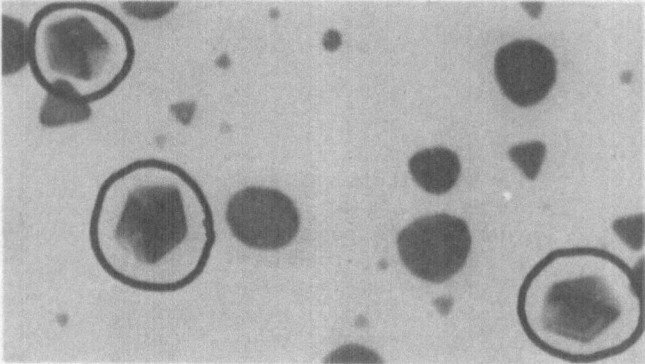

Fig. 36. Electron micrograph of several pentagonal gold crystals grown by evaporation of gold onto a heated mica substrate. (From Allpress and Sanders.[268])

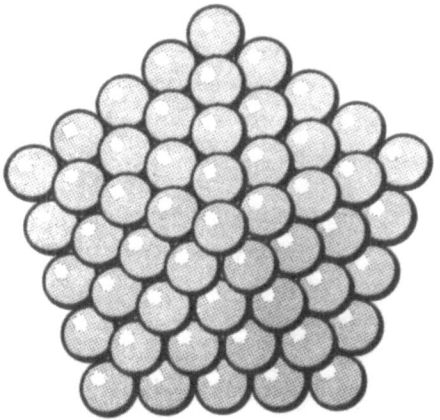

Fig. 37. Model of a fivefold symmetric crystal obtained by the continued packing of hard spheres on a seed of fivefold symmetry. Note that the exposed surfaces have close-packed (111) orientations. (From Bagley.[269])

the most stable geometry for a 13-atom cluster is not the fcc structure of Figure 38a, but is the fivefold symmetric icosahedral structure of Figure 38b. The 13-atom fcc structure is unstable, and would shear into the icosahedron if the fcc structure were somehow constructed. The

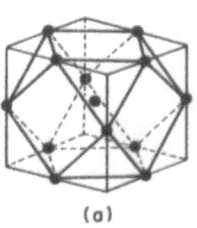
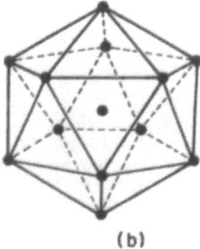

(a)   (b)

Fig. 38. (a) Thirteen-atom cubooctahedral cluster showing the nearest-neighbor environment corresponding to atoms packed in a face-centered cubic lattice. Note that the surface arrangement of atoms contains elements of both (100) and (111) surfaces. (b) Thirteen-atom icosahedral cluster. Note the axes of fivefold symmetry, as well as the fact that all of the exposed surfaces have (111) symmetry. (From Johnson and Messmer.[50])

next largest perfect icosahedron is a 55-atom cluster, which Burton reasons is the seed on which crystals exhibiting fivefold symmetry (Figure 36) have grown. It is also of interest to note that the surface planes of the 55-atom icosahedral clusters all have the structure of close packed (111) planes, whereas many planes in a 55-atom fcc structure are square (100) planes. Atoms in (111) planes have higher coordination numbers (more nearest neighbors) than those in (100) planes, so that elimination of (100) planes in rearrangement from the fcc to the icosahedral structure increases the number of nearest-neighbor bonds and lowers the energy of the system.

6.2.2. Quantum Chemistry of Small Particles

There have been a number of applications of molecular orbital theory to chemisorption and catalysis in the last few years. Reviews by Johnson and Messmer[50] and by Slater and Johnson[51] contain excellent overviews of the state of various molecular orbital approaches to catalysis problems. As discussed in Section 4, these authors have developed a technique called the "self-consistent-field X-alpha scattered wave" cluster method (SCF-Xα-SW). This approach has been applied to small Cu_8 and Ni_8 clusters having simple cubic geometry, the simplest structure which has been used to characterize small catalytic particles.[51] The SCF-Xα energy levels computed for Cu_8 and Ni_8 particles are shown in Figure 39. There is remarkable similarity between the electronic structure of the eight-atom clusters and the bulk solids, e.g., the complete set of Cu_8 energies can be characterized as a dense band of d levels bounded above and below by levels which are predominantly s- and p-like in character, but with some d admixture. The electronic structure of the Ni_8 cluster is similar to the spin-polarized band structure for ferromagnetic crystalline Ni, particularly with respect to the high density of d states around the Fermi level. However, the spin magneton number per atom (0.25) in paramagnetic Ni_8 is considerably less than the value for ferromagnetic crystalline Ni (0.54). This decrease is consistent with the paramagnetic susceptibility measurements of Carter and Sinfelt[272] for highly dispersed, silica-supported Ni. Carter and Sinfelt had concluded that the decrease in susceptibility was large when the metal particles become sufficiently small and significant fractions of the atoms are present on the surface.

SCF-Xα-SW calculations for 13-atom Li clusters[51] suggest that the icosahedral geometry is more stable than the cubooctahedral

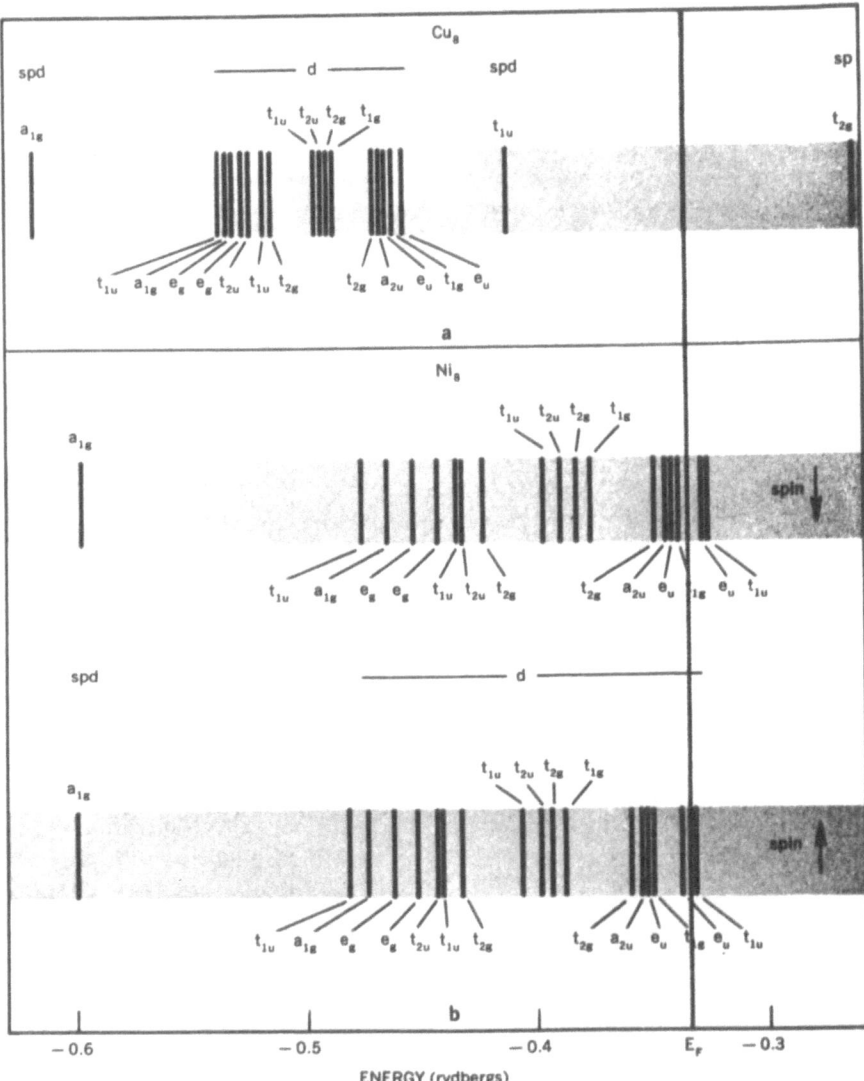

Fig. 39. Theoretical energy levels for copper and nickel clusters computed using the SCF-Xα method. Part (a) shows the SCF-Xα electronic energy levels for a simple cubic Cu_8 cluster; part (b) shows the spin–polarized SCF-Xα electronic energy levels of a simple cubic Ni_8 cluster. For each cluster, the results are shown for a nearest-neighbor internuclear distance equal to that in the corresponding bulk crystal. The levels are labeled according to the irreducible representations of the cubic (O_h) symmetry group. The "Fermi level" E_F separates the occupied levels from the unoccupied ones. (From Slater and Johnson.[51])

geometry, a result in good agreement with the pairwise interaction calculations (Section 6.2.1).

6.3. Stereochemical Methods as an Aid to the Determination of Catalytic Mechanisms

The determination of the structures of intermediate species in heterogeneous catalysis is an important driving force in catalytic research. Ideally, this objective also includes a specification of the structure of the catalytic site, which, as we have seen, is at present often an elusive goal. While studies of the kinetics of catalytic processes are an important aspect of the determination of catalytic mechanisms, kinetic arguments are, by themselves, usually not a completely definitive factor in selecting a model to describe the reaction. Often many dissimilar mechanisms give the same overall kinetics.[2] For this reason, in the case of complex molecules, stereochemical methods are often used to aid in the definition of the mechanism of a catalytic reaction, or in the definition of the geometry of adsorbed species involved in the reaction. A few examples of the usefulness of stereochemical arguments will be given below. Burwell[273] and Siegel[274] have discussed in detail the stereochemical evidence for various bonding modes in hydrocarbon adsorption and reaction. We will consider as a first example the catalytic reaction of a disubstituted acetylene (2-butyne) with hydrogen on a Pd catalyst. 2-Butyne is a linear molecule. It is well known that, in general, the *cis*-addition of two hydrogen atoms to the triple bond occurs, i.e., that both hydrogens add to the same side of the triple bond. In this case, *cis*-addition would therefore give an olefin in which the two methyl groups are on the side of the double bond opposite the H atoms. There is additional experimental evidence showing that deuterium addition yields the *cis*-dideutero species as a product. This further confirms that the two methyl groups do not interact with the surface. Thus, a proposed mechanism is

$$D_2 + 2Pd \rightarrow 2D\text{-}Pd$$

$$H_3CC{\equiv}CCH_3 + 2Pd \rightarrow \begin{array}{c} H_3C \\ \diagdown \\ \diagup \\ Pd \end{array} C{=}C \begin{array}{c} CH_3 \\ \diagup \\ \diagdown \\ Pd \end{array}$$

$$D{-}Pd + \begin{array}{c} CH_3 \\ \diagdown \\ \diagup \\ Pd \end{array} C{=}C \begin{array}{c} CH_3 \\ \diagup \\ \diagdown \\ Pd \end{array} \rightarrow \begin{array}{c} CH_3 \\ \diagdown \\ \diagup \\ D \end{array} C{=}C \begin{array}{c} CH_3 \\ \diagup \\ \diagdown \\ Pd \end{array} \rightarrow \begin{array}{c} CH_3 \\ \diagdown \\ \diagup \\ D \end{array} C{=}C \begin{array}{c} CH_3 \\ \diagup \\ \diagdown \\ D \end{array}$$

An alternative mechanism is that the 2-butyne reactant molecule collides with two sites containing adsorbed deuterium and that reaction occurs to yield the *cis*-dideutero species. Kinetic studies indicate that this alternative mechanism is not applicable since the process is roughly zeroth order in 2-butyne (i.e., almost pressure independent) and first-order in deuterium.[273]

As a second example of the utility of stereochemical reasoning, we consider, following Burwell,[273] the hydrogenation of molecules (olefins) containing a carbon–carbon double bond. Again, there is ample evidence that *cis*-addition of two hydrogens generally occurs, but the question arises as to whether this addition occurs from the gas phase to the topside of the adsorbed species, or from adsorbed hydrogen underneath the surface complex. Two general configurations for an olefin adsorbed on substrate atoms M are possible,

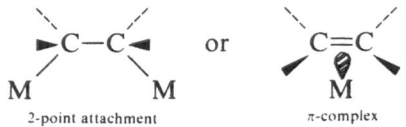

The two-point attachment would involve sp^3 hybridization of each C atom, whereas the π-complex could form with preservation of the planar geometry of the olefin. The discrimination between these two structures is, at present, difficult using chemical methods alone.[275]

In order to determine whether topside addition of hydrogen to an olefin is possible, Burwell[276] cited the work of others[277] dealing with the catalytic hydrogenation of large cyclic olefins. This work concerned the hydrogenation of *trans*-cyclononene, which contains a carbon–carbon double bond with one side protected by a chain of seven CH_2 groups connected *trans* across the double bond as shown schematically below:

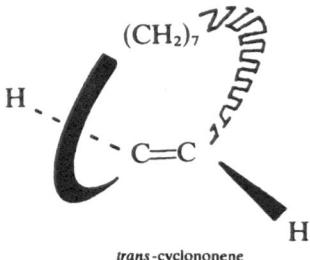

trans-cyclononene

This molecule is effectively hydrogenated catalytically, suggesting that H attack occurs *under* the molecule, involving electrons that were used to produce the chemisorption bond initially.

Fig. 40. Three stereochemical views of a chemisorbed olefin. Left figure: Carbon–carbon single bond with two-point attachment to surface atoms. Staggered conformation of sp^3 carbon orbitals. Central figure: Same as above, except eclipsed conformation of sp^3 carbon orbitals. Right figure: π-complex single-point attachment. Carbon is hybridized in sp^2 state. (From Burwell.[273])

We can ask even more detailed questions regarding the stereochemical nature of surface species involved in catalytic reactions. Burwell[273] has considered three possible stereochemical conformations for an adsorbed olefin as shown in Figure 40. The left-hand diagram represents a staggered form of a two-point attachment species; the central diagram represents an eclipsed form of a two-point attachment species; the right-hand diagram represents a π-complex. By freezing in either of the first two configurations using ring structures to orient C–H bond directions, it has been possible to show that the

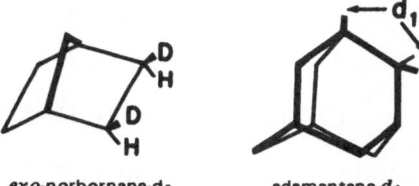

exo-norbornane-d_2 adamantane-d_1

Fig. 41. Stereochemical views of two cyclic hydrocarbon molecules in which the rigidity of the ring structure produces eclipsed or staggered C–H bonds. Left: exo-norbornane; eclipsed C–H bonds undergo double site deuterium exchange in one period of adsorption. Right: adamantane; staggered C–H bonds undergo single-site deuterium exchange in one period of adsorption. (From McMahon and Clarke[278] and Schrage and Burwell.[279])

eclipsed form is necessary for deuterium exchange to occur on the two C atoms together. Thus, as summarized in Figure 41, two deuterium atoms initially exchange with hydrogens in adsorbed norbornane[278]; the original C–H bonds here are held in the eclipsed conformation by the ring structure in the norbornane. In particular, the readily exchangeable hydrogens are in the equatorial position pointing outward, roughly in the plane of the ring. The interpretation was made using NMR analysis of exchanged products, and suggests that the molecule adsorbs on edge by means of two-point attachment. By contrast, adamantane (Figure 41) contains only staggered C–H bonds and exchanges only a single deuterium initially over Pd,[279] as deduced by mass spectrometry of the products.

As a final example of the influence of stereochemical factors on catalytic processes, we turn to a brief discussion of the stereoregular polymerization of propylene, as discovered by Ziegler and Natta, leading to the award of a Nobel Prize in Chemistry in 1963.[280] The basic discovery was that certain catalysts containing transition metals caused propylene to polymerize in such a way that the asymmetric carbon atoms in the polymer are all either right-handed (*d*-configuration) or left-handed (*l*-configuration). An asymmetric carbon atom is tetrahedrally bonded to four *different* substituents. This configuration of substituents implies the possibility of two conformations of the carbon atom and its substituents, which are mirror images. Such a polymer is termed isotactic and is schematically shown below, where the asterisk indicates an asymmetric C atom:

$$
\begin{array}{ccccccccccc}
& H & & H & H & & H & H & & H \\
& | & & | & | & & | & | & & | \\
-C^* & - & C & - & C^* & - & C & - & C^* & - & C- \\
& | & & | & | & & | & | & & | \\
& CH_3 & & H & CH_3 & & H & CH_3 & & H \\
\end{array}
$$

In a monumental series of three papers,[281] Cossee and Arlman have described the electronic and crystallographic factors responsible for olefin stereospecific polymerization on the surface of transition metal halides, specifically $TiCl_3$. The specific adsorption and polymerization of the parent olefin is envisioned to occur in the vicinity of a Ti^{+3} ion which is incompletely coordinated with Cl^- ions near the surface of an α-$TiCl_3$ crystal. The olefin is postulated to be bound to Ti^{3+} via a π-bond as shown in Figure 42, where ethylene is depicted as the adsorbed olefin. The Ti^{3+} ion, designated M, is octahedrally coordinated along the x, y, and z axes with either a Cl^- (designated X) or an alkyl group designated R as shown in Figure 43. It is seen that

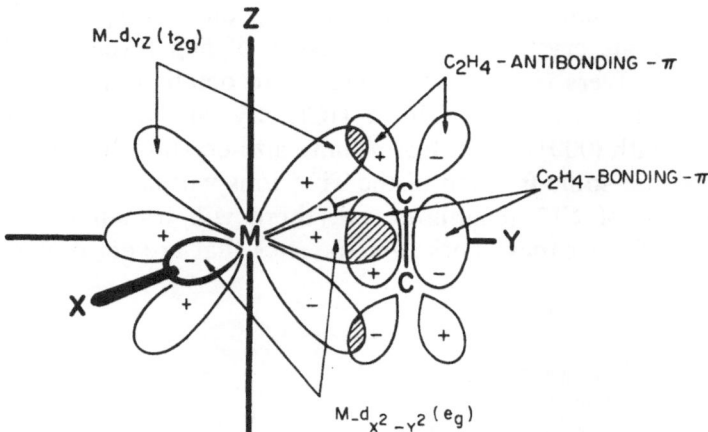

Fig. 42. Schematic view of π-bonded ethylene on a transition metal atom M. This complex is proposed as the initial species in the stereoregular polymerization of olefins by Ziegler–Natta catalysts. (From Cossee and Arlman.[281])

the olefin coordinates with the metal M via an octahedral site which is initially unfilled. The polymerization step, involving the linkage of R to one end of the olefin, is depicted in the third step of Figure 43. The final result is to generate an $R–CH_2–CH_2–$ group linked to M, plus a new vacant octahedral position at the point where R originally existed. The model of Cossee and Arlman now requires $R–CH_2–CH_2–$ to migrate back to the original site which contained R, so that adsorption of another olefin molecule can occur at the same site as before, since the model for production of isotactic polymer requires that the active polymerization site have identical stereochemical conformation for each polymerization step.

The basic idea formulated above has been woven by Cossee and Arlman[281] into a crystallographic model which seems appropriate

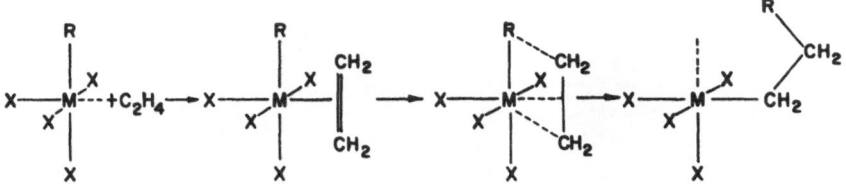

Fig. 43. Schematic view of polymerization step in Ziegler–Natta catalysis. R is an alkyl group which adds to the adsorbed olefin. In this model, $R–CH_2–CH_2–$ migrates back to the original R site ready for another polymerization step. $M = Ti^{3+}$; $X = Cl^-$. (From Cossee and Arlman.[281])

91

to explain recent observations regarding olefin polymerization on α-TiCl$_3$ single crystals. α-TiCl$_3$ consists of hcp layers of Cl$^-$ ions separating layers of Ti^{3+} ions, which are octahedrally coordinated with neighbor Cl$^-$ ions. The α-TiCl$_3$ crystals form as hexagonal platelets with (0001) faces. For a finite size crystal, charge neutrality requires that some fraction of the Ti^{3+} ions will have an incomplete octahedron of Cl$^-$ neighbors, i.e., there will be some Ti^{3+} sites having a Cl$^-$ vacancy. These are the sites that are active centers for

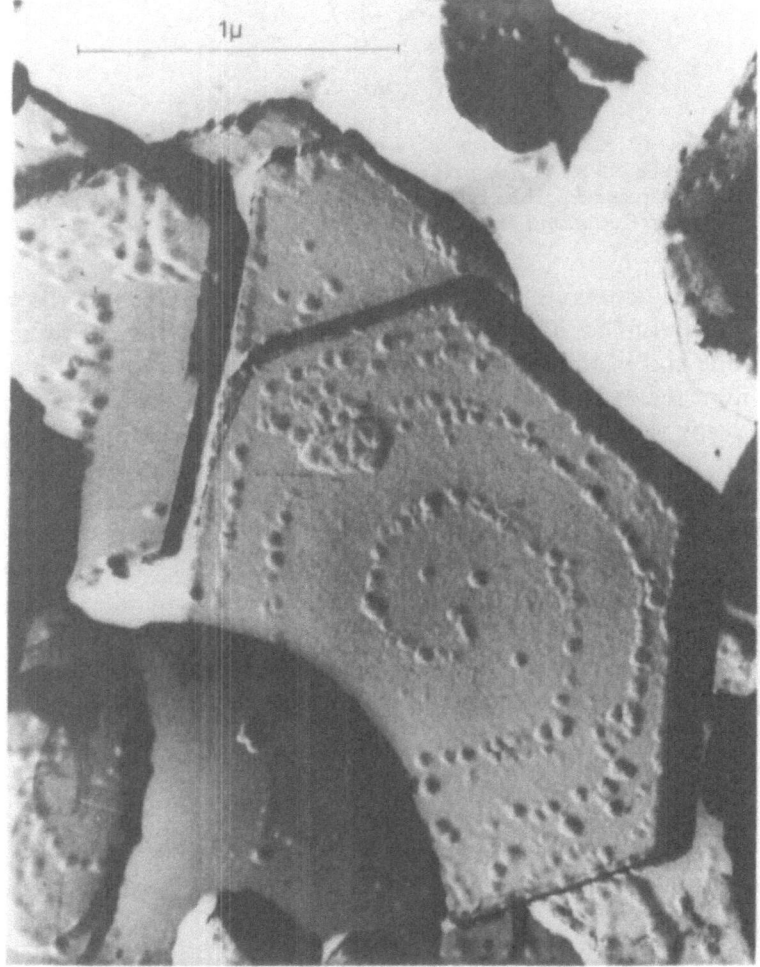

Fig. 44. Electron micrograph of α-TiCl$_3$ crystals containing polypropylene molecules at crystal edges and along spiral dislocation. (From work of Rodriguez and van Looy,[282] courtesy of Boudart.[103])

olefin adsorption and polymerization. Independent of the shape and size of the α-TiCl$_3$ crystal, the number of these vacancies is approximately equal to the number of Ti^{3+} ions located on the *edge* of the hexagonal platelets. Furthermore, electrostatic calculations indicate that Cl$^-$ ions are more likely to be absent from the crystal edge than from the (0001) faces. Thus, polymerization should occur specifically at edge sites. These basic conclusions, arrived at theoretically in 1963, have been confirmed by dramatic electron micrographs of α-TiCl$_3$ crystals following their use in propylene polymerization. As shown in Figure 44, Rodgriguez and Van Looy[282] have observed that polypropylene is selectively formed at the platelet edges (black), as well as along spiral dislocations (white) within the (0001) plane. Both of these regions would be expected to be Cl$^-$ deficient. Boudart[103] has pointed out that not every site in the spiral dislocation is active, perhaps because of the requirement of very specific geometry for a Cl$^-$-deficient center to be active for polymerization. Burwell[273] suggests that while the mechanism proposed above may be inaccurate in detail, it represents a historical landmark in heterogeneous catalysis because of the detailed understanding which has been made possible through consideration of stereochemical data.

7. Reduction and Enhancement of Catalytic Activity: Poisons and Promoters

Two factors of immense practical importance in industrial catalysis are (a) the degradation of catalyst activity as a function of time in use and (b) the enhancement of catalytic activity by the addition of promoter materials to the "pure" catalyst. Connecting catalyst deactivation and catalyst promotion is the fact that both of these processes proceed via alteration of the structural and/or electronic properties of the initially pure catalyst. Details of these electronic and structural modifications are explored in the following paragraphs.

7.1. Catalyst Deactivation

The decay in activity of a catalyst as a function of time in use may be due to a number of factors.[283] (a) Certain catalysts are poisoned by impurities in the feed gas, (b) others sinter and lose surface area under reaction conditions. (c) Under some conditions, catalysts may be self-poisoned via the deposition of by-products on the surface from decomposition of the reactants or products. In this case, the poisons can frequently be removed in a regeneration step in which the catalyst can be brought back to its original activity. (d) The surfaces of certain catalysts recrystallize under reaction conditions to expose facets having a symmetry different from the as-prepared catalyst. (e) During the catalytic process, the oxidation state of the catalyst may change. In the present discussion, the emphasis will be on poisoning by impurities, self-poisoning, and surface recrystallization. Sintering will be discussed in Section 7.2, since an important use of promoters is for prevention of sintering.

Catalyst poisoning by impurities in the reactants is a serious problem in a number of applications. One such problem of current concern for energy production is the catalytic methanation reaction[284] (the hydrogenation of carbon monoxide to methane), the crucial reaction in the manufacture of methane gas from coal. With the nickel catalysts presently employed in methanation reactors, it is usual to limit sulfur compounds in the $H_2 + CO$ feed gas to less than 1 ppm in order to avoid deactivation by sulfur poisoning. Despite the realization that sulfur adversely affects catalyst performance, there is little understanding of the mechanisms by which catalyst poisons act. As an example of the complexities involved in the understanding of poisoning processes, the most active methanation catalysts, Ni and Ru, are both highly susceptible to sulfur poisoning where the less active metals Mo and W are resistant to sulfur poisoning, and are frequently sulfided before use as methanation catalysts! In another example, the use of catalysts for the removal of CO, hydrocarbons, and NO from automotive exhaust, it was found that not only sulfur, lead, and halides from gasoline and gasoline additives but also phosphorus from motor oil additives poisoned the catalysts.

In spite of complexities of catalyst poisoning, answers are just starting to appear to such questions as: Does the poison inhibit the reaction by blocking active sites on a one-to-one basis, or does the poison operate via a long-range electronic interaction through the substrate (see Section 4.4)? Alternatively, does the presence of sulfur compounds result in a restructuring or recrystallization of the catalyst surface?

Because such small traces of impurities are able to poison catalysts, experimental studies are frequently limited by the sensitivity of chemical analytical methods. Several recent studies[285,286] of catalyst poisoning using Auger electron spectroscopy (AES) have demonstrated the utility of a specifically surface-sensitive technique

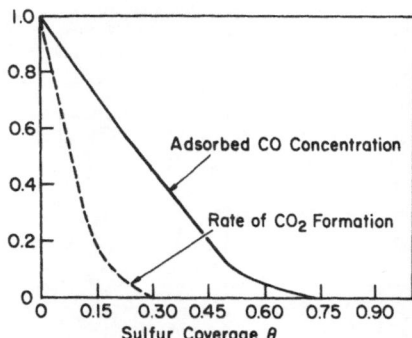

Fig. 45. Effect of sulfur as a poison in the oxidation of CO to form CO_2 on a (110) Pt surface. Composite plot of the normalized adsorbed CO concentration (expressed as a fraction of a monolayer) and normalized rate of CO_2 formation, both shown as a function of calibrated sulfur coverage (θ is the fraction of a monolayer of sulfur). (From Bonzel and Ku.[286])

for such investigations (as indicated in Section 5.2.1, AES is sensitive to the top few atomic layers of a solid). Bhasin[285] has shown that a surface concentration of 13 atoms of Pb per 100 surface atoms of Cu is sufficient to cause a dramatic reduction in the ability of powdered copper catalysts to catalyze the reaction of methyl chloride with silicon. Bonzel and Ku[286] used low-energy electron diffraction (LEED) and AES in a comprehensive study of the effect of sulfur on the catalytic oxidation of CO ($2CO + O_2 \rightarrow 2CO_2$) on a Pt(110) surface. They demonstrated that a very low sulfur coverage on Pt is needed to poison this reaction; as shown in Figure 45, the steady-state rate of CO_2 production decreases to less than 10% of its initial value at a sulfur coverage of ~ 0.20 monolayer. Figure 45 also shows that the concentration of adsorbed CO on the S-covered surface does not decrease as rapidly as the CO_2 production. This observation, when coupled with other kinetic evidence, led to the conclusion that the sulfur poisoning of CO oxidation on Pt(110) is correlated solely with the adsorption of oxygen on the partially S-covered surface. That is, the presence of sulfur limits the dissociative adsorption of oxygen. They reason that this is largely a site-blocking effect, but cannot eliminate the possibility of a longer range electronic effect.

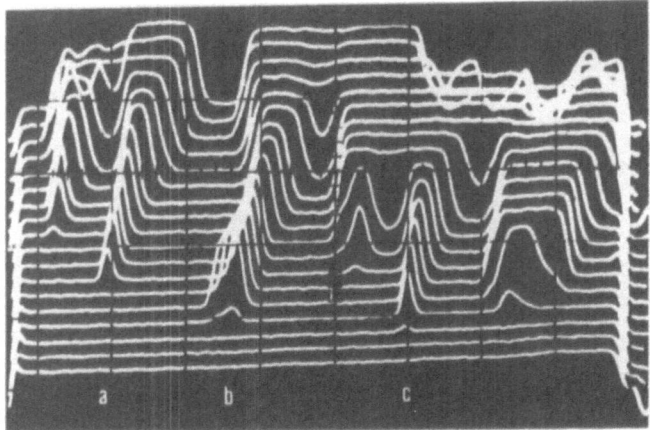

Fig. 46. Illustration of the non-uniform distribution of S left on an initially uniformly S-covered Pt(110) surface following reaction with O_2 at 2×10^{-7} Torr, $T_{(Pt)} = 350°C$. The curves are oscilloscope traces of the maximum in the sulfur Auger peak as a function of position on the Pt surface, and indicate that isolated patches of S are formed during the reaction. Note the ridges a, b, and c. (From Bonzel and Ku.[286])

Bonzel and Ku[286] also examined the interaction between adsorbed sulfur and oxygen on the Pt(110) surface, and found that SO_2 was the reaction product. For low sulfur coverage, $\theta < 0.28$, the reaction was quite fast, whereas for higher sulfur concentrations, $0.28 < \theta_s < 0.75$, the reaction became very slow. In this latter range, the reaction proceeded inhomogeneously, and resulted in a nonuniform distribution of adsorbed sulfur. Figure 46 shows several oscilloscope traces resulting from a two-dimensional scan of the sample surface using AES. These traces clearly indicate that the sulfur remains on the surface in discrete "patches" following the oxidation reaction and that under these conditions, the poisoned surface consists of a mixture of clean regions and sulfur-covered regions.

In addition to poisoning by impurities, catalysts are also susceptible to self-poisoning by the reactants, products, or both. In a careful study of self-poisoning, Hegedus and Petersen[287] have applied a method known as the single pellet diffusion reactor technique to the problem of poisoning of a supported Pt catalyst during the hydrogenation of cyclopropane to propane ($H_2 + C_3H_6 \rightarrow C_3H_8$). The technique allows one to discriminate among impurity poisoning, parallel self-poisoning, series self-poisoning, and triangular self-poisoning, whose poison precursors are, respectively, an impurity in the feed gas, the reactant itself, the product itself, or both the reactant and the product. By careful experimentation so as to eliminate impurities, and by comparison of data with the theoretical models, they concluded that the catalyst was poisoned by both the reactant cyclopropane (A) and product propane (B) (triangular self-poisoning) at temperatures between 50 and 75°C according to the mechanism

where X is the poison structure on the catalyst surface. The exact nature of X is not well-defined, but it is clearly a carbonaceous residue due to decomposition of A and B.

Another example of self-poisoning is the fouling of a methanation catalyst due to carbon deposition. Mills and Steffgen[284] have tabulated the effects of temperature, pressure, and H_2/CO ratios on the degree of carbon deposition. In this case, equilibrium thermodynamics was used to define the boundaries above and below which extensive carbon deposition occurs.

As indicated previously, the effects of self-poisoning are often reversible, and such poisoned catalysts may frequently be returned to their original activity by regeneration steps such as changing feed stock ratios, catalyst temperature, etc.

Somorjai[9] has pointed out that deposition of a carbonaceous overlayer on the surface of a catalytically active metal does not always result in self-poisoning. An *ordered* carbonaceous overlayer on a stepped Pt single-crystal surface is necessary for high activity in the catalytic dehydrocyclization of *n*-heptane to toluene. In contrast, however, a *disordered* carbon-containing layer will inhibit the same reaction. Thus, the real catalyst in this reaction is not the clean Pt metal, but rather the metal surface covered in a special way with a well-defined carbonaceous layer.

Another possible mechanism of catalyst poisoning involves faceting, i.e., changes in surface structure, in the presence of impurities adsorbed on the surface. There is good evidence that in some cases, Pt and Pt alloy surfaces develop crystal planes having (100) orientation in the presence of sulfur.[288] As Somorjai[289] has pointed out, adsorbed impurities change the free energy of a surface. If the foreign atoms alter the surface free energy of various crystal planes by different amounts, they can induce the rearrangement of the surface structure to form crystal planes that have lower surface free energy in the presence of the adsorbed impurity than in the absence of the impurity.

If this model of poisoning is correct, then "structure-sensitive" reactions should be affected by impurities that induce recrystallization, whereas "structure-insensitive" reactions should be relatively unaffected. To date, there does not appear to be enough evidence to draw a firm conclusion concerning the role of surface recrystallization in catalyst poisoning, but it is a promising area of investigation.

7.2. Enhancement of Catalytic Activity: Promoters

It has long been recognized that the activity of catalysts for certain reactions may be enhanced by the incorporation of small quantities of specific additives (promoters) during the preparation of the catalyst. Two well-studied examples of catalyst promotion involve the use of oxide promoters in iron catalysts for ammonia synthesis, and the role of Ca promotion in the Ag-catalyzed oxidation of ethylene.

In a pioneering series of experiments, Emmett and Brunauer[290] showed that the addition of promoter oxides such as Al_2O_3, K_2O, and Na_2O increases considerably the total surface area per gram of a reduced iron synthetic ammonia catalyst. By using gas adsorption techniques to determine surface areas, they found that iron synthetic ammonia catalysts promoted with small amounts of oxides ($\sim 1\%$) have $\sim 60\%$ of their surface covered with promoter. More recently, Solbakken et al.[291] have verified this earlier result using an isotope exchange method, and further showed that the promoter molecules (in this case, Al_2O_3 in Fe) are present at the surface in a single atomic layer.

Alumina incorporated in iron catalysts is known to be a textural promoter which prevents sintering; sintering of unpromoted iron particles occurs quite easily, whereas singly promoted (one promoter oxide) catalysts have a much better resistivity toward sintering. Thus, at least part of the mechanism for the promotion of catalytic activity is the fact that the surface area of the catalyst is not reduced by sintering during the course of reaction.

Topsoe et al.[292] have challenged the traditional view that textural promotion of Fe by Al_2O_3 is purely a surface effect. They recognized, as have others, that a major fraction of the promoter must remain inside the iron crystallites, even though the surface has a high coverage of promoter. Using Mössbauer spectroscopy, they found that the Al_2O_3 is present in the iron in the form of ~ 30-Å inclusions of $FeAl_2O_4$ in the incompletely reduced catalyst and of

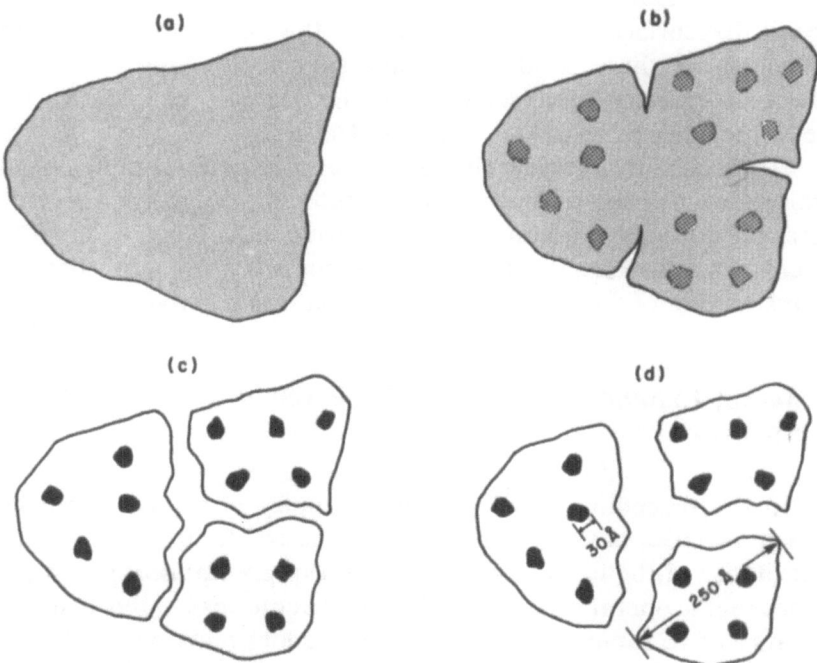

Fig. 47. Model of the effect of a textural promotor (Al_2O_3) on the reduction of an iron synthetic ammonia catalyst. (a) Unreduced large catalyst particle with the promotor distributed homogeneously. (b) Catalyst after a short reduction; aluminum-rich regions appear. (c) Catalyst after further reduction consists of pure α-Fe and $FeAl_2O_4$ inclusions. (d) Fully reduced catalyst. Consists of small α-Fe particles with Al_2O_3 inclusions. (From Topsoe *et al.*[292])

Al_2O_3 itself after thorough reduction. They suggest that the existence of strain in the iron particles created by the occluded Al_2O_3 phase will shift the equilibrium particle size to smaller particles. Thus, small catalyst particles are stabilized against sintering by the occluded promoter. Their model of the effect of Al_2O_3 on the reduction process of an iron catalyst is shown in Figure 47.

The remarkable selectivity of silver catalysts in the oxidation of ethylene to ethylene oxide has also been a subject of intense investigation. Kilty and Sachtler[293] have reviewed the literature concerning this reaction, and conclude that the formation of diatomic adsorbed oxygen (O_2^-) on the catalyst surface is the key to the selective oxidation reaction. Furthermore, the selectivity is enhanced by a CaO promoter in technical Ag catalysts. Forsalti *et al.*[294] found an increase in selectivity upon irradiating a supported Ag catalyst by γ-rays, and found, using ESCA (electron spectroscopy for chemical analysis, a specifically surface-sensitive technique), that Ca appeared on the surface under these conditions. They suggest that the correlation between selectivity enhancement and the presence of surface Ca is due to the adsorption of O_2^- on these Ca sites.

In conclusion, it is clear that understanding of the complexities of catalyst deactivation and promotion is receiving a new impetus from the application of modern surface-sensitive techniques to catalytic systems. It is anticipated that we may soon have a more detailed understanding of factors which cause a catalyst's activity to vary.

8. Recent Examples of Catalytic Research on Well-Defined Surfaces

In the preceding pages, several examples of catalysis on well-defined surfaces have already been discussed. In many cases, the insensitivity of the catalyst to the method of preparation provides a surface that is operationally well-defined. In some cases, this behavior appears due to rapid solid-state diffusion under reaction conditions. Sometimes, a degree of confidence as to the nature of the catalytic surface may be gained by crushing crystals to obtain fresh surface area of the same composition as the bulk. This method is particularly suitable for the oxide catalysts discussed in Section 5. Where applicable, the use of the UHV techniques of surface physics is of much advantage, and a degree of characterization can then be obtained by LEED, AES, and other spectroscopies. The presence of steps and of some point defects is often still difficult to detect by these methods. Even so, high-index faces of Pt have been well-characterized by these methods, as discussed in Section 6. In the present section, some additional examples are discussed to illustrate the use of UHV methods in catalytic research.

8.1. Catalytic Decomposition of Ammonia on Tungsten Single Crystals

An interesting experiment to accurately compare the catalytic activity of different single-crystal surfaces is the recent work of McAllister and Hansen.[295] These workers used standard ultrahigh procedures to prepare clean tungsten crystals; the catalyst crystals were then exposed to NH_3 at pressures up to 10^{-1} Torr. The crystals were heated with a focused beam of light to temperatures in the range

800–970°K and the kinetics of catalytic decomposition was studied. It was shown that NH_3 decomposes stoichiometrically on all three crystal faces into N_2 and H_2 following the rate expression

$$\text{rate} = A + BP_{NH_3}^{2/3}$$

No influence of N_2 or H_2 on the decomposition kinetics was detected over a wide partial pressure range in accordance with the above expression. A significant retardation of the decomposition rate on W(111) was observed when ND_3 was substituted for NH_3 ($B_{NH_3} = 1.47 B_{ND_3}$), indicating that the breaking of an N–H bond is involved in the rate-limiting step at higher NH_3 pressures, where the B term becomes dominant. It is proposed that the "B" process occurs on a substantially complete nitrogen adlayer of stoichiometry W–N, with complex intermediates being formed which eventually liberate N_2 and H_2, leaving behind W–N.

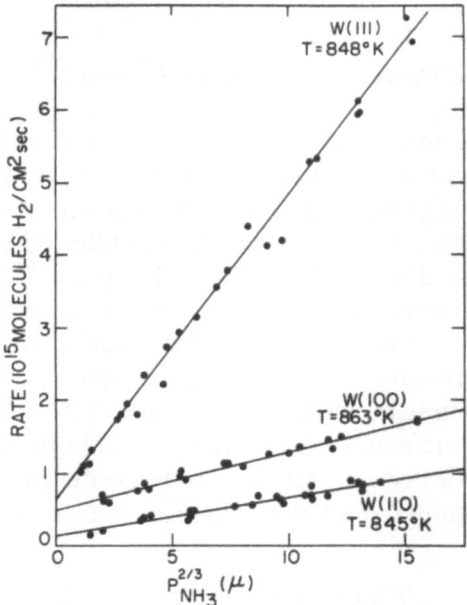

Fig. 48. Comparison of NH_3 decomposition rates on three tungsten single crystals at similar temperatures. The tungsten (111) crystal is the most effective catalyst of the three. The pressure is in units of 10^{-3} Torr. (From McAllister and Hansen.[295])

At low NH_3 pressures, the "A" process contributes significantly to the overall rate, and the model devised suggests that a surface process

$$2W\text{--}N \longrightarrow W_2N + \tfrac{1}{2}N_2(g)$$

is rate limiting. The rate of N_2 liberation observed at low NH_3 pressures (i.e., the value of A) is of the same order of magnitude as that observed in thermal desorption experiments from pure nitrogen adlayers on polycrystalline surfaces, thereby giving support to the model proposed.

In Figure 48, a comparison is given of the rates of decomposition of NH_3 on the three tungsten crystals. Although the crystal temperatures differ slightly in the figure, it can be seen that the W(111) plane is significantly more active than either W(100) or W(110). No explanation is offered for the enhanced activity of the W(111) plane. This impressive measurement thus forms a sound basis for future theoretical considerations of the effect of surface structure and perhaps electronic factors on the rate of catalytic decomposition of NH_3.

8.2. Oxidation of Carbon Monoxide by Platinum

One of the most thoroughly investigated catalytic reactions through the years has been the oxidation of CO by Pt. This reaction has been both intriguing and vexing, starting with the classic work of Langmuir[296] and ranging to the recent studies on single crystals of Pt using modulated molecular beam techniques.[297] The same basic questions are as important today as they were over 50 years ago, namely, is the basic mechanism of the reaction unique, or does it depend on experimental boundary conditions? Two points of view concerning mechanism have been advanced.[298] In the so-called "Langmuir–Hinshelwood" (LH) mechanism, both reactants have to be adsorbed on the surface of the catalyst in order to form the reaction product. For example, the LH mechanism in CO oxidation is

$$CO(a) + O(a) \longrightarrow CO_2(g) + 2M$$

where M represents a bare site on the surface. In the "Rideal–Eley" (RE) mechanism, only one reactant has to be firmly adsorbed on the catalyst while the other has to strike the adsorbate from the gas phase in order to form a bond between them. For example, a possible RE

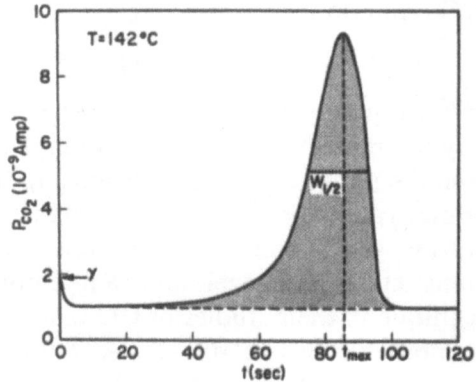

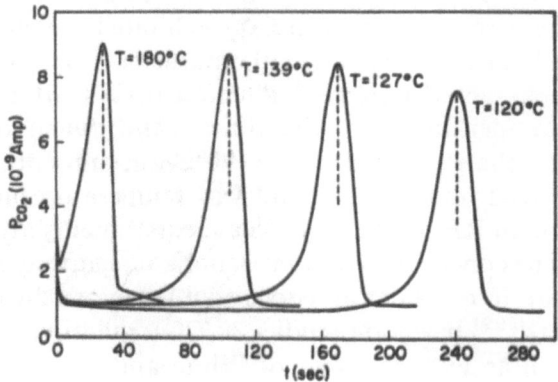

Fig. 49. *Upper figure*: Langmuir–Hinshelwood (LH) mechanism in CO oxidation. Plot of the partial pressure of CO_2 as a function of time, with the Pt(110) crystal at 142°C in an O_2 atmosphere. At time $t < 0$, the gas phase contains CO + O_2 as well as product CO_2; at $t = 0$, the partial pressure of CO is suddenly decreased. Since CO_2 formation increases and goes through a maximum at t_{max}, it follows that *only the CO present on the surface can contribute to the CO_2 formation*. Also, the strong temperature dependence of t_{max} suggests that oxygen must replace CO adsorbed on the surface in order for the reaction to proceed. This means that CO_2 is formed by a reaction between adsorbed CO and adsorbed oxygen (LH). *Lower figure*: Theoretical plots of CO_2 evolution curves based on the LH mechanism for CO_2 formation. Both the time at which the maximum occurs and the width of the peak are functions of the sample temperature. Note the correspondence between the upper experimental plot and these curves. (From Bonzel and Ku.[298])

mechanism in CO oxidation is

$$O(a) + CO(g) \rightarrow CO_2(g) + M$$

Langmuir[296] originally proposed that "oxygen is removed from the surface not by the reaction of adjacent CO molecules, but by the action of CO from the gas phase" (RE mechanism). In a recent series of experiments, Bonzel and Ku[298] have reported that the two different reaction mechanisms were found to prevail depending on the experimental conditions chosen. Bonzel and Ku employed a kinetic perturbation technique in their studies of CO oxidation on a Pt(110) surface. In the temperature range $100 < T < 220°C$, where CO was preadsorbed on the Pt surface, the subsequent adsorption of O_2 was dependent on desorption of part of the CO as shown in Figure 49, and the resultant reaction to form CO_2 exhibited the characteristics of an LH mechanism. In this case, the onset of CO_2 formation was delayed by a characteristic time (induction period) which depended strongly on temperature. On the other hand, when oxygen was preadsorbed on the Pt surface at $T > 90°C$, the subsequent reaction with CO occurred immediately, and was temperature independent, consistent with an RE mechanism. These experiments are illustrated in Figure 50. The conclusions of Bonzel and Ku regarding the reaction mechanisms are in excellent agreement with the conclusions of Ertl and co-workers[299] based on studies of CO oxidation on Pd single crystals, as well as with the data of White and co-workers[300] for polycrystalline Pd. Unanimity regarding the mechanism of CO oxidation on Pt has not yet been reached, however; in a recent molecular beam study, Palmer and Smith[297] have concluded that *all* of their CO oxidation results are consistent with an LH mechanism.

8.3. Decomposition of Formic Acid by Metals

Studies of the catalytic decomposition of formic acid (HCOOH) have long fascinated research workers for a variety of reasons, including the simplicity of the major products (H_2, CO, H_2O, CO_2) and the existence of different reaction paths.[301,302] Formic acid can decompose along three different pathways:

$$HCOOH \left\{ \begin{array}{ll} \rightarrow H_2 + CO_2 & (1) \\ \rightarrow H_2O + CO & (2) \\ \rightarrow H_2O + CO_2 + H_2CO & (3) \end{array} \right.$$

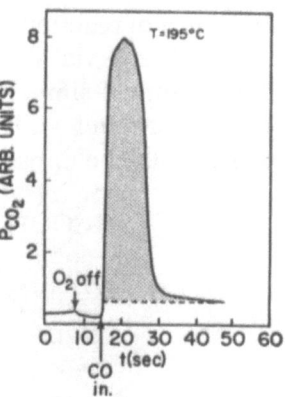

Fig. 50. Rideal–Eley (RE) mechanism in CO oxidation. Plot of partial pressure of CO_2 as function of time with the sample in a CO atmosphere. At $t < 0$, the sample is exposed to O_2, and the oxygen flow is stopped at 8 sec. At $t = 15$ sec, CO is admitted to the sample and, immediately, CO_2 is evolved. The CO_2 evolution rate is dependent on CO pressure but independent of temperature at $T > 90°C$, indicating a RE mechanism involving reaction of O(a) and CO(g) to make CO_2. (From Bonzel and Ku.[298])

Reaction (3) is not observed on metals; the dehydrogenation reaction (1) is considered to be the major reaction on metal catalysts, and the dehydration reaction (2) seems to be specific to certain oxide catalysts. In the following paragraphs, we will attempt to summarize what is known about this reaction based on two types of experiments:

(a) Studies of decomposition of HCOOH on finely dispersed metals, powders, films, and wires using kinetic methods combined with infrared absorption spectroscopy; HCOOH pressures were in the range greater than a few millitorr in these studies.

(b) Studies of decomposition of HCOOH on single crystals of Ni under ultrahigh vacuum conditions, using LEED, AES, and flash desorption mass spectrometry.

One of the primary objectives of the studies of decomposition of HCOOH has been to identify the catalytic intermediate and to correlate the nature of the intermediate with the specificity of the reaction. Kinetic methods, coupled with isotope substitution techniques, provided the first information concerning the mechanism of decomposition. It was speculated rather early (1918) that the formation of metal formates might play a role, but the application of infrared spectroscopy (1959–1960) to the study of HCOOH on metal catalysts greatly clarified the picture; the background of the kinetic and infrared studies is reviewed by Mars *et al.*[301] In a combination kinetic and infrared study of the decomposition of HCOOH on a Ni-on-silica catalyst, Fahrenfort and Sachtler[17] found that (a) formate ions (HCOO⁻) exist on the surface during the decomposition reaction, and (b) decomposition of the formate is the rate-controlling step in the formation of reaction products. Further evidence that the decomposition proceeds via a formate reaction intermediate is based on the observation of similarities between the stability and decomposition of bulk formates and the catalytic activity of the relevant metal or oxide. Thus, the apparent mechanism is

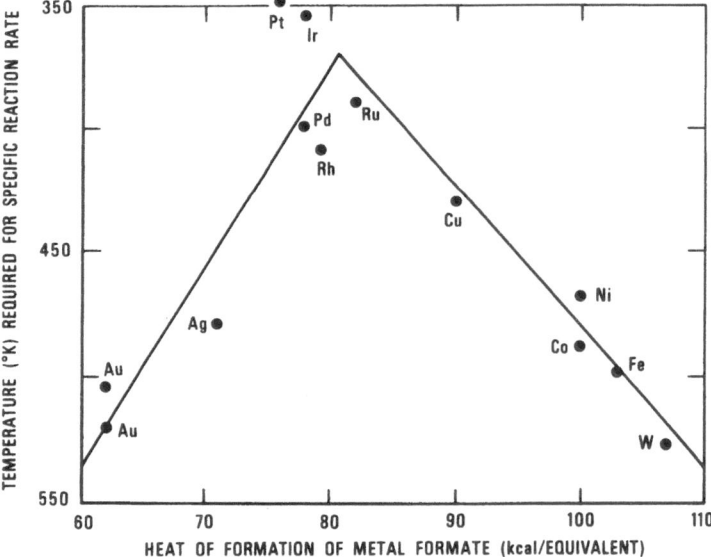

where the structure of the intermediate is only schematically illustrated. Dependent on experimental conditions of temperature and pressure, the rate-controlling step shifts from step A to step B or C on different metal catalysts. Fahrenfort and Sachtler[17] have plotted a "volcano curve" in which a relationship between the catalytic activity of heavy metals and the heats of formation of their formates is

Fig. 51. Correlation between catalysis and thermochemistry: relationship between the catalytic activity of different metals for the decomposition of formic acid (HCOOH) and the corresponding heat of formation for the bulk metal formate. The temperature required for a specific rate of decomposition on a metal is plotted vs. the heat of formate formation. A low value of temperature implies a low activation energy for the catalytic decomposition. (From Sachtler and Fahrenfort.[17])

observed (Figure 51). They concluded that the noble metals like Au and Ag exhibit a high activation energy for step A and a relatively unstable formate, whereas on other metals (Ni, Co, Fe, W) the stability of the formate is so high that step B or C governs the overall reaction rate. The best catalysts are those metals (Pt, Ir, etc.) whose formate stabilities are between the two extremes, so that neither the formation nor the decomposition of the formate intermediate will severely limit the reaction rate.

It should be cautioned that the present treatment of this complex problem is necessarily superficial, and that a closer look at the details of formation, structure, and decomposition of the intermediate reveals a host of difficulties. This has also been demonstrated recently in an elegant series of experiments in which Madix and his colleagues[303-305] have examined the decomposition of HCOOH on both clean and carbon-covered Ni(110) single-crystal surfaces using LEED, AES, and thermal desorption mass spectrometry.

One of the first surprises to come from this work was the observation that desorption of adsorbed HCOOH on Ni(110) yielded four

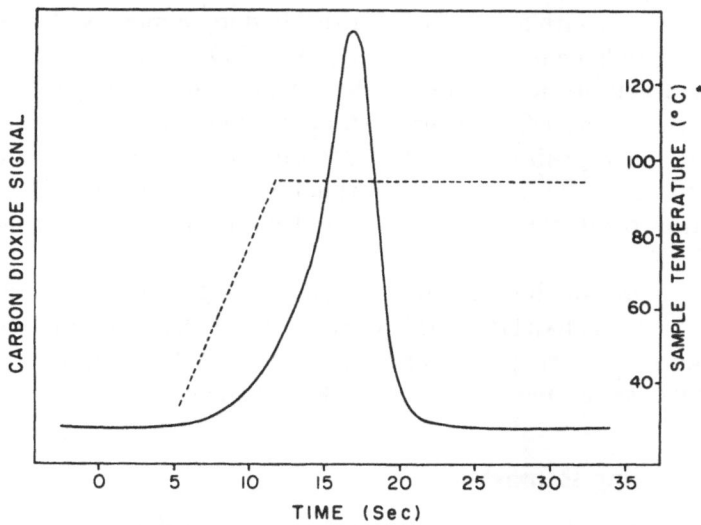

Fig. 52. Autocatalytic decomposition of HCOOH on a Ni(100) surface. The HCOOH-covered surface was heated in vacuum (dashed curve indicates sample temperature vs. time) and decomposition was monitored by observing the evolution of CO_2 from the surface. At constant temperature, the reaction rate *increased* as surface coverage *decreased*, evidence of explosive decomposition. (From Falconer et al.[304])

decomposition products (H_2, CO_2, H_2O, and CO) whose concentration is a function of the initial coverage.[303] At a coverage of one monolayer, the ratio CO_2:CO was found to be 1:1. The previously reported selectivity of Ni for the dehydrogenation reaction is not seen when the Ni substrate is atomically clean prior to adsorption.

Second, the kinetics of decomposition of the adsorbed species cannot be ascribed to simple desorption kinetics; the desorption peaks of H_2, CO_2, and CO are much too narrow for desorption to be the rate-controlling process in their liberation (Figure 52). Falconer *et al.*[304] have shown that the narrow thermal desorption peaks were the result of a self-accelerating surface reaction; i.e., a surface explosion. To verify this, they observed the decomposition of adsorbed HCOOH at constant temperature by following the desorption of CO_2 from the surface. As shown in Figure 41, the HCOOH-covered sample was heated in vacuum to a temperature below the thermal desorption temperature and held constant. At constant *T*, the rate of reaction *increased* as surface coverage *decreased*, clear evidence for an "explosive" autocatalytic reaction. The details of this process are currently under study.

Finally, these workers[305] observed that contamination of the Ni(110) surface with a carbide layer resulted in an increased specificity for the dehydrogenation reaction (CO_2:CO $\sim$ 10:1), but a much decreased activity for HCOOH decomposition ($\sim$100 times slower than on the clean surface at certain temperatures). They also concluded that the decomposition of DCOOH on the carbide surface clearly indicated that the reaction intermediate was DCOO (dominant desorption products were H_2 at $\sim$290°K, followed by D_2 and CO_2 at $\sim$450°K).

One can conclude from this type of experiment that even an idealized system, like HCOOH decomposition on a well-characterized single crystal, is very complicated, and that the simplicity of the mechanism suggested by earlier work is illusory.

9. Concluding Remarks

This review has attempted to bring together the major streams of thought in the field of heterogeneous catalysis. One is derived from our experience with practical catalysts of high surface area; another is from the recent injection of modern methods of surface characterization into this field. In addition, we have shown how increasingly sophisticated methods of solid-state chemistry and theoretical chemistry are being brought into the field, which has thereby become one of the outstanding examples of interdisciplinary science. Since heterogeneous catalysis is essentially concerned with the enhancement of the rate of desirable reactions, we have chosen, where possible, to use kinetic measurements as a bridge between these two broad areas of research activity. In addition, it is apparent that linkages exist via other routes, such as the various spectroscopies, coordination chemistry, and stereochemistry. All of these ways of thinking about surface processes play a central role in recent developments within the field.

It seems clear that we are entering a period of intense experimental activity where much will be learned about the fundamental nature of surface processes. The problem of understanding heterogeneous catalysis is particularly difficult since even with the new surface measurement techniques, it is often necessary to reason indirectly about mechanisms and intermediates which are of catalytic import-ance. These difficulties have, so far, prevented the formulation of predictive theories of catalytic activity, although a number of semi-empirical correlations are now evident. It is anticipated that as we learn more of the atomic nature of surfaces and surface processes, we will stimulate theoretical activity directed to an understanding of the factors influencing the rate of catalytic reactions. This is the frontier of an exciting field of intellectual endeavor of immense importance to mankind.

Acknowledgment

The portion of this work completed at the National Bureau of Standards was supported in part by the Office of Naval Research.

References

1. J. T. Yates, Jr., Catalysis, *Chem. Eng. News*, **52**, 19–29 (1974).
2. R. L. Burwell, Jr., Heterogeneous catalysis, in *Catalysis—Progress in Research* (F. Basolo and R. L. Burwell, Jr., eds.), Plenum Press, London and New York (1973), pp. 51–74.
3. H. S. Taylor, A theory of the catalytic surface, *Proc. Roy. Soc. A* **108**, 105–111 (1925).
4. J. M. Thomas, Enhanced reactivity at dislocations in solids, *Advan. Catal.* **19**, 293–400 (1969).
5. J. M. Thomas and W. J. Thomas, *Introduction to the Principles of Heterogeneous Catalysis*, Academic Press, London and New York (1967).

6. G. C. Bond, *Catalysis by Metals*, Academic Press, London and New York (1962).

7. A. A. Balandin, Modern state of the multiplet theory of heterogeneous catalysis, *Advan. Catal.* **19**, 1–210 (1969).

8. M. Boudart, Catalysis by supported metals, *Advan. Catal.* **20**, 153–166 (1969).

9. G. A. Somorjai, The surface structure of and catalysis by platinum single crystal surfaces, *Catal. Rev.* **7**, 87–120 (1972).

10. D. W. Griffiths and M. L. Bender, Cycloamyloses as catalysts, *Advan. Catal.* **23**, 209–261 (1973).

11. M. Boudart, Pauling's theory of metals in catalysis, *J. Am. Chem. Soc.* **72**, 1040 (1950).

12. O. Beeck, Hydrogenation catalysts, *Disc. Faraday Soc.* **8**, 118–128 (1950).

13. L. Pauling, A resonating-valence-bond theory of metals and intermetallic compounds, *Proc. Roy. Soc. A* **196**, 343–362 (1949).

14. J. H. Sinfelt, Catalysis by metals, *Catal. Rev.* **9**, 147–168 (1974).

15. A. T. Gwathmey and R. E. Cunningham, The influence of crystal face in catalysis, *Advan Catal.* **10**, 57–95 (1958).

16. R. Van Hardeveld and F. Hartog, Influence of metal particle size on nickel-on-aerosil catalysts on surface site distribution, catalysis activity and selectivity, *Advan. Catal.* **22**, 75–114 (1972).

17. W. M. H. Sachtler and J. Fahrenfort, The catalytic decomposition of formic acid vapour on metals, in *Actes du Deuxième Congrès International de Catalyse*, Editions Technip, Paris (1960), pp. 831–852.

18. A. A. Holscher, Adsorption studies with the field-emission and field-ion microscope, Thesis, Leiden (1967).

19. Th. Wolkenstein, Effect of small quantities of impurities on the catalytic activity of ionic catalysts, *J. Phys. Chem. USSR* **24**, 1068–82 (1950).

20. Th. Wolkenstein, The electron theory of catalysis, *Advan. Catal.* **12**, 189–264 (1960).

21. K. Hauffe and H. J. Engell, The mechanism of chemical sorption from the standpoint of the disorder theory, *Z. Elektrochem.* **56**, 366–73 (1952).

22. K. Hauffe, in *Semiconductor Surface Physics* (R. H. Kingston, ed.), pp. 259–282, Univ. Pennsylvania Press, Philadelphia (1957).

23. P. B. Weisz, Electronic barrier layer phenomena in chemisorption and catalysis, *J. Chem. Phys.* **20**, 1483–84 (1952).

24. C. Wagner and K. Hauffe, The stationary state of catalysts in heterogeneous reactions, I., *Z. Electrochem.* **44**, 172–8 (1938).

25. F. S. Stone, in *Chemistry of the Solid State* (W. E. Garner, ed.), pp. 367–404, Academic Press, New York (1955).

26. C. Wagner, The mechanism of the decomposition of nitrous oxide on zinc oxide as catalyst, *J. Chem. Phys.* **18**, 69–71 (1950).

27. W. E. Garner, T. J. Gray, and F. S. Stone, The oxidation of copper and the reactions of hydrogen and carbon monoxide with copper oxide, *Proc. Roy. Soc. A* **197**, 294–314 (1949).

28. G. Parravano, The catalytic oxidation of carbon monoxide on nickel oxide I. Pure nickel oxide, II. Nickel oxide containing foreign ions, *J. Amer. Chem. Soc.* **75**, 1448–1454 (1953).

29. G. M. Schwab and J. Block, Über die Oxydation von Kohlenmonoxyd und den Zerfall von Distickstoffoxyd an definiert halbleitenden Oxygen, *Z. Physik Chem. N.F. (Frankfurt)* **1**, 42–62 (1954).

30. G. Parravano and M. Boudart, Chemisorption and catalysis on oxide semi-conductors, *Advan. Catal.* **7**, 50–74 (1955).
31. C. G. B. Garrett, Quantitative considerations concerning catalysis at a semi-conductor surface, *J. Chem. Phys.* **33**, 966–979 (1960).
32. F. Garcia-Moliner, The band picture in the electronic theories of chemisorption on semiconductors, *Catal. Rev.* **2**, 1–66 (1969).
33. H. Chon and C. D. Prater, Hall effect studies of carbon monoxide oxidation over doped zinc oxide catalysts, *Disc. Faraday Soc.* **41**, 380–393 (1966).
34. D. A. Dowden, N. Mackenzie, and B. M. W. Trapnell, The catalysis of H_2–D_2 exchange by oxides, *Proc. Roy. Soc. London* **237A**, 245–254 (1956).
35. D. A. Dowden, Crystal and ligand field models of solid catalysts, *Catal. Rev.* **5**, 1–31 (1972).
36. R. J. H. Voorhoeve and L. E. Trimble, Reduction of nitric oxide with carbon monoxide and hydrogen over ruthenium catalysts, *J. Catal.* **38**, 80–91 (1975).
37. R. W. Joyner, B. Lang, and G. A. Somorjai, Low pressure studies of dehydro-cyclization of *n*-heptane on platinum crystal surfaces using mass spectrometry, Auger electron spectroscopy, and low energy electron diffraction, *J. Catal.* **27**, 405–415 (1972).
38. R. J. H. Voorhoeve, J. P. Remeika, and L. E. Trimble, Defect chemistry and catalysis in oxidation and reduction over perovskite-type oxides, *Ann. N.Y. Acad. Sci.* (1976).
39. R. L. Banks and G. C. Bailey, Olefin disproportionation, *I&EC Product Res. Develop.* **3**, 170–173 (1963).
40. F. D. Mango and J. H. Schàchtschneider, in *Transition Metals in Homogeneous Catalysis* (G. N. Schrauzer, ed.), pp. 223–296, Marcel Dekker, New York (1971).
41. F. D. Mango, Molecular orbital symmetry conservation in transition metal catalysis, *Advan. Catal.* **20**, 291–326 (1969).
42. P. Mars and D. W. Van Krevelen, Oxidations carried out by means of vanadium oxide catalysts, *Chem. Eng. Sci. Suppl.* **3**, 41–59 (1954).
43. G. W. Keulks, The mechanism of oxygen atom incorporation into the products of propylene oxidation over bismuth molybdate, *J. Catal.* **19**, 232–235 (1970).
44. D. J. Hucknall, *Selective Oxidation of Hydrocarbons*, Academic Press, London (1974).
45. Y. Morooka and A. Ozaki, Regularities in catalytic properties of metal oxides in propylene oxidation, *J. Catal.* **5**, 116–124 (1966).
46. Y. Morooka, Y. Morikawa, and A. Ozaki, Regularity in the catalytic properties of metal oxides in hydrocarbon oxidation, *J. Catal.* **7**, 23–32 (1967).
47. O. A. Hougen and K. M. Watson, *Chemical Process Principles*, Part III, *Kinetics and Catalysis*, Wiley, New York (1947).
48. M. Boudart and R. L. Burwell, Jr., Mechanism in heterogeneous catalysis, in *Techniques of Chemistry*, Vol. 6, Pt. 1 (E. S. Lewis, ed.), John Wiley and Sons (1974), pp. 693–740.
49. J. E. Demuth and D. E. Eastman, Photoemission observations of π–d bonding and surface reactions of adsorbed hydrocarbons on Ni(111), *Phys. Rev. Letters* **32**, 1123–1127 (1974).
50. K. H. Johnson and R. P. Messmer, Clusters, chemisorption and catalysis, *J. Vac. Sci. Technol.* **11**, 236–242 (1974).
51. J. C. Slater and K. H. Johnson, Quantum chemistry and catalysis, *Physics Today* **1974** (October), 34–41.

52. C. L. Thomas, *Catalytic Processes and Proven Catalysts*. Academic Press, New York (1970), Ch. 1.
53. P. B. Weisz and C. D. Prater, Interpretation of measurements in experimental catalysis, *Advan. Catal.* **6**, 143–196 (1954).
54. P. H. Emmett (ed.), *Catalysis*, Vols. III–V, Reinhold, New York (1955, 1956, 1957).
55. F. G. Dwyer, Catalysis for control of automotive emissions, *Catal. Rev.* **6**, 261–292 (1972).
56. E. G. Rochow, *An Introduction to the Chemistry of the Silicones*. 2nd ed., Wiley, New York (1951).
57. R. J. H. Voorhoeve, *Organohalosilanes. Precursors to Silicones*. Elsevier, Amsterdam (1967).
58. F. G. Ciapetta and C. J. Plank, Catalysis preparation, in *Catalysis*. Vol. I, *Fundamental Principles* (P. H. Emmett, ed.), Reinhold, New York (1954), pp. 315–352.
59. H. Adkins and H. R. Billica, Preparation of Raney nickel catalysts and their use under conditions comparable with those for platinum and palladium catalysts, *J. Amer. Chem. Soc.* **70**, 695 (1948).
60. V. Haensel and R. Burwell, Catalysis, *Sci. Am.* **225**(6), 46–58 (1971).
61. V. Haensel, Catalytic cracking of pure hydrocarbons, *Advan. Catal.* **3**, 179–198 (1951).
62. G. C. A. Schuit and B. C. Gates, Chemistry and engineering of catalytic hydrodesulfurization, *AIChE J.* **19**, 417–438 (1973).
63. M. N. Berger, G. Boocock, and R. N. Hawarr, The polymerization of olefins by Ziegler catalysts, *Advan. Catal.* **19**, 211–241 (1969).
64. W. R. Kleckner and R. C. Sutter, in *Encyclopedia of Chemical Technology* Vol. II, 2nd ed., pp. 307–337, Interscience, New York (1966).
65. M. Leva, *Fluidization*. McGraw-Hill, New York (1959).
66. O. T. Zimmerman and I. Lavine, *Chemical Engineering Laboratory Equipment Design, Construction, Operation*. Industrial Research Service, Dover, N.H. (1955).
67. R. E. Kirk and D. F. Othmer (eds.), *Encyclopedia of Chemical Technology*. Interscience, New York (1963–70).
68. P. H. Emmett, Measurement of surface area, in *Catalysis*, Vol. I, Reinhold, New York (1965), pp. 63–67.
69. S. Brunauer, P. H. Emmett, and E. Teller, Adsorption of gases in multimolecular layers, *J. Amer. Chem. Soc.* **60**, 309–19 (1938).
70. J. E. Benson and M. Boudart, Hydrogen–oxygen titration method for the measurement of supported platinum surface areas, *J. Catal.* **4**, 704–710 (1965).
71. D. M. Young and A. D. Crowell, *Physical Adsorption of Gases*, Butterworths, London (1962).
72. J. H. Sinfelt, Catalytic hydrogenolysis over supported metals, *Catal. Rev.* **3**, 175–203 (1970).
73. J. E. Benson, H. S. Hwang, and M. Boudart, Hydrogen–oxygen titration method for the measurement of supported palladium surface areas, *J. Catal.* **30**, 146–153 (1973).
74. T. E. Whyte, Jr., Metal particle size determination of supported metal catalysts, *Catal. Rev.* **8**, 117–134 (1974).
75. H. Chon, R. A. Fisher, E. Tomeszko, and J. G. Aston, Chemisorption of hydrogen and oxygen on platinum black at low temperatures, in *Actes Congr. Intern. Catalyse 2e*, Paris 1960, Vol. 1, pp. 217–224.

76. E. L. Kugler, A. B. Kadet, and J. W. Gryder, The nature of NO adsorption on chromia, *J. Catal.* (to be published).

77. P. Stonehart and P. H. Zucks, Sintering and recrystallization of small metal particles. Loss of surface area by platinum-black fuel cell electro-catalysts, *Electrochim. Acta* **17**, 2333–2351 (1972).

78. R. P. Eischens, S. H. Francis, and W. A. Pliskin, The effect of surface coverage on the spectra of chemisorbed CO, *J. Phys. Chem.* **60**, 194–201 (1956).

79. J. C. Tracy and P. W. Palmberg, Structural influences on adsorbate binding energy. I. CO on (100) palladium, *J. Chem. Phys.* **51**, 4852–4862 (1969).

80. T. E. Madey, Adsorption of oxygen on tungsten (100): Adsorption kinetics and electron stimulated desorption, *Surface Sci.* **33**, 355–376 (1972); Chemisorption of H_2 on W(100), *Surface Sci.* **36**, 281–294 (1973).

81. S. E. Voltz, C. R. Morgan, D. Liederman, and S. M. Jacob, Kinetic study of carbon monoxide and propylene oxidation on platinum catalysts, *Ind. Eng. Chem. Prod. Res. Develop.* **12**, 294–301 (1973).

82. C. P. C. Bradshaw, E. J. Howman, and L. Turner, Olefin dismutation: reactions of olefins on cobalt oxide–molybdenum oxide–alumina, *J. Catal.* **7**, 269–276 (1967).

83. D. R. Kahn, E. E. Petersen, and G. A. Somorjai, The hydrogenolysis of cyclopropane on a platinum stepped single crystal at atmospheric pressure, *J. Catal.* **34**, 294–306 (1974).

84. Y. F. Yu Yao, The oxidation of hydrocarbons and CO over metal oxides IV. Perovskite-type oxides, *J. Catal.* **36**, 266–275 (1975).

85. P. K. Gallagher, D. W. Johnson, Jr., J. P. Remeika, F. Schrey, L. E. Trimble, E. M. Vogel and R. J. H. Voorhoeve, The activity of $La_{0.7}Sr_{0.3}MnO_3$ without Pt and $La_{0.7}Pb_{0.3}MnO_3$ with varying Pt contents for the catalytic oxidation of CO, *Mater. Res. Bull.* **10**, 529–538 (1975).

86. V. A. Dris'ko, Specific activity of metallic catalysts, *Russ. Chem. Rev.* **43**, 435–451 (1974).

87. P. B. Weisz and C. D. Prater, Interpretation of measurements in experimental catalysis, *Advan. Catal.* **6**, 143–196 (1954).

88. R. J. Madix and M. Boudart, Sticking probabilities by an effusive beam technique, the germanium–oxygen system, *J. Catal.* **7**, 240–251 (1967).

89. R. J. Madix and A. Susu, Reactive scattering of halogen molecules from ⟨111⟩ surfaces of silicon and germanium: Comparison with oxygenic species, *J. Catal.* **28**, 316–321 (1973).

90. S. L. Bernasek and G. A. Somorjai, Molecular beam study of the mechanism of catalyzed hydrogen–deuterium exchange on platinum single crystal surfaces, *J. Chem. Phys.* **62**, 3149–3161 (1975).

91. S. L. Bernasek and G. A. Somorjai, Small molecule reactions on stepped single crystal platinum surfaces, *Surface Sci.* **48**, 204–213 (1975).

92. D. O. Hayward, D. A. King, and F. C. Tompkins, Sticking probabilities, heats of adsorption and redistribution processes of nitrogen on tungsten films at 195 and 290°K, *Proc. Roy. Soc.* **297A**, 305–320 (1967).

93. A. A. Bell and R. Gomer, Adsorption of carbon monoxide on tungsten. Abundances, dipole moments, and sticking coefficients, *J. Chem. Phys.* **44**, 1065–1080 (1966).

94. I. Langmuir, Chemical reactions on surfaces, *Trans. Faraday Soc.* **17**, 607–675 (1922).

95. E. H. Taylor, The effects of ionizing radiation on catalysts, *Advan. Catal.* **18**, 111–258 (1968).

96. H. E. Farnsworth and R. F. Woodcock, Effects of radiation quenching, ion bombardment and annealing on catalytic activity of pure Ni and platinum surfaces, *Advan. Catal.* **9**, 123–130 (1957).

97. I. Yasumori, T. Kabe, and Y. Inoue, Radiochemical study of active sites on palladium *J. Phys. Chem.* **78**, 583–588 (1974).

98. J. W. Hall and H. F. Rase, Dislocations and catalysis, *Nature* **4893**, 858 (1963).

99. H. M. C. Sosnovsky, The catalytic activity of silver crystals of various orientations after bombardment with positive ions, *J. Phys. Chem. Solids* **10**, 304–310 (1959).

100. I. Matsuura and G. C. A. Schuit, Adsorption and reaction of adsorbed species on Bi_2MoO_6 catalysts, *J. Catal.* **25**, 314–325 (1972).

101. K. Aykan, A. W. Sleight, and D. B. Rogers, Defect control in oxidation catalysts, *J. Catal.* **29**, 185–187 (1973).

102. R. van Hardeveld and A. Van Montfoort, Infrared spectra of nitrogen adsorbed on nickel-on-aerosil catalysts. Effects of intermolecular interaction and isotopic substitution, *Surface Sci.* **17**, 90–124 (1969).

103. M. Boudart, Four decades of active centers, *Am. Scientist* **57**, 1, 97–111 (1969).

104. R. J. Kokes, in *Experimental Methods in Catalytic Research* (R. B. Anderson, ed.), pp. 436–476, Academic Press, New York (1968).

105. R. J. H. Voorhoeve, Electron spin resonance study of active centers in nickel–tungsten sulfide hydrogenation catalysts, *J. Catal.* **23**, 236–242 (1971).

106. R. J. H. Voorhoeve and H. B. M. Wolters, Tungsten sulfides obtained by decomposition of ammonium tetrathiotungstate, *Z. Anorg. Allgem. Chem.* **376**, 165–179 1970; **380**, 326 (1971).

107. R. J. H. Voorhoeve and J. C. M. Stuiver, The mechanism of the hydrogenation of cyclohexene and benzene on nickel–tungsten sulfide catalysts, *J. Catal.* **23**, 243–252 (1971).

108. A. L. Farragher and P. Cossee, Catalytic chemistry of molybdenum and tungsten sulfides and related ternary compounds, in *Proc. Fifth Internat. Congr. Catal.* (J. W. Hightower, ed.), pp. 1301–1318, North-Holland, Amsterdam (1973).

109. A. Aoshima and H. Wise, Hydrodesulfurization activity and electronic properties of molybdenum sulfide catalyst, *J. Catal.* **34**, 145–151 (1974).

110. G. C. A. Schuit, Structure of hydrodesulfurization catalysts, *Ann. N.Y. Acad. Sci.* (1976).

111. V. H. J. DeBeer, Some structural aspects of the $CoO-MoO_3-\gamma-Al_2O_3$ hydrodesulfurization catalyst, Thesis, Eindhoven (Netherlands), 1975.

112. J. H. Lunsford and T. W. Leland, Effects of neutron and ultraviolet irradiation on the catalytic activity of magnesium oxide, *J. Phys. Chem.* **66**, 2591–7 (1962).

113. C. G. Harkins, W. W. Shang, and T. W. Leland, Relation of the catalytic activity of MgO to its electron energy states, *J. Phys. Chem.* **73**, 130–141 (1969).

114. J. H. Lunsford, A study of irradiation-induced active sites on magnesium oxide using electron paramagnetic resonance, *J. Phys. Chem.* **68**, 2312–2316 (1964).

115. M. Boudart, A. Delbouille, E. G. Derouane, V. Indovina, and A. B. Walters, Activation of hydrogen at 78°K on paramagnetic centers of magnesium oxide, *J. Am. Chem. Soc.* **94**, 6622–30 (1972).

116. R. P. Eischens, W. A. Pliskin, and M. J. D. Low, The infrared spectrum of hydrogen chemisorbed on ZnO, *J. Catal.* **1**, 180–191 (1962).

117. R. J. Kokes, A. L. Dent, C. C. Chang, and L. T. Dixon, Infrared studies of isotope effects for hydrogen adsorption on zinc oxide, *J. Am. Chem. Soc.* **94**, 4429–4436 (1972).

118. R. E. Allen, G. P. Alldredge, and F. W. de Wette, Studies of vibrational surface modes. I. General formulation, *Phys. Rev. B* **4**, 1648–1660 (1971).

119. R. E. Allen, G. P. Alldredge, and F. W. de Wette, Studies of vibrational surface modes. II. Monoatomic fcc crystals, *Phys. Rev. B* **4**, 1661–1681 (1971).

120. G. P. Alldredge, R. E. Allen, and F. W. de Wette, Studies of vibrational surface modes. III. Effect of an adsorbed layer, *Phys. Rev. B* **4**, 1682–1697 (1971).

121. T. Wolfram and R. E. De Wames, Surface dynamics of magnetic materials, in *Progress in Surface Science*, Vol. 2, Part 4, pp. 233–330 (1972).

122. H. Suhl, Magnetic effects in surface reactions, in *AIP Conference Proceedings*, No. 18, Part 1, pp. 33–47 (1974).

123. P. Kumar, Magnetic phase transition at a surface: mean field theory, *Phys. Rev. B* **10**, 2928–2933 (1974).

124. P. Kumar and K. Maki, Surface spin fluctuations in paramagnetic phase, unpublished (1975).

125. K. L. Kliewer and R. Fuchs, Theory of dynamical properties of dielectric surfaces, *Advan. Chem. Phys.* **27**, 355–541 (1974).

126. J. A. Appelbaum and D. R. Hamann, Surface states and surface bonds of Si(111), *Phys. Rev. Lett.* **31**, 106–9 (1973).

127. J. A. Appelbaum and D. R. Hamann, Surface potential, charge density, and ionization potential for Si(111)—a self-consistent calculation, *Phys. Rev. Lett.* **32**, 225–228 (1974).

128. J. A. Appelbaum and D. R. Hamann, Electronic structure of Si and Ge surfaces, in *Proc. 12th International Conference on the Physics of Semiconductors*, Stuttgart, Germany, July 15–19 (1974).

129. J. A. Appelbaum and D. R. Hamann, Self-consistent quantum theory of chemisorption: H on Si(111), *Phys. Rev. Lett.* **34**, 806–809 (1975).

130. P. W. Anderson, Localized magnetic states in metals, *Phys. Rev.* **124**, 41–53 (1961).

131. D. Newns, Self-consistent model of hydrogen chemisorption, *Phys. Rev.* **178**, 1123–1135 (1969).

132. J. R. Smith, S. C. Ying, and W. Kohn, Charge densities and binding energies in hydrogen chemisorption, *Phys. Rev. Letters* **30**, 610–613 (1973).

133. J. R. Schrieffer and R. Gomer, Induced covalent bond mechanism of chemisorption, *Surface Sci.* **25**, 315–320 (1971).

134. T. B. Grimley, Chemisorption of some small molecules on transition metals, in *Molecular Processes on Solid Surfaces* (E. Drauglis, R. Gretz, and R. Jaffee, eds.), McGraw-Hill, New York (1969), pp. 299–316.

135. J. R. Schrieffer, Theory of chemisorption, *J. Vac. Sci. Technol.* **9**, 561–568 (1972).

136. J. R. Schrieffer and P. Soven, Surface physics: Theory of the electronic structure, *Physics Today* **1975** (April), 24–30.

137. A. Madhukar, Chemisorption on transition-metal surfaces: Electronic structure, *Phys. Rev. B* **8**, 4458–4463 (1973).

138. R. H. Paulson and J. R. Schrieffer, Induced covalent bond theory of chemisorption of hydrogen on a metal surface, *Surface Sci.* **48**, 329–352 (1975).

139. B. J. Thorpe, Chemisorption theory and the surface molecule, *Surface Sci.* **33**, 306–325 (1972).

140. T. L. Einstein, Short-chain model of chemisorption: Exact and approximate results, *Phys. Rev. B* **11**, 577–587 (1975).

141. A. P. Mortola, C. Geller, C. Melius, J. W. Moskowitz, and M. B. Bailee, A Theoretical model for chemisorption, *Abstracts ACS Mexico City Meeting*, (November 1975).

142. A. Van der Avoird, Some model calculations for adsorption on transition metals, *Surface Sci.* **18**, 159–177 (1969).

143. K. H. Johnson, J. G. Norman, Jr., and J. W. D. Connolly, in *Computational Methods for Large Molecules and Localized States in Solids* (F. Herman, A. D. McLean, and R. K. Nesbet, eds.), pp. 161–201, Plenum, New York (1973).

144. S. J. Niemczyk, A SCF-Xα-SW investigation of chemisorption bonding of chalcogens on nickel (001), *J. Vac. Sci. Technol.* **12**, 246–248 (1975).

145. I. P. Batra and O. Robaux, Electronic energy level structure calculations of chemisorbed CO on Ni(100), *J. Vac. Sci. Technol.* **12**, 242–245 (1975).

146. I. P. Batra and P. S. Bagus, Interpretation of the photoemission spectrum of chemisorbed carbon monoxide on Ni(100), *Solid State Commun.* **16**, 1097–1100 (1975).

147. J. T. Waber, H. Adachi, F. W. Averill, and D. E. Ellis, Molecular cluster study of the approach of CO to a Ni(100) surface, *Japan J. Appl. Phys.* (Suppl. 2, Pt. 2) **1974**, 695–697.

148. D. E. Eastman and J. K. Cashion, Photoemission energy-level measurements of chemisorbed CO and O on Ni, *Phys. Rev. Lett.* **27**, 1520–1523 (1971).

149. W. G. Klemperer and K. H. Johnson, *J. Chem. Phys.* (to be published).

150. W. H. Weinberg, The bond-energy bond-order (BEBO) model of chemisorption, *J. Vac. Sci. Technol.* **10**, 89–94 (1973).

151. W. H. Weinberg, H. A. Deans, and R. P. Merrill, The structure and chemistry of ethylene adsorbed on platinum (111), *Surface Sci.* **41**, 312–336 (1974).

152. R. Hoffmann, An extended Hückel theory. I. Hydrocarbons, *J. Chem. Phys.* **39**, 1397–1412 (1963).

153. C. J. Ballhausen and H. B. Gray, The electronic structure of the vanadyl ion, *Inorg. Chem.* **1**, 111–122 (1962).

154. R. F. Fenske, K. G. Caulton, D. D. Radtke, and C. C. Sweeny, A method for molecular orbital calculations for metal complexes, *Inorg. Chem.* **5**, 951–964 (1966).

155. R. Hoffman and R. B. Woodward, Selection Rules for Concerted Cycloaddition Reactions, *J. Amer. Chem. Soc.* **87**, 2046–2048 (1965).

156. F. D. Mango and J. H. Schachtschneider, Orbital symmetry restraints to transition metal catalyzed [2 + 2] cycloaddition reactions, *J. Am. Chem. Soc.* **93**, 1123–1130 (1971).

157. H. E. Simmons and J. F. Bunnett (eds.), *Orbital Symmetry Papers*, American Chemical Society, Washington, D.C. (1974).

158. J. Halpern, Catalysis of the rearrangements of strained hydrocarbons by transition metal compounds, *Proc. 14th Internat. Conference on Coordination Chemistry, Toronto*, pp. 698–703 (1972).

159. R. B. Woodward and R. Hoffman, Stereochemistry of electrocyclic reactions, *J. Am. Chem. Soc.* **87**, 395–397 (1965).

160. R. G. Pearson, Symmetry rules for chemical reactions, *Accts. Chem. Res.* **4**, 152–160 (1971).

161. V. I. Labunskaya, A. D. Shebaldova, and M. L. Khidekal', Catalysis of symmetry-disallowed reactions, *Russ. Chem. Rev.* **43**, 1–16 (1974).

162. L. Kleinman, Orbital-symmetry rules for chemisorption, *Phys. Rev. B* **9**, 1989–1992 (1974).

163. A. B. Anderson and R. Hoffmann, Molecular orbital studies of dissociative chemisorption of first period diatomic molecules and ethylene on (100) W and Ni surfaces, *J. Chem. Phys.* **61**, 4545–4559 (1974).

164. D. Menzel, Investigations of adsorption on metal surfaces by photoelectron spectroscopy, combined with other methods, *J. Vac. Sci. Technol.* **12**, 313–323 (1975).

165. L. Hedin, in *Electrons in Crystalline Solids, Proceedings of the Seventh International Course at Trieste, 1972*, pp. 665–731, International Atomic Energy Agency, Vienna (1973).

166. R. P. Eischens, W. A. Pliskin, and S. A. Frances, Infrared spectra of chemisorbed carbon monoxide, *J. Chem. Phys.* **22**, 1786–1787 (1954).

167. Extensive references to the work of A. Terenin are contained in Ref. 168.

168. L. H. Little, *Infrared Spectra of Adsorbed Species*, Academic Press, London (1966).

169. M. L. Hair, *Infrared Spectroscopy in Surface Chemistry*, Marcel Dekker, New York (1967).

170. A. C. Yang and C. W. Garland, Infrared studies of carbon monoxide chemisorbed on rhodium, *J. Phys. Chem.* **61**, 1504–1512 (1957).

171. L. F. Dahl, C. Martell, and D. L. Wampler, Structure of and metal–metal bonding in $Rh(CO)_2Cl$, *J. Amer. Chem. Soc.* **83**, 1761–1762 (1961).

172. T. L. Brown and D. J. Darensbourg, Intensities of CO stretching of adsorbed CO and metal carbonyls, *Inorg. Chem.* **6**, 971–977 (1967).

173. J. T. Yates, Jr., R. G. Greenler, I. Ratajczkowa, and D. A. King, Reflection-absorption infrared spectrum of α-CO chemisorbed on polycrystalline tungsten, *Surface Sci.* **36**, 739–755 (1973).

174. J. Pritchard, T. Catterick, and R. K. Gupta, Infrared spectroscopy of chemisorbed carbon monoxide on copper, to be published.

175. G. Blyholder, Infrared spectrum of CO chemisorbed on iron, *J. Chem. Phys.* **36**, 2036–2039 (1962).

176. D. E. Eastman and J. E. Demuth, Photoemission studies of inorganic and organic adsorbates on Ni(111) and surface reactions, *Japan J. Appl. Phys.* (Suppl. 2, Pt. 2) **1974**, 827–836.

177. G. Blyholder, CNDO model and interpretation of the photoelectron spectrum of CO chemisorbed on Ni, *J. Vac. Sci. Technol.* **11**, 865–868 (1974).

178. T. Gustafsson, E. W. Plummer, D. E. Eastman, and J. L. Freeouf, Identification of the energy levels of molecularly adsorbed CO on Ni and Pd by means of $h\omega$ dependent photoemission, *Bull. Am. Phys. Soc.* **1975** (March), 304.

179. J. Fuggle, T. E. Madey, M. Steinkilberg, and D. Menzel, X-Ray photoelectron spectroscopy (XPS) of adsorbate valence bands, *Phys. Letters* **51A**, 163 (1975).

180. K. Yu, I. Lindau, P. Pianetta, and W. E. Spicer, Ultraviolet photoemission studies of O_2, CO, C_2H_2 and C_2H_4 chemisorbed on copper, *Bull. Am. Phys. Soc.* **1975** (March), 359.

181. G. A. Somorjai and F. J. Szalkowski, Simple rules to predict the structure of adsorbed gases on crystal surfaces, *J. Chem. Phys.* **54**, 389–399 (1971).

182. R. J. H. Voorhoeve, Molecular beam deposition of solids on surfaces: ultra-thin films, Chapter 4, this Volume of the Treatise.

183. J. J. Lander, in *Progress in Solid State Chemistry*, Volume 2, 26 (H. Reiss, ed.), Pergamon, Oxford (1965).

184. D. Haneman, Surface structures and properties of diamond-structure semiconductors, *Phys. Rev.* **121**, 1093–1100 (1961).

185. E. Tosatti and P. W. Anderson, Two-dimensional excitonic insulators: Si and Ge(111) surfaces, *Solid State Commun.* **14**, 773–777 (1974).

186. E. Tosatti and P. W. Anderson, Charge and spin density waves on semiconductor surfaces, *Japan J. Appl. Phys.* (Suppl. 2, Pt. 2) **1974**, 381–388.

187. J. Koutecky, A contribution to the molecular orbital theory of chemisorption, *Trans. Faraday Soc.* **54**, 1038–1052 (1958).

188. T. B. Grimley, The indirect interaction between atoms or molecules adsorbed on metals, *Proc. Phys. Soc.* **90**, 751–764 (1967).

189. T. B. Grimley, The electron density in a metal near a chemisorbed atom or molecule, *Proc. Phys. Soc.* **92**, 776–782 (1967).

190. T. B. Grimley and S. M. Walker, Interactions between adatoms on metals and their effects on the heat of adsorption at low surface coverage, *Surface Sci.* **14**, 395–406 (1969).

191. T. L. Einstein and J. R. Schrieffer, Indirect interactions between adatoms on a tight-binding solid, *Phys. Rev. B* **7**, 3629–3648 (1973).

192. J. T. Yates, Jr. and T. E. Madey, Interactions between chemisorbed species: H_2 and CO on (100) tungsten, *J. Chem. Phys.* **54**, 4969–4978 (1971).

193. M. M. Siddiqui and F. C. Tompkins, Surface potential of mixed adsorbates on metal films, *Proc. Roy. Soc.* (*London*) *A* **268**, 452–473 (1962).

194. D. O. Hayward and B. M. W. Trapnell, *Chemisorption*, Butterworths, Washington (1964).

195. A. Clark, *The Chemisorptive Bond*, Academic Press, New York and London (1974).

196. J. R. Anderson and B. G. Baker, Catalytic reactions on metal films, in *Chemisorption and Reactions on Metallic Films*, Vol. 2 (J. R. Anderson, ed.), Academic Press, London and New York (1971).

197. E. G. Allison and G. C. Bond, The structure and catalytic properties of Pd–Ag and Pd–Au alloys, *Catal. Rev.* **7**, 233–289 (1972).

198. R. L. Moss and L. Whalley, Adsorption and catalysis on evaporated alloy films, *Advan. Catal.* **22**, 115–189 (1972).

199. N. F. Mott and H. Jones, *The Theory of the Properties of Metals and Alloys*, Clarendon Press, Oxford (1936).

200. A. Couper and D. D. Eley, The parahydrogen conversion on palladium gold alloys, *Disc. Faraday Soc.* **8**, 172–184 (1950).

201. S. Hufner, G. K. Wertheim, and J. H. Wernick, X-ray photoelectron spectra of the valence bonds of some transition metals and alloys, *Phys. Rev. B* **8**, 4511–4524 (1973).

202. W. E. Spicer, in *Band Structure Spectroscopy of Metals and Alloys* (D. J. Fabian and L. M. Watson, eds.), pp. 7–53, Academic, London (1973).

203. W. M. H. Sachtler and R. Jongepier, The surface of copper–nickel alloy films, *J. Catalysis* **4**, 665–671 (1965).

204. P. van der Plank and W. M. H. Sachtler, Interaction of benzene and protium and deuterium on copper–nickel films with known surface composition, *J. Catalysis* **12**, 35–44 (1968).

205. M. Ono, Y. Takasu, K. Nakayama, and T. Yamashina, Auger spectroscopy study on Cu–Ni alloy surfaces related to catalysis, *Surface Sci.* **26**, 313 (1971).

206. K. Nagayama, M. Ono, and H. Shimizu, Auger spectroscopy study on the surface composition of copper–nickel alloys after annealing and sputtering, *J. Vac. Sci. Technol.* **9**, 749–751 (1972).

207. C. R. Helms, Observation of the segregation of Cu to the surface of a clean, annealed 50% Cu/50% Ni alloy by Auger electron spectroscopy, *J. Catalysis* **36**, 114–117 (1975).

208. K. Y. Yu, C. R. Helms, and W. Spicer, UPS Measurements of the surface electronic structure and surface composition of the Cu–Ni alloys, *Phys. Rev.*, to be published.

209. F. L. Williams and D. Nason, Binary alloy surface compositions from bulk alloy thermodynamic data, *Surface Sci.* **45**, 377–408 (1974).

210. R. A. Van Santen and W. M. H. Sachtler, A theory of surface enrichment in ordered alloys, *J. Catalysis* **33**, 202–209 (1974).

211. Y. Soma-Noto and W. M. H. Sachtler, Infrared spectra of CO adsorbed on supported Pd and Pd–Ag alloys, *J. Catalysis* **32**, 315–324 (1974).

212. Y. Soma-Noto and W. M. H. Sachtler, Infrared spectra of CO adsorbed on silica-supported Ni–Cu alloys, *J. Catalysis* **34**, 162–165 (1974).

213. F. L. Williams and M. Boudart, Surface composition of nickel–gold alloys, *J. Catalysis* **30**, 438–443 (1973).

214. J. H. Sinfelt, Supported 'bimetallic cluster' catalysts, *J. Catalysis* **29**, 308–315 (1973).

215. E. Ruckenstein, Phase stability of supported bimetallic catalysts, *J. Catalysis* **35**, 441 (1974).

216. R. B. Levy and M. Boudart, Platinum-like behavior of tungsten carbide in surface catalysis, *Science* **181**, 547–549 (1973).

217. L. H. Bennett, J. R. Cuthill, A. J. McAlister, N. E. Erickson, and R. E. Watson, Electronic structure and catalytic behavior of tungsten carbide, *Science* **184**, 563–565 (1974).

218. J. E. Houston, G. E. Laramore, and R. L. Park, Surface electronic properties of tungsten, tungsten carbide, and platinum, *Science* **185**, 258–259 (1974).

219. L. Fiermans, R. Hoogewijs, and J. Vennik, Electron spectroscopy of transition metal oxide surfaces, *Surface Sci.* **47**, 1–40 (1975).

220. G. Parravano, Ferroelectric transitions and heterogeneous catalysis, *J. Chem. Phys.* **20**, 342–343 (1952).

221. G. Parravano, Catalytic activity of lanthanum and strontium manganite, *J. Am. Chem. Soc.* **75**, 1497–8 (1953).

222. F. F. Vol'kenshtein, *The Electronic Theory of Catalysis on Semiconductors*, Pergamon/Macmillan, New York (1963).

223. V. J. Lee, Electrons and holes as energy-transport agents in catalysis on semiconductors—Part I, *J. Catalysis* **17**, 178–189 (1970).

224. V. J. Lee, Catalysis on wide-band-gap semiconductors, *J. Chem. Phys.* **55**, 2905–2913 (1971).

225. O. V. Krylov, *Catalysis by Non-Metals*. Academic Press, New York (1970).

226. A. Many, Y. Goldstein, and N. B. Grover, *Semiconductor Surfaces*, Chapter 4, North-Holland, Amsterdam (1965).

227. P. C. Gravelle, G. El Shobaky, and S. J. Teichner, Determination a l'aide du microcalorimetre calvet du mechanisme de l'oxydation catalytique de l'oxyde de carbone au contact de l'oxyde de nickel VII. Oxyde de nickel contenant des ions lithium, *J. Chim. Phys.* **66**, 1953–1965 (1969).

228. G. El Shobaky, P. C. Gravelle, and T. J. Teichner, Influence of the surface structure of a nickel oxide catalyst on the mechanism of the room temperature oxidation of carbon monoxide, *J. Catal.* **14**, 4–22 (1969).

229. J. Deren, Z. Guzik, and J. Sloczynski, Catalytic oxidation of carbon monoxide on monocrystalline nickel oxide catalysts, *Bull. Acad. Pol. Sci.. Ser. Sci. Chim.* **20**, 361–368 (1972).

230. D. Adler, The imperfect solid—transport properties, Vol. 2, pp. 237–332 of this Treatise.

231. C. Wagner, Reduction of the rate of carbon monoxide combustion in nickel oxide (NiO) through catalyst doping with gallium oxide (Ga_2O_3), *Ber. Bunsenges. Phys. Chem.* **74**, 1270–3 (1970).

232. J. Deren and A. Kowalska, The changes of adsorptive properties of nickel oxide in the course of adsorption-desorption cycles, *Bull. Acad. Pol. Sci.. Ser. Sci. Chem.* **21**, 131–136 (1973).

233. F. Morin, in *Semiconductors* (N. B. Hannay, ed.), Chapter 14, pp. 600–633, Reinhold, New York (1959).

234. J. Bardeen, Surface states and rectification at a metal semiconductor contact, *Phys. Rev.* **71**, 717–727 (1947).

235. T. J. Gray and N. Lowery, Surface electronic structure of oxides as established by thermally stimulated electron current measurements, *Disc. Faraday Soc.* **52**, 132–139 (1971).

236. A. Cimino, E. Molinari, F. Cramarossa, and G. Ghersini, Hydrogen chemisorption and electrical conductivity of zinc oxide semiconductors, *J. Catal.* **1**, 275–292 (1962).

237. Ph. Coussel and S. J. Teichner, Electron localization and oxygen transfer reactions on zinc oxide, *Catal. Rev.* **6**, 133–260 (1972).

238. E. Arijs and F. Cardon, The influence of surface donor states on the chemisorption kinetics of oxygen at the surface of ZnO single crystals, *J. Solid State Chem.* **6**, 310–326 (1973).

239. R. J. Kokes and A. L. Dent, Hydrogenation and isomerization over zinc oxide, *Advan. Catal.* **22**, 1–50 (1972).

240. E. R. S. Winter, Adsorption upon pure and lithium-doped nickel oxide, *J. Catal.* **6**, 35–49 (1966).

241. L. A. Peterman, in *Adsorption-Desorption Phenomena* (F. Ricca, ed.), pp. 227–243, Academic, London (1972).

242. Y. Morooka and A. Ozaki, Regularities in catalytic properties of metal oxides in propylene oxidation, *J. Catal.* **5**, 116–124 (1966).

243. Y. Morooka, Y. Morikawa, and A. Ozaki, Regularity in the catalytic properties of metal oxides in hydrocarbon oxidation, *J. Catal.* **7**, 23–32 (1967).

244. W. M. H. Sachtler, G. J. H. Dorgelo, J. Fahrenfort, and R. J. H. Voorhoeve, Correlations between catalytic and thermodynamic parameters of transition metal oxides, *Rec. Trav. Chim.* **89**, 460–480 (1970).

245. R. J. H. Voorhoeve, J. P. Remeika, and L. E. Trimble, Nitric oxide and perovskite-type catalysts: Solid state and catalytic chemistry, in *Proc. Symp. Chemistry of Nitrogen Oxides* (R. L. Klimisch and J. G. Larson, eds.), pp. 215–233, Plenum, New York (1975).

246. F. Basolo and R. G. Pearson, *Mechanism of Inorganic Reactions*. Wiley, New York (1967).

247. A. Cimino, Catalytic activity of transition metal ions in an oxide matrix, *Chim. Ind. (Milan)* **56**(1), 27–38 (1974).

248. C. C. Chao and J. H. Lunsford, An infrared and electron paramagnetic resonance study of some silver–nitric oxide complexes in Y type zeolites, *J. Phys. Chem.* **78**, 1174–1177 (1974).

249. K. Klier, R. Kellerman, and P. J. Hutta, Spectra of synthetic zeolites containing transition metal ions. V. π Complexes of olefins and acetylene with Co(II) A molecular sieve, *J. Chem. Phys.* **61**, 4224–4234 (1974).

250. P. A. Batist, P. C. van der Heyden, and G. C. A. Schuit, Isomerization of butenes on bismuth molybdate—recirculating and pulse reactions of butenes on koechlinite, *J. Catal.* **22**, 411–418 (1971).

251. I. Matsuura and G. C. A. Schuit, Adsorption equilibria and rates of reactions of adsorbed compounds on reduced and oxidized Bi–Mo catalysts, *J. Catal.* **20**, 19–39 (1971).

252. A. W. Sleight and W. J. Linn, Olefin oxidation over oxide catalysts with the scheelite structure, *Ann. New York Acad. Sci.* (1976).

253. K. Aykan, D. Halvorson, A. W. Sleight, and D. B. Rogers, Olefin oxidation and ammoxidation on studies over molybdate, tungstate and vanadate catalysts having point defects, *J. Catal.* **35**, 401–406 (1974).

254. A. W. Sleight, K. Aykan, and D. B. Rogers, New nonstoichiometric molybdate, tungstate and vanadate catalysts with the scheelite-type structure, *J. Solid State Chem.* **13**, 231–236 (1975).

255. R. J. H. Voorhoeve, J. P. Remeika, L. E. Trimble, A. S. Cooper, F. J. DiSalvo, and P. K. Gallagher, Perovskite-like $La_{1-x}K_xMnO_3$ and related compounds: Solid state chemistry and the catalysis of the reduction of NO by CO and H_2, *J. Solid State Chem.* **14**, 395–406 (1975).

256. H. C. Yao and M. Shelef, Nitric oxide and carbon monoxide chemisorption on cobalt-containing spinels, *J. Phys. Chem.* **78**, 2490–2496 (1974).

257. M. Boudart, A. Aldag, J. E. Benson, N. A. Dougharty, and C. G. Harkins, On the specific activity of platinum catalysts, *J. Catal.* **6**, 92–99 (1966).

258. G. A. Somorjai, The mechanism of hydrocarbon catalysis on platinum single crystal surfaces, Battelle Conf. on The Physical Basis for Heterogenous Catalysis, Gstaad, Switzerland, Sept. 1974; also see Ref. 83.

259. H. Kral, Freie und aktive Oberfläche von Pd/C-Katalysatoren, *Chemiker-Zeitung, Chem. Apparatur* **91**, 41–47 (1967).

260. J. T. Yates, Jr., T. E. Madey, and M. J. Dresser, Adsorption and decomposition of formaldehyde on tungsten (100) and (111) crystal planes, *J. Catal.* **30**, 260–275 (1973).

261. M. Boudart, Chemisorption during catalytic reaction on metal surfaces, *J. Vac. Sci. Technol.* **12**, 329–332 (1975).

262. S. L. Bernasek, W. J. Siekhaus, and G. A. Somorjai, Molecular-beam study of hydrogen–deuterium exchange on low- and high-Miller-index platinum single-crystal surfaces, *Phys. Rev. Letters* **30**, 1202–1204 (1973).

263. G. A. Somorjai and S. B. Brumbach, The interaction of molecular beams with, solid surfaces, *CRC Critical Reviews in Solid State Sciences*. pp. 429–453 (May 1974).

264. R. R. Rye and K. E. Lu, Equilibrium of H_2 and D_2 on single crystal surfaces of platinum, *J. Vac. Sci. Technol.* **12**, 334–337 (1975).

265. R. W. Joyner, B. Lang, and G. A. Somorjai, Low pressure studies of dehydrocyclization of *n*-heptane on platinum crystal surfaces using mass spectrometry, Auger electron spectroscopy, and low energy electron diffraction, *J. Catal.* **27**, 405–415 (1972).

266. G. A. Somorjai, private communication.

267. T. E. Whyte, Jr., Metal particle size determination of supported metal catalysts, *Catal. Rev.* **8**, 117–134 (1973).

268. J. G. Allpress and J. V. Sanders, The structure and orientation of crystals in deposits of metals on mica, *Surface Sci.* **7**, 1–25 (1967).

269. B. G. Bagley, Five-fold pseudosymmetry, *Nature* **225**, 1040–1041 (1970).

270. M. R. Hoare and P. Pal, Statistics and stability of small assembles of atoms, *J. Crystal Growth* **17**, 77–96 (1972).

271. J. J. Burton, Structure and properties of microcrystalline catalysts, *Catal. Rev.—Sci. Eng.* **9**, 209–222 (1974).

272. J. L. Carter and J. H. Sinfelt, The paramagnetic susceptibility of supported nickel, *J. Catal.* **10**, 134–139 (1968).

273. R. L. Burwell, Jr., The use of stereochemistry in heterogeneous catalysis, *Intra-Science Chem. Rept.* **6**, 135–149 (1973).

274. S. Siegel, Stereochemistry and the mechanism of hydrogenation of unsaturated hydrocarbons, *Adv. Catal.* **16**, 123–177 (1966).

275. R. L. Burwell, Jr., Deuterium as a tracer in reactions of hydrocarbons on metallic catalysts, *Accounts Chem. Res.* **2**, 289–296 (1969).

276. R. L. Burwell, Jr., Stereochemistry and heterogeneous catalysis, *Chem. Rev.* **57**, 895–934 (1957).

277. A. T. Bloomquist, L. H. Liu, and J. C. Bohrer, Many membered carbon rings. VI. Unsaturated nine-membered cyclic hydrocarbons, *J. Amer. Chem. Soc.* **74**, 3643–3647 (1952).

278. E. McMahon and J. K. A. Clarke, The exchange of norbornane with deuterium on a palladium surface, *Tetrahedron Letters* **18**, 1413–1414 (1971).

279. K. Schrage and R. L. Burwell, Jr., The mechanism of the isotopic exchange between deuterium and cycloalkanes on palladium catalysts, *J. Amer. Chem. Soc.* **88**, 4555–4560 (1966).

280. *Nobel Lectures. Chemistry*. 1963–70, Elsevier Publishing Co., (1972), pp. 1–61.

281. P. Cossee and E. J. Arlman, Ziegler–Natta catalysis, I, II, and III, *J. Catal.* **3**, 80–104 (1964).

282. L. A. M. Rodriquez and H. M. Van Looy, Studies on Ziegler–Natta catalysts, Part V. Stereospecificity of the active center, *J. Polymer Sci.* A-1 **4**, 1971–1992 (1966).

283. B. W. Wojceichowski, A theoretical treatment of catalyst decay, *Can. J. Chem. Eng.* **46**, 48–52 (1968); The kinetic foundations and the practical application of the time on stream theory of catalyst decay, *Catal. Rev.* **9**, 79–113 (1974).

284. G. A. Mills and F. W. Steffgen, Catalytic methanation, *Catal. Rev.* **8**(2), 159–210 (1973).

285. M. M. Bhasin, Auger electron spectroscopic study of the surface poisoning of copper catalysts, *J. Catal.* **34**, 356–359 (1974).
286. H. P. Bonzel and R. Ku, Adsorbate interactions on a Pt(110) surface. II. Effect of sulphur on the catalytic CO oxidation, *J. Chem. Phys.* **59**, 1641–1651 (1973).
287. L. L. Hegedus and E. E. Petersen, Experimental study of the poisoning of a single catalyst pellet in a diffusion reactor, *Chem. Eng. Sci.* **28**, 345–356 (1973).
288. L. D. Schmidt and D. Luss, Physical and chemical characterization of platinum–rhodium gauze catalysts, *J. Catal.* **22**, 269–279 (1971).
289. G. A. Somorjai, On the mechanism of sulphur poisoning of platinum catalysts, *J. Catal.* **27**, 453–456 (1972).
290. P. H. Emmett and S. Brunauer, Accumulation of alkali promotors on surfaces of iron synthetic ammonia catalysts, *J. Am. Chem. Soc.* **59**, 310–315 (1937).
291. V. Solbakken, A. Solbakken, and P. H. Emmett, The exchange of $H_2\,^{18}O$ with the oxygen of promoters on the surface of iron catalysts, *J. Catal.* **15**, 90–98 (1969).
292. H. Topsoe, J. A. Dumesic, and M. Boudart, Alumina as a textural promoter of iron synthetic ammonia catalysts, *J. Catal.* **28**, 477–488 (1973).
293. P. A. Kilty and W. M. H. Sachtler, The mechanism of the selective oxidation of ethylene to ethylene oxide, *Catal. Rev.—Sci. Eng.* **10**, 1–16 (1974).
294. P. Forsalti, E. Martinez, G. C. Kuczynski, and J. J. Carberry, Influence of γ-irradiation upon catalytic selectivity. II. The role of Ca in the Ag catalyzed oxidation of ethylene, *J. Catal.* **28**, 455–458 (1973).
295. J. McAllister and R. S. Hansen, Catalytic decomposition of ammonia on tungsten (100), (110), and (111) crystal faces, *J. Chem. Phys.* **59**, 414–422 (1973).
296. I. Langmuir, The mechanism of the catalytic action of platinum in the reactions $2CO + O_2 = 2CO_2$ and $2H_2 + O_2 = 2H_2O$, *Trans. Faraday Soc.* **17**, 621–654, 672–673 (1922).
297. R. L. Palmer and J. N. Smith, Molecular beam study of CO oxidation on a (111) Pt surface, *J. Chem. Phys.* **60**, 1453–1463 (1974).
298. H. P. Bonzel and R. Ku, Mechanisms of the catalytic carbon monoxide oxidation on Pt(110), *Surface Sci.* **33**, 91–106 (1972).
299. G. Ertl and J. Koch, Adsorption studies with a Pd(111) surface, in *Adsorption-Desorption Phenomena* (F. Ricca, ed.), Adademic Press, London and New York (1972), pp. 345–357.
300. T. Matsushima and J. M. White, On the mechanism and kinetics of the CO oxidation reaction on polycrystalline Pd. I. The reaction paths, to be published.
301. P. Mars, J. J. F. Scholten, and P. Zwietering, The catalytic decomposition of formic acid, *Advan. Catal.* **14**, 35–113 (1963).
302. S. J. Thomson and G. Webb, *Heterogeneous Catalysis*, Oliver and Boyd, Edinburgh and London (1968), pp. 184–188.
303. J. McCarty, J. Falconer, and R. J. Madix, Decomposition of formic acid on Ni(110), *J. Catal.* **30**, 235–249 (1973).
304. J. L. Falconer, J. G. McCarty, and R. J. Madix, Surface explosion: HCOOH on Ni(110), *Surface Sci.* **46**, 473–504 (1974).
305. J. McCarty and R. J. Madix, A study of the kinetics and mechanism of the decomposition of formic acid on carburized and graphetized Ni(110) using AES, LEED and flash desorption, *Surface Sci.* submitted for publication.
306. T. A. Dorling and R. L. Moss, The structure and activity of supported metal catalysts, *J. Catal.* **5**, 111–115 (1966).

307. P. C. Aben, J. C. Platteeuw, and B. Stouthamer, The hydrogenation of benzene over supported platinum, palladium, and nickel catalysts, *Rec. Trav. Chim.* **89**, 449–459 (1970).
308. V. Nikolajenko, V. Bosacek, and Vl. Dames, Investigation of properties of the metallic nickel surface in mixed Ni–MgO catalysts, *J. Catal.* **2**, 127–130 (1963).
309. F. M. Dautzenberg and J. C. Platteeuw, Isomerization and dehydrocyclization of hexanes over monofunctional supported platinum catalysts, *J. Catal.* **19**, 41–48 (1970).
310. O. M. Poltorak and V. S. Boronin, Chemisorption and catalysis on platinized silica gels. II. Specific catalytic activity, *Zh. Fiz. Khim.* **39**, 2491–2498 (1965).
311. V. S. Chesalova and G. K. Boreskov, The specific catalytic activity of metals. I. Oxidation of sulfur dioxide on platinum catalysts, *Zh. Fiz. Khim.* **30**, 2560 (1956).
312. D. J. C. Yates, W. F. Taylor, and J. H. Sinfelt, Catalysis over supported metals, I. Kinetics of ethane hydrogenolysis over nickel surfaces of known area, *J. Amer. Chem. Soc.* **86**, 2996–3001 (1964).
313. G. K. Boreskov, M. G. Slin'ko, and V. S. Chesalova, The specific catalytic activity of platinum. II. The reaction between hydrogen and oxygen, *Zh. Fiz. Khim.* **30**, 2787–2797 (1956).

Ion Implantation and Channeling

Fred H. Eisen
Science Center, Rockwell International, Thousand Oaks, California
and
James W. Mayer
California Institute of Technology, Pasadena, California

1. Introduction

Ion implantation is the introduction of atoms into a solid substrate by bombardment of the solid with ions in the keV-to-MeV energy range. The effects of implantation may be due to the properties of the implanted species or to the damage produced by the energetic ions. Damage will always be present following implantation of energetic ions. If one desires to make use of the properties of the implanted atoms, the damage effects must be minimized by suitable thermal treatment. In some cases implantation techniques may be employed only to utilize the damage effects. However, in this review we are concerned primarily with effects due to the properties of the implanted atoms. In fact, we direct our attention principally to implantation in semiconductors, where there has been the most effort devoted toward the characterization of the properties of the implanted layer.

Ion implantation provides the possibility of introducing a wide selection of atomic species and of externally controlling both the number of atoms per cm² introduced into the host substrate and the depth distribution of the implanted species. This is in contrast to diffusion processes where the temperature governs the surface concentration of the diffusing species and the time and temperature determine the depth distribution (profile). The ability to control the

125

number of implanted dopant species per cm^2 has made a major impact on silicon technology. It has been possible to control the number of dopant atoms per cm^2 to about 1 % over the surface of a wafer and to control the number per cm^2 within 3 % from wafer to wafer. Also, very simple masking techniques can be used to define the geometric area of the implanted region. It is also possible to introduce large doses of the implanted species so that the composition and structure of the host substrate can be changed. In this latter case one must consider sputtering effects, which can lead to pronounced erosion of the implanted layer.

There are limitations and disadvantages to the implantation process. With commercial implantation systems (energies less than ~ 300 keV), the range of the implanted ion is typically less than 1 μm and is often of the order of a few tenths of a micron. To achieve greater depths, the implantation process can be followed with a thermal diffusion step. The most significant drawback to the implantation doping process in semiconductors is the amount of lattice disorder and the number of radiation-induced defects that are caused by the implanted ion as it comes to rest in the crystal. Although implantation is considered as a room-temperature process, it is necessary to thermally anneal the samples after implantation. Often these anneal temperatures are comparable to those used in diffusion.

The solid-state aspects of ion implantation are particularly broad because of the range of physical properties that are affected by trace amounts of foreign atoms in the surface layers of solid substrates. In recent years there have been investigations of the influence of implantation on mechanical, electrical, optical, magnetic, and superconducting properties of materials. The basic concepts of ion implantation were discussed around 1970 in several review articles and books.[1-4] There have been several conferences concerned with implantation since that time.[5-9] In this review we will concentrate on the developments in ion implantation in the last few years. Over this period the predominant emphasis has been on implantation in semiconductors, in particular silicon and GaAs, and the review will reflect this emphasis. There is a growing interest in implantation in a wider range of compound semiconductors and in creation of structural changes in solids. We will try to indicate some of the results obtained in these areas.

We introduce this review with a discussion of channeling effect measurements. The development of this technique and that of ion implantation have been linked together since early 1965. Both involve the penetration of energetic ions in solids. Ion implantation is used to create changes within the first micron of the sample surface, while channeling effect measurements are used to analyze lattice disorder and the lattice site location of the implanted atoms. Much of the early insight into implantation processes in semiconductors was obtained by channeling measurements. On the other hand, the channeling measurements themselves were developed and refined in the course of analysis of implanted samples.

2. Channeling Effect Measurements

Channeling effect measurements provide a relatively simple method of studying the lattice location of implanted dopants and of determining the amount of lattice disorder produced during their implantation. Channeling measurements have contributed much of our understanding of ion implantation in semiconductors. Many of these results will be referred to in the discussion of implantation. We present a brief review of the principles of the channeling effect technique to provide a basis for the understanding and evaluation of these results. More detailed discussions of the technique may be found in books[4,10,11] and conference proceedings.[12-15] The most recent reviews of the applications of the channeling effect technique can be found in Refs. 11 and 15a.

2.1. General Principles

When a beam of energetic ions is incident on a crystalline sample at a sufficiently small angle to a symmetry axis of the crystal (i.e., an angle smaller than the "critical angle" for channeling) most of the ions in the beam will be steered by a series of small-angle deflections.[16] The energetic ions do not approach closer than about 0.1 Å to the atoms in the rows of the crystal. This is the so-called channeling phenomenon. It provides a method of determining whether or not atoms are located in the atomic rows along a particular crystallographic direction. In order to apply channeling, it is necessary to be able to monitor a small impact parameter reaction between the incident ions and the atoms in the crystal. The possibilities include ion induced X rays, nuclear reactions, and nuclear backscattering. By counting the number of reaction events as the beam of energetic ions is aligned along a symmetry axis, it is possible to reach conclusions concerning the lattice location of the implanted atoms and the

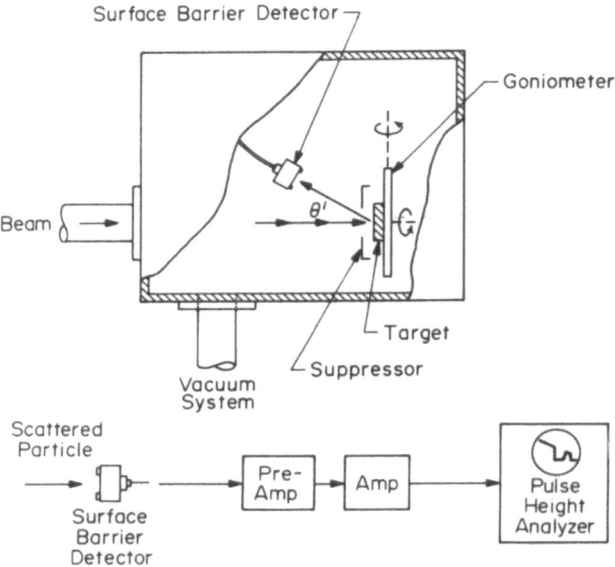

Fig. 1. Schematic diagram of scattering chamber used for observing backscattered particles in channeling effect measurements. The electronic system is indicated at the bottom of the figure.

amount of disorder introduced in the crystal during the implantation process.[4,11] Channeling is also observed when the incident beam is aligned with low-index symmetry planes of the crystal. Therefore planar orientations can also be used in studying disorder and lattice location, although the use of axial alignments is more common.

Backscattering has been the most widely applied technique in implantation investigations. This is in part because of the ease of observing nuclear backscattering and also because it is possible to extract depth information from the energy spectra of the back-scattered particles. A typical experimental setup for observing back-scattering of channeled ions is indicated schematically in Fig. 1. Silicon surface barrier detectors are commonly used to detect the backscattered ions. The pulses from these detectors are amplified using standard electronics and the pulse height spectra are recorded using a multichannel analyzer.

Figure 2 shows backscattering spectra for protons and helium ions incident on a silicon sample which has been implanted with antimony. Spectra are shown for the beam incident along a ⟨110⟩ axis and incident away from any symmetry axis or plane, so that no

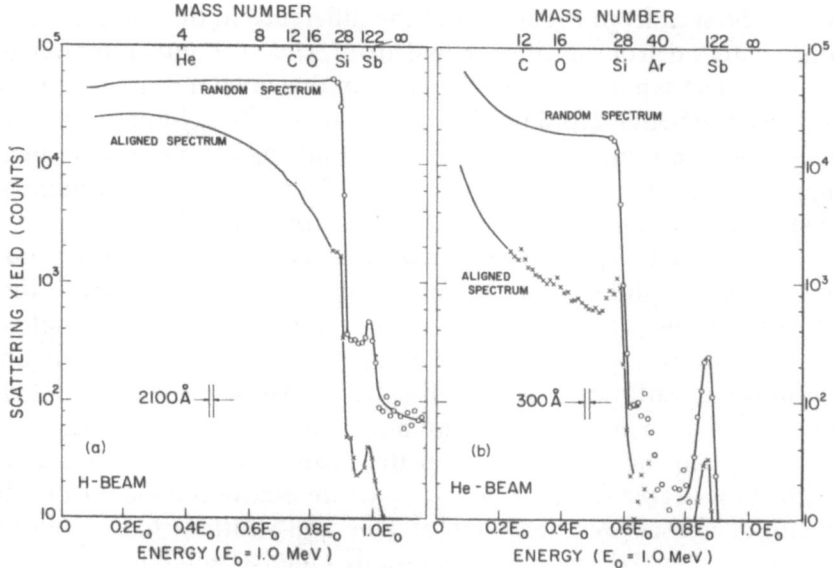

Fig. 2. Comparison of backscattering spectra obtained with (a) 1.0-MeV protons and (b) 1.0-MeV helium ions. The target in both cases was a silicon crystal implanted with 3×10^{14} 30-keV antimony ions/cm^2. Both sets of spectra were obtained with the same electronic gain and the same total dose. Mass and depth resolution scales are also included (from Davies *et al.*[16a]).

steering occurs (random direction). The energy of the protons or helium ions scattered from the silicon atoms on the surface of the crystal is somewhat less than the energy of the particles scattered from the implanted antimony, which lies close to the surface of the crystal. This is because the energy of the particles after a back-scattering event is a function of the mass of the target atom. For 180° backscattering the relation between the initial energy of the ion E_1 and its energy after the backscattering event E_2 is as follows:

$$E_2 = KE_1 = \left(\frac{M_2 - M_1}{M_2 + M_1}\right) E_1$$

where M_1 is the mass of the ions in the analyzing beam and M_2 is the mass of the target atom. If the backscattering angle is less than 180°, K is also a function of the scattering angle but it is independent of E_1. This relation between the incident and scattered energies of the ions in the analyzing beam means that it is possible by energy discrimination to detect scattering from impurity atoms with a mass greater than that of the substrate in which they are implanted. It can readily be seen from Figure 2 that the difference in the energies of the ions scattered from antimony and from silicon is much greater for helium ions than for protons. Because of this greater mass resolution and also the better depth resolution obtained with helium (see below), helium ions are more often used in channeling effect measurements than protons. The general application of backscattering techniques to surface layer analysis has recently been reviewed.[17]

The silicon yield shown in Figure 2 (at about $0.5E_0$) for the helium beam incident along the $\langle 110 \rangle$ direction is a factor of 40 lower than when the beam was incident in a random direction. An example of the angular dependence of the scattering yield from silicon and antimony in antimony-implanted silicon is shown in Figure 3. (The difference between the silicon and antimony curves will be discussed below.) It can be seen that when the beam is aligned within a few tenths of a degree of the symmetry axis the scattering yield from the silicon decreased by a large factor. The half-width at half-minimum $\psi_{1/2}$ of this angular yield curve is usually taken as equal to the critical

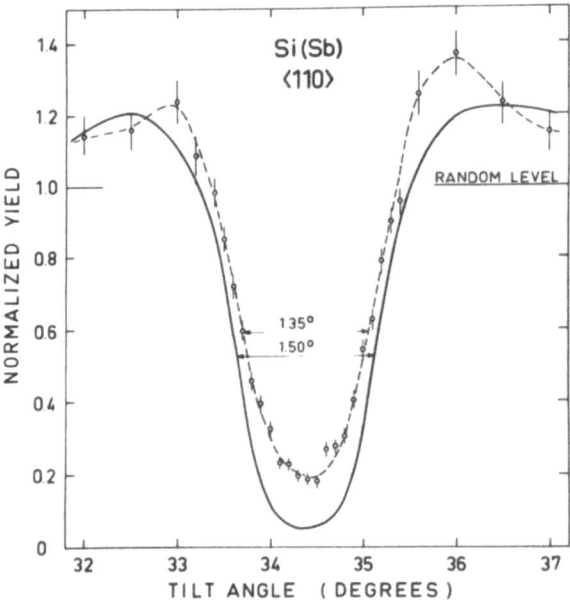

Fig. 3. The angular distribution for 1.0-MeV helium ions scattered from a silicon crystal implanted with 6×10^{14} antimony ions/cm². These curves were taken by tilting the crystal about the $\langle 110 \rangle$ axis (from Sigurd and Björkqvist[25]).

angle for channeling, and is given approximately by the relation

$$\psi_{1/2} = \left(\frac{2Z_1 Z_2 e^2}{Ed}\right)^{1/2}$$

where Z_1 is the atomic number of the projectile atoms and Z_2 the atomic number of the target atoms, e is the electronic charge, E is the energy of the ions, and d is the spacing between the atoms along the atomic rows.[16] This relation has been verified experimentally in a number of diamond lattice semiconductors.[18]

2.2. Lattice Location of Implanted Atoms

If the implanted dopant atoms occupy substitutional sites in the lattice, well-channeled ions will not make close impact collisions with the dopant atoms. Consequently the yield of particles backscattered from substitutional dopants will exhibit the same normalized angular yield curve as that for the host crystal. When such equivalence between angular yield curves for the dopant and the host is observed for a particular crystal axis it indicates the dopants are located along the corresponding crystal rows. Because the beam can be aligned with only one of several equivalent axes (for example, one of six $\langle 110 \rangle$ axes), there is a further constraint on the position of the dopant atoms. They must lie at the intersection of these axes. In the fcc lattice these intersections occur at substitutional sites for $\langle 110 \rangle$ axes, whereas they occur at both tetrahedral interstitial sites and substitutional sites for $\langle 111 \rangle$ axes.[19]

If the impurities are not substitutionally located, the interpretation of the channeling effect data is not as straightforward. When the impurity is located in the channel, i.e., not along the atomic rows, the scattering yield from it will depend upon the probability that a channeled ion will pass through the point in the channel at which the impurity is located. In other words, it will depend upon the spatial distribution of the channeled ions within the channel. This spatial distribution is not uniform over the channel. It is possible to have a flux of channeled ions at the center of the channel which is appreciably higher, by a factor of 2–4, than the random flux.[11,20,21] In this case the scattering yield from interstitial impurities located at the center of the channel would be greater than the random yield. This was first observed experimentally for ytterbium implanted into silicon.[20] The angular dependence of the scattering yield from the silicon and the ytterbium for a helium beam incident near the $\langle 110 \rangle$, $\langle 111 \rangle$, and

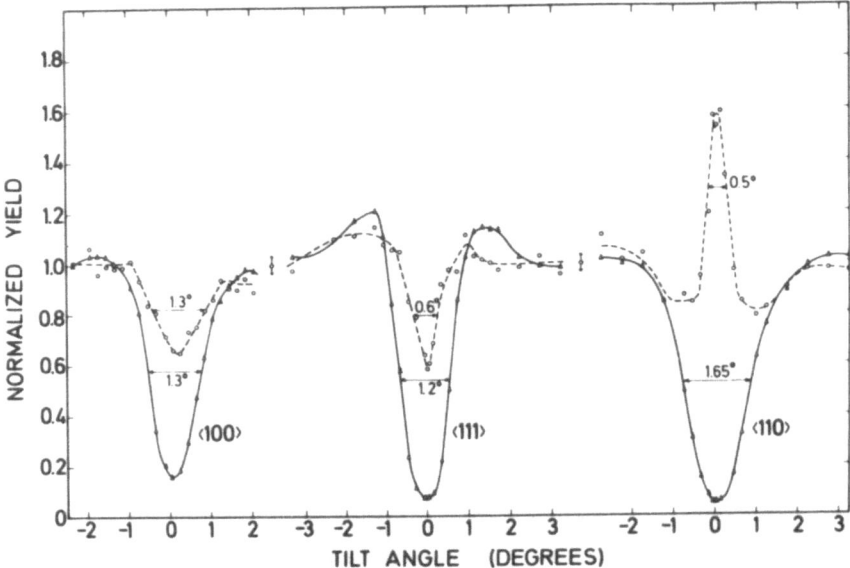

Fig. 4. Angular distributions for 1.0-MeV helium ions scattered from a silicon crystal implanted with 5×10^{14} ytterbium ions/cm². The scattering yield from the silicon atoms (△) and from the ytterbium atoms (○) are shown as a function of the angle between the incident beam and various low-index axes (from Andersen *et al.*[20]).

⟨110⟩ axes is shown in Figure 4. The scattering yield from the ytterbium exhibits a very narrow peak about the ⟨110⟩ direction which rises to a value nearly twice that of the random yield.

In order to understand flux peaking, it is necessary to introduce several concepts from Lindhard's theory of channeling.[16] He has shown that the energy of motion of the channeled ions in the plane transverse to the channeling axis (the transverse energy) is a constant of the motion of the channeled ions. When the channeled beam is incident exactly along a symmetry axis the transverse energy of a particular ion is determined by the potential at the point at which it enters the channel. In order to compute this potential, it is necessary to consider the interaction between the channeled ions and the atomic rows surrounding the channel. Lindhard has shown that this interaction can be approximated by a continuum potential. The potential at any point within a channel can be calculated by summing the values of the continuum potential at that point for the individual rows surrounding the channel. This has been calculated for some specific cases. An example of the resultant equipotential curves for helium ions in the ⟨100⟩ axial channel of copper is shown in Figure 5.[21] The

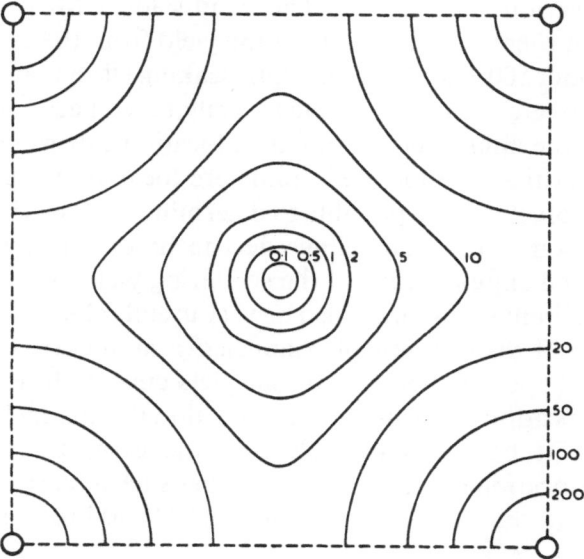

Fig. 5. The equipotential curves for helium ions in the
⟨100⟩ channel of copper (from Van Vliet[21]).

potential is nearly flat in the center of the channel and rises steeply close to the atomic rows. Conservation of the transverse energy of channeled ions means, for example, that an ion entering the channel at a point where the potential is 10 eV may only move within that part of the channel that lies within the 10-eV equipotential curve. It is then easy to see that the flux of channeled ions will be highest near the center of the channel and will be lowest near the rows of atoms surrounding the channel.

There are several scattering processes which will tend to increase the transverse energy of channeled ions and so decrease the flux peaking effect by smearing out the spatial distribution of the channeled ions. Because of the shape of the potential, these scattering processes can produce large reductions in the maximum flux at the center of the channel, but the changes in the flux distribution close to the strings will be small. The tendency is to approach a uniform flux across the channel. Van Vliet,[21] in his theoretical treatment of the spatial distribution of channeled ions, has estimated that flux peaking effects should be observable to depths of several thousand angstroms in silicon crystals. He has further predicted that there should be an oscillation with depth of the flux peaking phenomena. This oscillatory effect has been observed in experiments with ytterbium implanted into silicon at various energies.[22] The results for 2-MeV helium ions showed that there is a maximum in the yield from the ytterbium at a depth of about 500 Å and also that flux peaking effects can be observed at depths as great as 1500 Å. Such oscillatory effects only serve to complicate the interpretation of lattice location experiments.

If impurities or implanted atoms are located at a unique interstitial position, it may be possible to determine this position by a careful application of the channeling effect technique. It is necessary to make detailed angular scans of the scattering yield near several symmetry axes. Planar scans may also be very useful. The theoretical work on the spatial distribution of channeled ions makes it possible to predict the shape of the backscattering yield curve to be expected for a particular location of the dopant atom within the channel. Using such predictions, it has been possible in some cases to suggest fairly exactly the location of interstitial impurities such as carbon implanted in iron,[23] ytterbium implanted in silicon,[20] and bromine implanted in iron.[24]

In cases where the minimum axial yield from the dopant atoms is about as low as that from the atoms of the host crystal, it is often found that the width of the angular yield curve for the dopant is lower than the corresponding angular width for the host atoms.[25,26] An example of this is shown in Figure 3 for antimony implanted in silicon.[25] The width of the antimony yield curve is about 10% smaller than that for silicon. Such differences in angular width may be due to small equilibrium displacements of the dopant atoms from lattice sites. This may be related to the high concentration of implanted dopant atoms present in the sample.

To summarize, the application of the channeling effect technique to lattice location studies is simplest for substitutional species. The measurement of the angular yield curve for the ⟨110⟩ axial direction can suffice in this case. In fact, a simple measurement of the ⟨110⟩ aligned and random yields can give strong evidence for a high substitutional component. The measurements must be much more detailed for dopants which are not located on substitutional lattice sites.

2.3. Disorder

Channeling effect measurements will of course indicate when the atoms of the host crystal are not located on lattice sites. The technique can therefore be used to study the disorder introduced during ion

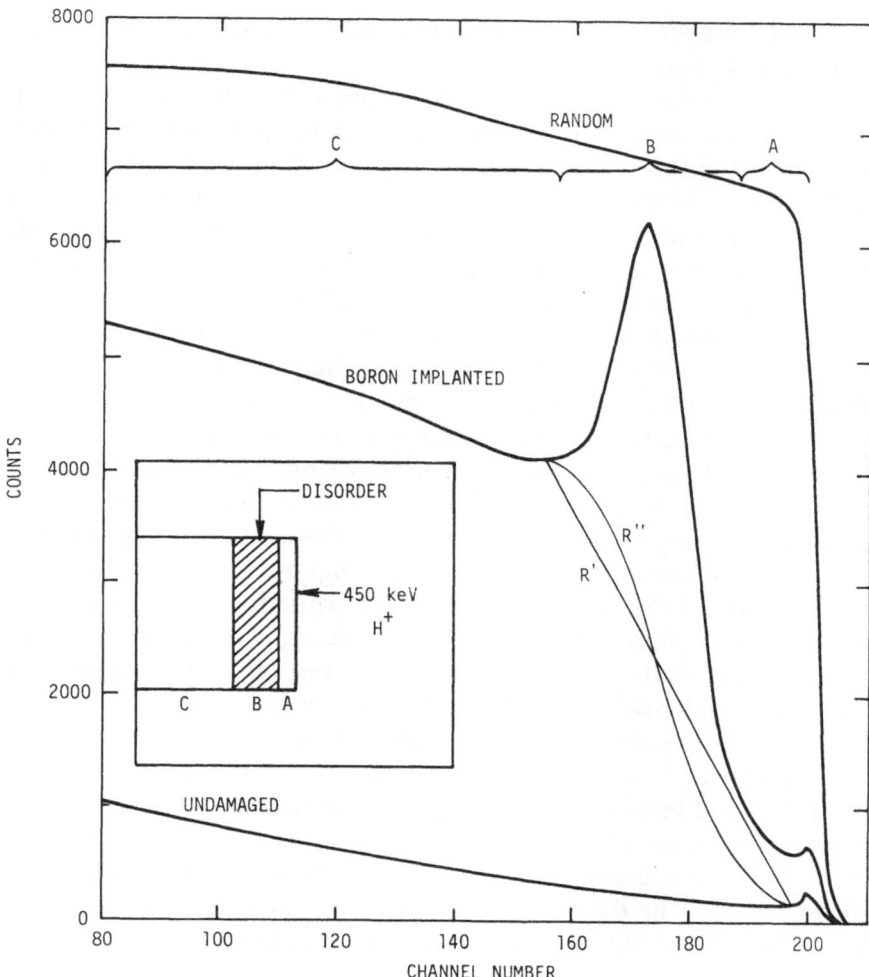

Fig. 6. Aligned ($\langle 110 \rangle$) and random backscattering spectra for 450-keV protons incident on silicon. The spectrum showing a damage peak was obtained from a sample implanted with 8×10^{15} boron ions/cm². The lines labeled R' and R'' are two approximations to the scattering yield from the random component of the analyzing beam. The labeled regions of the spectrum for the boron-implanted sample correspond to the regions with the same lable in the schematic disorder distribution shown in the inset (adapted from Eisen and Welch[27]).

implantation. A typical backscattering curve from an implanted sample is shown in Fig. 6, together with the energy spectrum from an undamaged crystal.[27] The data for the implanted sample exhibit a peak in the backscattered yield, and at lower energies, that is, lower channel numbers, the scattering yield falls to a value which is somewhat higher than the yield at the corresponding energy from the undamaged crystal.

It is convenient at this point to introduce the concept of an aligned and a random component of the analyzing beam. The aligned component is the portion of the analyzing beam that is channeled. The random component of the beam is composed of all the other ions that are incident in the analyzing beam. It includes those ions that were not channeled on entry into the sample and those that were later scattered from the aligned beam (dechanneled).

The higher yield at depths below the disorder peak, shown in Figure 6, is due to the fact that some of the channeled ions have become dechanneled as a result of the scattering or spreading of the channeled beam as it passed through the damaged layer. These dechanneled ions may then interact with and be scattered from all the atoms of the crystal. One of the problems of applying the channeling effect technique to the study of disorder is to separate the scattering of the aligned beam by displaced atoms from the scattering of the random beam by all the atoms of the crystal. A useful approximation is to assume that the random beam rises linearly from a value equal to the aligned yield at the surface of the undamaged crystal to the value of the scattering yield observed behind the damage peak, as illustrated by the line labeled R' in Figure 6. Generally there is little difference between the straight line approximation and the more exact treatment, which is labeled R'' in Figure 6. The observed backscattering yield and the estimated yield from the random component of the analyzing beam (straight line in Figure 6) differ due to the contribution from displaced silicon atoms. The disorder distribution and total disorder can then be calculated directly. This procedure and other methods of estimating the dechanneled fraction of the beam have been described in several publications.[28-31]

The energy of a backscattered particle depends on the mass of the atom it is scattered from and the depth in the crystal at which it is scattered. The conversion between the energy scale of the backscattering spectra and the depth scale is given by

$$\Delta t = \Delta E / \{KS(E_1) + [S(E_2)(\cos\theta_1)/(\cos\theta_2)]\} \tag{1}$$

where Δt is the change in depth corresponding to a change in energy ΔE; $S(E_1)$ is the stopping power of the analyzing ions at energy E_1 before they are backscattered; $S(E_2)$ is the stopping power after backscattering at energy E_2; and θ_1 and θ_2 are the angles between the normal to the sample surface and the incident and backscattered beam directions, respectively.[28] For thin layers E_1 is the incident energy and $E_2 = E_1 - \Delta E$. $S(E_1)$ is the stopping power of channeled ions. The accuracy with which the depth scale can be calculated from Eq. (1) depends on the accuracy with which $S(E_1)$ and $S(E_2)$ are known. Measurements of channeled and random stopping have only been made for a limited number of projectile–target combinations. Since proton stopping powers are appreciably lower than helium-ion stopping powers in a given material, the depth resolution in channeling effect measurements is generally much better using helium rather than protons. A comparison at 1 MeV in silicon is shown in Figure 2.

The random stopping power is sometimes used as an approximation to $S(E_1)$. This may result in depth errors as large as 20%, since observed stopping powers for channeled ions are lower than random stopping powers. A second approximation which has been made is to use one-half the random stopping power for $S(E_1)$. While this may result in a smaller error than using the random stopping power, it may not be the best choice for $S(E_1)$, since experiments on helium stopping in silicon have shown that the ratio of channeled stopping power to random stopping power is a function of energy varying for $\langle 110 \rangle$ between 0.3 and 0.7.[32] If the appropriate channeled energy loss spectra are available, there may still be a question about how to obtain a value for $S(E_1)$, since there can be a relatively wide spread in the width of such spectra. Some recent experiments have indicated that if the crystal is not heavily disordered, the stopping power which would be derived from the most probable (rather than the minimum) energy loss of channeled ions in a perfect crystal should be used.[32,33]

The backscattering energy spectra from disordered samples give no direct information concerning the nature of the defects producing the scattering. It has often been assumed that all atoms which are more than 0.1 Å off lattice sites will be detected by the channeling effect technique. In lightly damaged samples the flux of channeled ions near the atomic rows should be appreciably smaller than the flux near the center of the channel. Therefore the contribution to the backscattering events from an atom which is only 0.1 Å or so off its lattice site will be appreciably smaller than the contribution from atoms which lie well toward the center of the channel (i.e., what might be termed true displaced atoms). It therefore seems that the defects which are detected by the channeling effect technique are primarily interstitials and interstitial clusters, and that the contribution from atoms which are slightly displaced due to the strain around such defects or around vacancies or vacancy clusters is relatively small.

Because of the uncertainty in identifying the nature of the center responsible for producing a contribution to the aligned yield, these displaced atoms are often called "scattering centers."

A problem which has been associated with the use of the channeling effect technique to determine the total amount of disorder introduced during ion implantation is that the results of the measurements do not agree with theoretical predictions.[4] Such predictions are based upon the theoretical work of Kinchin and Pease.[34] The modifications proposed by Sigmund[35] still contain an inverse proportionality of the total number of displaced atoms to the threshold displacement energy. It was early recognized that room-temperature measurements gave twice the predicted number of displaced atoms.[4,36] Recent measurements[37] made at 25°K (in order to reduce the annealing effects) gave about ten times the disorder predicted from Sigmund's result. It has sometimes been suggested that this may be due to the sensitivity of the channeling effect technique to very small displacements of atoms off their lattice sites because of strain.[36] We have argued above that such effects should be small compared to the scattering from interstitials and interstitial clusters. Perhaps a more satisfactory explanation is to be found in the suggestion that the effective threshold energy for damage produced by ion bombardment (where large clusters of displacements may be produced) is lower than the threshold energy which is determined from electron bombardment experiments.[38] The point is that when one or two of the neighbors of an atom have already been displaced the energy required to displace that atom should be significantly smaller than the value derived from electron irradiation experiments in which only single displacements are produced.

When the aligned scattering yield from an implanted sample is equal to the random scattering yield this is taken as an indication that an amorphous layer has been formed.[4] The random and aligned yields should be equal if an amorphous layer has been formed but their equality does not necessarily indicate the formation of such a layer.[39] It will be seen later that the presence or absence of an amorphous layer is crucial in some situations. Therefore we discuss here some problems connected with the use of the channeling effect technique to determine the presence of such a layer. If the width of the damaged region is small compared to the depth resolution of the channeling effect measurement [this can be calculated from Eq. (1)], then the height of the damage peak in the aligned spectrum will not be indicative of the actual maximum amount of disorder in the crystal.

Under this condition the height of the damage peak will reach a saturation value which is less than the random yield. In compound semiconductors the aligned yield from a given constituent of the compound should actually be compared with the random yield from that constituent. For a compound with a small mass difference between the two constituents such as GaAs, helium-ion beam energies of the order 2.5–3 MeV are required to resolve the scattering from the gallium and the arsenic.[40]

2.4. Beam Effects and Sensitivity

There are a number of cases in which the analyzing beam itself has influenced the amount of lattice disorder or the lattice location of dopant atoms. For example, beam annealing of disorder has been found to be caused by both proton and helium-ion beams in silicon samples implanted with boron ions at low temperature[30] or with carbon or antimony ions at room temperature.[41,42] Bøgh and co-workers have observed an opposite effect, namely an increase in the amount of damage, particularly in germanium.[41] Beam-induced motion of dopant atoms off lattice sites has been found for boron[4] and thallium[43] in silicon and also in samples of silicon heavily diffused with arsenic.[44,45] The arsenic-in-silicon case has been studied in the most detail. The fraction of arsenic which comes off lattice sites has been found to saturate at high helium-ion doses and to be a function of the arsenic doping of the sample. A detailed mechanism has not yet been suggested for these phenomena. However, their occurrence suggests that interpretation of channeling effect data which indicate dopant location off lattice sites should be undertaken with caution.

The sensitivity of the channeling effect technique in lattice location experiments depends upon the mass of the dopant which is being studied. When backscattering is used to detect the interaction of the channeled beam with these dopants the sensitivity will be greater the higher the mass or atomic number of the dopant, reaching levels of the order about 10^{-4} atomic fraction.[19] This sensitivity is sufficient for many implantation situations in which a relatively high atomic fraction of the dopant is introduced in the thin layer near the sample surface. For dopants with a mass lower than that of the substrate into which they are implanted nuclear reactions or ion-introduced X rays may be employed in some cases. However, extensive use has not been made of nuclear reactions (except for boron in silicon) and work has only recently begun to evaluate the sensitivity of ion-induced X rays in lattice location experiments.

In measuring disorder the sensitivity is determined by the magnitude of the minimum yield observed from an undamaged crystal of the material of interest. This is usually of the order of 1%, so that the channeling effect technique can only be applied in cases where approximately 1% or more of the atoms in the damaged layer have been displaced. Double alignment techniques provide a method of decreasing the minimum yield from the undamaged sample.[11] This is accomplished by aligning the detector with a symmetry axis in the crystal as well as aligning the beam with such an axis. Double alignment may produce an order of magnitude or more reduction in the aligned yield from the undamaged sample, thus lowering the background which limits the sensitivity in detecting disorder. However, double alignment techniques must be used with care because they require appreciably higher doses of the analyzing beam in order to perform the measurement. This greatly increases the possibility that the results will be affected by the type of beam-induced effects discussed in the preceding paragraph.

3. Implantation in Silicon

Silicon provides an ideal host for studying the basic parameters and concepts involved in implantation processes. In order to obtain a feeling for the trend in implantation research, it is necessary to put the work in silicon in some perspective. Ion implantation is a complicated process with many parameters, such as implantation temperature, anneal temperature, total dose of ions per cm^2, dose rate, beam energy, and beam species. All of these parameters must be considered when evaluating implantation behavior in any solid. An often overlooked factor is that silicon technology is so advanced that reproducible, well-characterized, large-area, samples of silicon can be obtained in large quantities. Another advantage provided by silicon technology is that contact and measurement techniques have been developed so that electrical evaluation can be made in a standardized fashion. Then, too, the solubilities of dopants in silicon have been well established so that one can make comparisons between implantation, diffusion, and bulk doping.

In this section we will consider range and disorder distributions, the nature of lattice disorder, and lattice location and electrical activity of the implanted species. We will treat situations where the concentration of the implantation species is less than 1 at. %. This allows us to consider the more basic aspects without becoming involved in composition changes, precipitation of dopant atoms, sputtering, and other problems associated with high dose implants.

3.1. Range Distributions

Over the past years there have been extensive studies, both experimental and theoretical, on the penetration of energetic particles in matter. Lindhard *et al.*[46] have developed "unified" range-energy relations (generally referred to as LSS theory) which have been very successful in predicting the range of implanted ions in semiconductors and other materials. They distinguish between two different mechanisms of energy loss: nuclear collisions, in which energy is transferred to the target atom as a whole, and electronic collisions, in which the atomic electrons are excited or ejected. Large energy transfers in nuclear collisions lead to displacement of lattice atoms. Nuclear energy loss is the more important process at low energy. It reaches a maximum value at energies of 3, 17, 73, and 180 keV for B, P, As, and Sb in Si.[4] For energies generally of interest in ion implantation electronic energy loss increases linearly with velocity and becomes dominant at energies of 17, 140, 800, and 2000 keV for B, P, As, and Sb in Si.[4] Values for nuclear stopping can be calculated in a straightforward fashion. However, electronic stopping powers have been observed to oscillate as a function of atomic number about the values predicted from LSS theory.[4,11,47] Accurate values of electronic stopping power (energy loss per unit path length due to electronic collisions) are not readily available. This results in some uncertainty in predicting the range of low-mass ions in the energy range where electronic stopping is dominant.

For amorphous material the depth distribution is roughly Gaussian and can be characterized by a mean projected range R_p and a standard deviation ΔR_p about the mean. Approximations and rules of thumb for ion ranges and range straggling have been given by Schiøtt.[48] Detailed calculations based on the theoretical concepts developed by Lindhard and co-workers have been presented by Winterbon.[49] Johnson and Gibbons[50] utilized the LSS theory to tabulate values of R_p and ΔR_p for application to a wide variety of semiconductors.

Figure 7 shows the depth distribution at several energies of boron implanted in Si.[51] With increasing energy the penetration depth increases and the distribution broadens. From the dashed

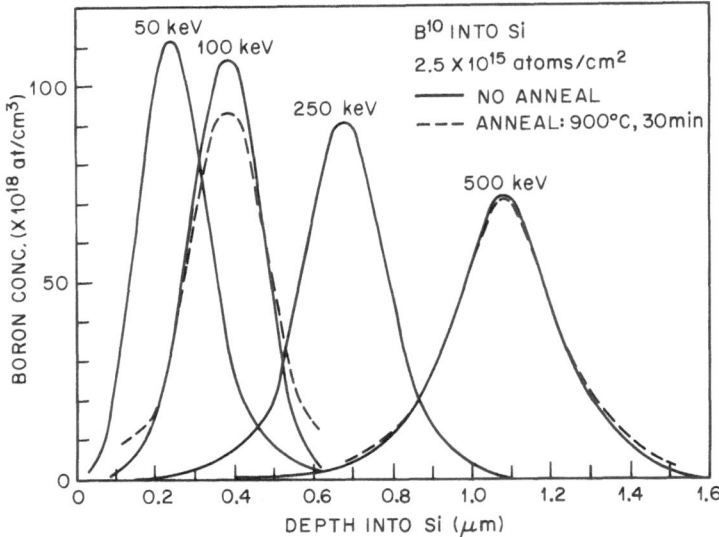

Fig. 7. Depth distribution of boron ions implanted into silicon at room temperature (from Ziegler et al.[51]).

curves one can see that there is no major change in the distribution following anneal at 900°C. This behavior is not always observed, as we will see later. The data in Figure 7 were obtained by thermal neutron reaction techniques. There have been a variety of other boron range measurements in Si[9,52,53] which agree with each other within about 10 to 20%. At present we must accept this uncertainty because a calculation of the boron range is not more accurate, due to the uncertainty in the value of the electronic stopping power.

For beam energies less than 1 MeV the penetration depth for most ion species is less than 1 μm. Crowder[54] has compared the measured values of the projected range of implanted species in amorphous silicon with the LSS predictions. His values are shown in Table 1. The close agreement between the measured values and the theoretical predictions indicates that the LSS theory provides reasonable estimates for the projected range. An early survey of range measurements was given by Gibbons.[55]

Measured values for projected range straggling are also listed in Table 1. Several authors have given theoretical estimates for ΔR_p and simple rules for making such estimates have been given by Schiøtt.[48] Johnson and Gibbons[50] have tabulated values for range straggling; however, there are systematic errors in these tabulated values. This is discussed in more detail in Ref. 54, which also suggests

TABLE 1

A Comparison of Range Parameters Observed in Amorphous Silicon with Lindhard Theory Calculations*

Ion	Energy, keV	Range parameters, μm		ΔR_p(obs)†
		R_p(obs)†	R_p(LSS)	
^{11}B	60	0.23	0.244	0.07
	120	0.40	0.467	0.10
^{27}Al	200	0.26	0.297	0.08
^{31}P	100	0.12	0.123	0.048
	140	0.18	0.175	0.07
	200	0.28	0.254	0.074
	280	0.38	0.359	0.091
^{71}Ga	140	0.09	0.083	0.035
	280	0.175	0.163	0.07
^{75}As	80	0.06	0.048	0.025
	140	0.095	0.079	0.040
	280	0.18	0.154	0.07
^{121}Sb	120	0.06	0.053	0.016
^{123}Sb	260	0.105	0.102	0.03
^{209}Bi	240	0.08	0.076	0.02

* LSS calculations from Johnson and Gibbons' tabulation.[50]
† The experimental errors in these parameters are approximately ±10%.

methods for using the tabulated values. Detailed calculations have been made by Winterbon[49] and others[56] that indicate agreement between experimental and theoretical values for range straggling.

The preceding discussion treated silicon as essentially an amorphous lattice and neglected the steering effects of the crystalline rows and planes. Channeling effects can play a large role in determining the depth distribution of implanted ions.[4] This process has been described in detail in a number of review articles.[55,57,58] The essential point is that the implanted atom can penetrate significantly deeper into the lattice. Figure 8 shows a comparison[59] between range distributions for arsenic atoms implanted in silicon along a $\langle 110 \rangle$ axis and those implanted in a direction which does not coincide with a major axis or plane (random direction). In this example the peak in the channeling distribution is a factor of ten deeper than the value of R_p for random incidence.

The maximum range to which the incoming energetic particles can penetrate is determined by the orientation of the silicon and the

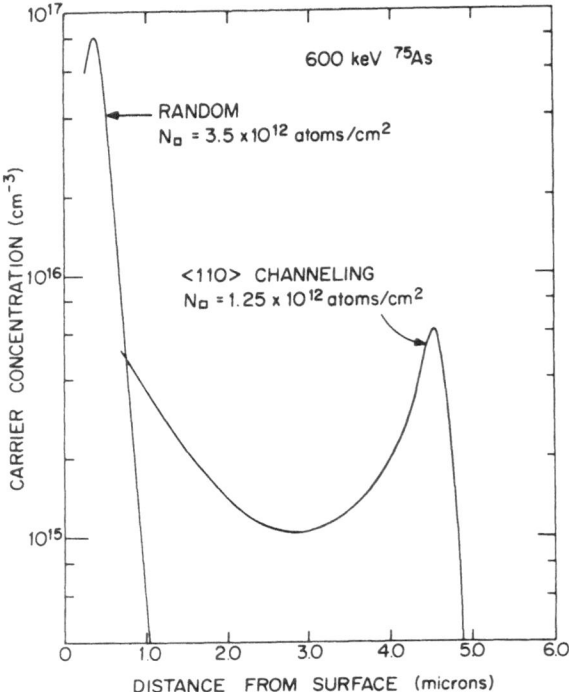

Fig. 8. Carrier concentration profiles resulting from 600-keV arsenic-ion implantation into $\langle 110 \rangle$ and random orientations of silicon (from Reddi and Yu[59]).

incident ion. For channeled phosphorus implants in silicon, the most studied case, good agreement for the range–energy relations has been obtained in experiments in different laboratories.[60-63]

The conditions are very stringent for the incident ion to remain channeled throughout its trajectory. Increases in target temperature, in lattice disorder, or in the amount of oxide layer on the surface produce deflections in the path of the energetic ion sufficient to cause the ion to become dechanneled as it travels through the crystal.[60,64,65] The shape of the depth distribution is strongly influenced by temperature and disorder. Figure 9 shows the depth distribution of phorphorus ions in silicon as a function of orientation off the $\langle 111 \rangle$ axis.[64] The beam fraction that channels deeply in the crystal is only slightly affected by small misorientations, but a larger misorientation causes a marked reduction in the number of particles that can penetrate deeply into the crystal. Even then, there is a tail in the distribution that can extend far in the crystal.

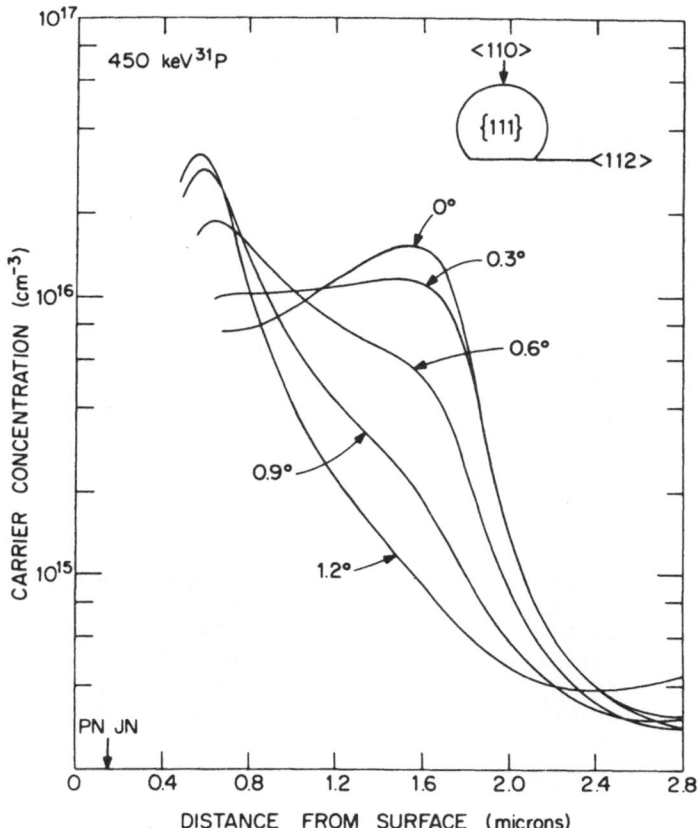

Fig. 9. Carrier concentration profiles resulting from 450-keV phosphorus implantation into silicon for tilt angles 0–1.2° away from ⟨111⟩ orientation (from Reddi and Sansbury[64]).

The general role of channeling in silicon is reasonably well understood. Under certain conditions it is very hard to distinguish whether a tail in the implanted distribution is caused by channeling of the incident ions or to interstitial diffusion. Experiments are still underway to resolve these two effects.[66]

Diffusion processes can result in a major distortion of the profile of implanted species.[4] Early work on enhanced diffusion of dopants in silicon under proton bombardment suggested that diffusion of dopants might occur during implantation. Indeed, such diffusion effects were observed for phosphorus implantations in silicon when the substrate temperature was raised above 600°C. Recently work on proton-enhanced diffusion has been carried out in implanted structures.[67]

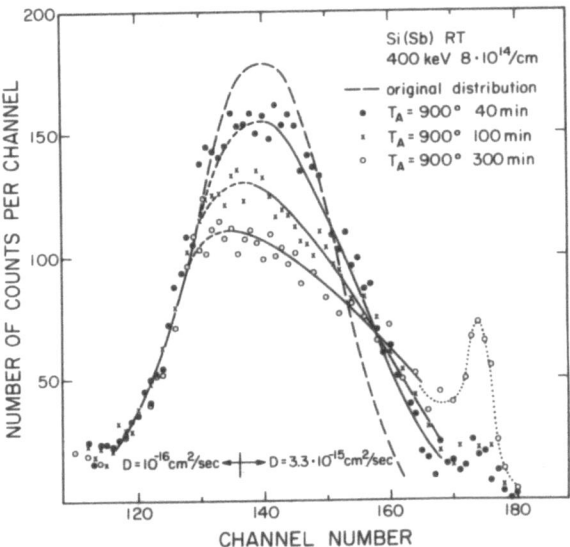

Fig. 10. The depth distribution of antimony implanted into silicon at room temperature to a dose of 8×10^{14} ions/cm^2. The depth profile was obtained by backscattering measurements. The peak at channel number 175 corresponds to antimony on the surface, and lower channel numbers correspond to greater depths in the crystal (from Meyer and Mayer[68]).

Enhanced diffusion effects have also been noted during the anneal of implanted layers. These effects generally take the form of diffusion toward the surface and may be due to the defects along the track of the implanted atoms. Figure 10 shows the depth distribution of antimony implanted into silicon and annealed at 900°C for successively longer times.[68] In this case the diffusion coefficient toward the surface was enhanced by a factor of 30.

Enhanced diffusion has been noted for other species.[69] Great caution must be used in attempting to extract diffusion coefficients from the dopant profile in implanted and annealed samples. At the present time we are not aware of any experiments which have shown agreement between diffusion coefficients extracted from implanted samples and those from bulk diffused samples. However, implantation techniques have been used to introduce dopants which are then driven deeper into the sample during a high-temperature diffusion anneal.

3.2. Disorder Distributions

As an energetic ion slows down and stops in a crystal it makes a number of collisions with the lattice atoms. The kinetic energy given to a recoiling target atom in these collisions leads to a distribution of lattice disorder in the region around the primary ion track. Again, as in the case of range distributions, there have been extensive theoretical studies of the depth distribution of lattice disorder. Experimentally, a number of techniques have been used to determine lattice disorder.[4,5] Perhaps the most direct technique applicable to a single-crystal target is channeling measurements. This technique, as described in Section 2, provides a direct measure of the depth distribution of target atoms moved off lattice sites.

Calculation of disorder involves both the depth distribution of energy lost in nuclear collisions and the range distribution of recoiling target atoms. For high-energy ions where the primary ion range is much greater than that of the recoils the distribution of lattice disorder can be estimated from the amount of energy lost in nuclear collisions by the primary ion. As an example, Figure 11 shows the depth distribution of nuclear energy losses for 40-, 100-, 200-, 300-, and 400-keV boron ions in silicon.[70] Experimental measurements of disorder are in reasonable agreement with energy loss predictions for boron[30] and oxygen[71–74] in silicon. For low-energy heavy ions the range of

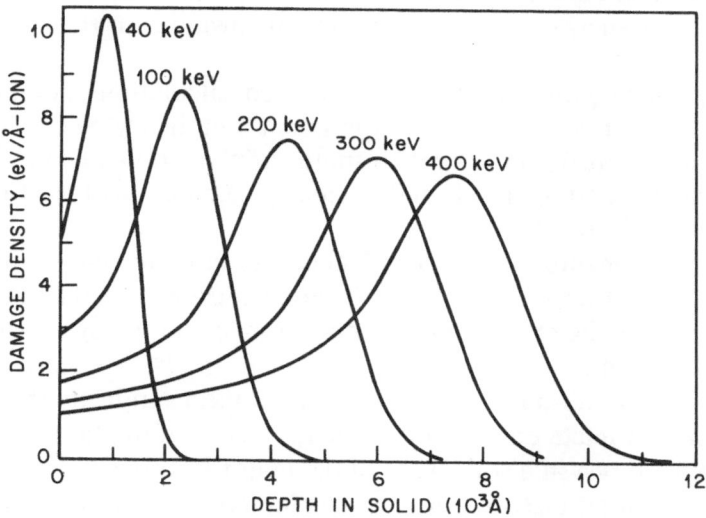

Fig. 11. Depth distribution of energy deposited into atomic processes for ^{11}B incident on an amorphous silicon target (from Brice[70]).

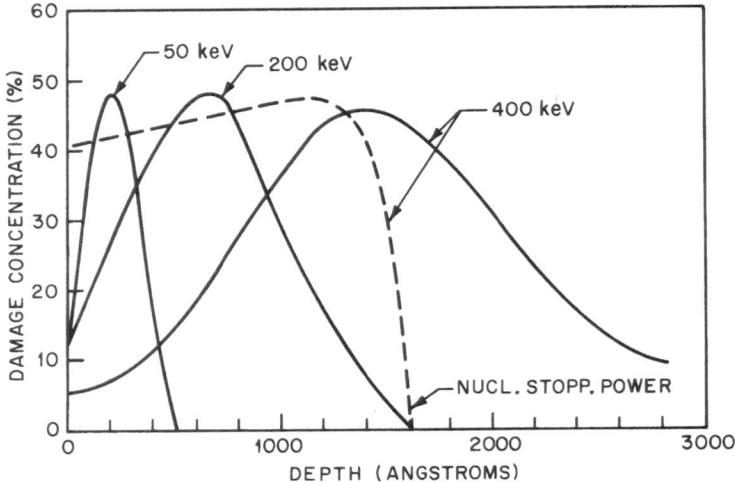

Fig. 12. Heavy-ion damage distributions in silicon. Dashed line is the specific nuclear energy loss versus distance traveled, normalized to give same total damage (area under curve) as the 400-keV distribution (from Bøgh et al.[41]).

the recoils may be great enough that the damage distribution will deviate significantly from the energy loss distribution. Such an example is shown in Figure 12, where the disorder distribution extends deeper into the crystal than that predicted on the basis of nuclear collisions alone (dashed curve).[41] It points out the importance of taking into account the energy transported by recoils away from their point of origin.

By utilizing the results of more detailed calculations, good agreement can be obtained between experiment and theory. Such a case is shown in Figure 13 for 50-keV bismuth implanted in silicon.[75] The experimental data of Bøgh match closely calculations based on the work of Winterbon.[49]

Ion channeling can also influence the amount and depth distribution of lattice disorder.[41,76] There is a pronounced decrease in disorder since the channeled ions do not make large-angle collisions with lattice ions.

The data on range and disorder distributions indicate that there is a fairly complete description of energy losses of implanted ions as they come to rest in a solid. One of the major features of comparison between disorder distributions and implanted ion range distributions is that the peak in the disorder distribution lies at somewhat smaller depths in the crystal than that of the range distribution. Generally,

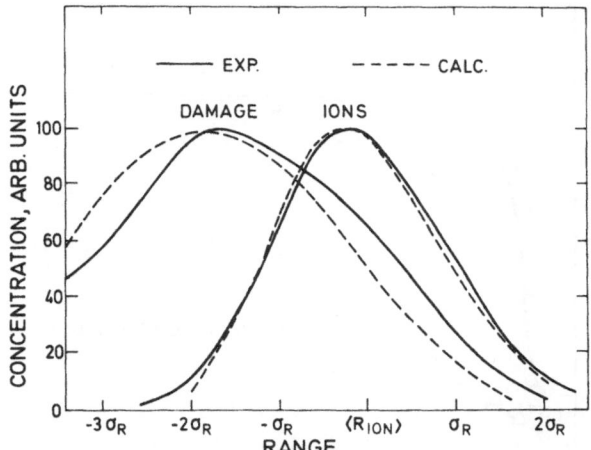

Fig. 13. Comparison of the ion and damage depth distributions for 50-keV bismuth implanted in silicon. The calculations are from Winterbon[49] (from Bøgh[75]).

the effects of implantation are confined to a region within 1 μm of the target surface. Ion channeling can lead to significantly deeper penetration.

3.3. Production and Annealing of Disorder

Over the past three or four years one of the major areas of study in implantation has been with regard to characterization of the amount and nature of lattice disorder produced by energetic ions. There have been a wide range of experimental techniques utilized[4]: channeling techniques, electron paramagnetic resonance (EPR), electron diffraction in either the transmission or reflection mode, optical absorption and reflection, and internal strain techniques.[77,78] The results of these investigations indicated that there are a large number of experimental variables which influence the amount of lattice disorder produced. These variables include ion mass, substrate temperature during implantation, dose rate, and dose. The studies form a fascinating and complex field, and a comprehensive understanding of the detailed mechanisms has not yet been achieved. However, it has been shown that there are strong correlations of defect species between fast neutron-irradiated and ion-implanted silicon.[73,79]

Figure 14 shows a comparison of channeling effect measurements of the disorder produced by implanting antimony in silicon at room

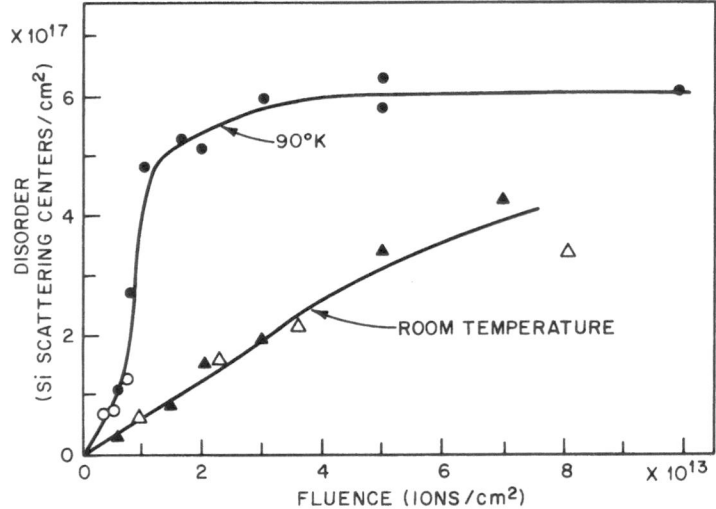

Fig. 14. Silicon lattice disorder vs. implantation dose for 90°K and room-temperature antimony implantations (from Picraux *et al.*[80]).

temperature and at 90° K.[80] The difference between these two curves indicates that some annealing of disorder takes place during room-temperature implantation. The amount of annealing depends upon the ion type, with more annealing occurring for damage produced by boron than antimony. In order to make comparisons of damage produced by different ions and at different energies, it is desirable to make measurements at low temperatures. Recent investigations at 27° K have suggested that some annealing of disorder may occur below 90° K.[81] However, at present we only have a significant amount of data for implantations carried out at 90° K and must limit our discussion to these results.

The average disorder per unit volume produced by implanting 100-keV oxygen or 200-keV antimony into silicon at 87° K is plotted vs. the average energy into atomic process per unit volume in Figure 15.[82] This shows that at this temperature disorder production depends only on the energy into atomic processes and is independent of the ion mass. This result is to be expected from theoretical considerations. However, it is possible for annealing effects to obscure this result.

The annealing of lattice disorder produced at low temperature appears to be independent of ion type. For example, Figure 16 (upper) shows a comparison between annealing results obtained in silicon implanted with boron or with antimony at low temperatures.[80]

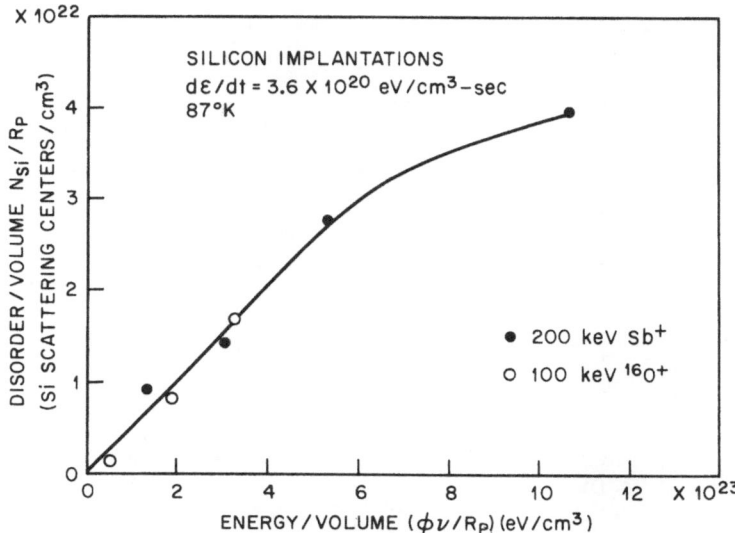

Fig. 15. The average disorder per unit volume vs. the average energy into atomic processes per unit volume for 100-keV oxygen and 200-keV antimony implants in silicon performed at 87°K (from Picraux and Vook[82]).

The lower part of Figure 16 shows the growth in divacancy concentration which occurs during the annealing of the gross disorder.[83] This divacancy formation is thought to take place because neutral vacancies are released from more complicated defect configurations. These measurements of divacancy concentration at low temperature are the only case in which defect structures have been identified following irradiation at low temperature. Consequently, it is not possible to identify the configurations of the defects from which the neutral vacancies are released. On the other hand, there has been identification of vacancy-associated defects following implantation at room temperature. Optical absorption[79] and EPR[84-86] measurements have been utilized to give an inventory of defects. Isotopic doping by implantation has been employed in an elegant fashion in the case of the vacancy–oxygen center.[85] The ^{17}O hyperfine spectrum is ordinarily very weak and unresolved from the fine structure and ^{29}Si hyperfine spectrum. This difficulty was overcome by implanting ^{17}O into silicon samples and thus enhancing the intensity of the hyperfine spectrum.

In order to identify these isolated defects, it is necessary that they reside in a crystalline region. The volume of crystalline material in the implanted layer decreases with an increase in dose. Consequently,

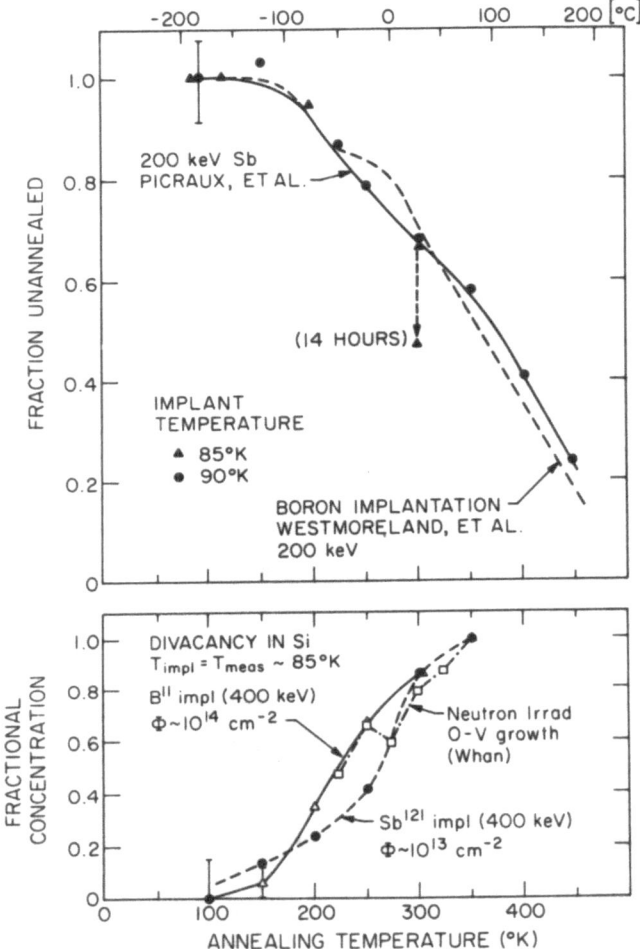

Fig. 16. Upper—Channeling effect measurements of disorder anneal behavior for low-temperature 200-keV antimony and boron implants in silicon (from Picraux *et al.*[80]). Lower— Fraction of divacancies vs. annealing temperature for 400-kev boron- and antimony-ion implantation into silicon at 85°K. Also shown is the vacancy–oxygen center growth following neutron irradiation (from Vook and Stein[83]).

the signal strength from isolated defects first increases with dose and then decreases. Eventually an amorphous layer is formed[79] at high doses, when the average amount of energy into atomic processes approaches $\sim 10^{21}$ keV/cm^3.[87]

The properties of the amorphous layer have been studied by Raman spectroscopy[88] and X-ray and electron diffraction.[4] During anneal at temperatures of about 550°C this amorphous layer re-crystallizes epitaxially on the underlying crystalline material.[4]

Following anneal of an ion-implanted sample to 800 or 900°C it is usually not possible to detect disorder with channeling effect techniques irrespective of whether or not an amorphous layer was formed. Hall effect measurements also indicate good electrical activity and the absence of compensating defect centers. However, other techniques indicate the presence of defects. Transmission electron micrographs show the presence of dislocations.[39,89,90] The detailed morphology of these dislocations depends on implantation parameters, as illustrated in Figure 17 for two samples implanted with 10^{15} boron ions/cm^2. Both samples were annealed at 800°C, but the sample shown on the left in Figure 17 was implanted at low temperature and that on the right in Figure 17 at high temperature. These dislocations in the implanted layer act as precipitation centers and can getter mobile impurities such as Cu and Au.[91] It has been shown that the depth distribution of Cu precipitates follows the initial depth distribution of disorder produced during implantation. Finally we comment that attempts to grow epitaxial layers on top of an annealed amorphous layer have not been successful.[92]

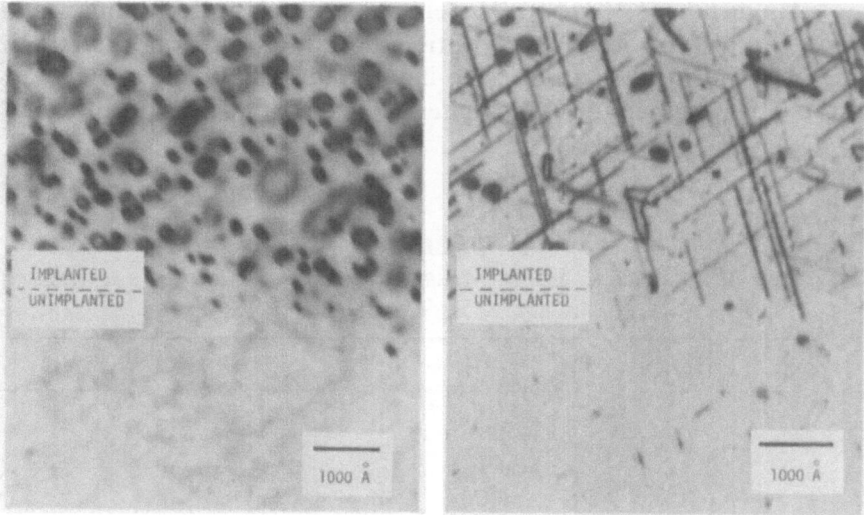

Fig. 17. Electron micrographs showing the difference between a silicon sample irradiated at low temperature and annealed to 800°C (left) and at room temperature followed by the same annealing treatment (right) (from Chadderton and Eisen[39]).

3.4. Lattice Location of Implanted Atoms

The lattice locations of implanted atoms in silicon have been studied primarily by the channeling effect technique.[4] Some measurements for arsenic and antimony have been made using EPR spectroscopy.[93] For low-dose room-temperature implants all species studied have exhibited substitutional behavior. This is true for low-mass elements (boron) and high-mass elements such as bismuth and thallium. It has been suggested that the implanted atom is readily able to find a substitutional site because of the large number of vacancies created along its track.[4] For higher dose implants, where the implanted atom comes to rest in a completely disordered region, the concept of lattice location loses its meaning.

After annealing, or under hot-substrate implant conditions, the lattice location of the implanted atom becomes species and temperature dependent. Table 2 gives a survey of the lattice location of implanted species in silicon under these conditions.[4] The data indicate that the behavior of a particular atom is correlated with its group in the chemical periodic table. Elements from groups IV, V, and VI are substitutional and their concentration may exceed equilibrium solubilities.[4] Note, however, that these high substitutional concentrations are not maintained at high anneal temperatures where appreciable diffusion, and hence precipitation, can occur.

The behavior of group III elements is much more complex, and they can exhibit both substitutional and tetrahedral interstitial characteristics. Furthermore, the interstitial component can be surpressed by implantations in material heavily doped with group V

TABLE 2
Summary of Channeling Effect Measurements of the Lattice Location of Implanted Species in Silicon

Period	Group:	I*	II†	III†‡	IV‡	V‡	VI‡
2		—	—	B	—	—	—
3		—	—	—	—	P	—
4		—	Zn	Ga	Ge	As	Se
5		Rb	Cd	In	Sn	Sb	Te
6		Cs	Hg	Tl	Pb	Bi	—

* Randomly located off substitutional sites.
† Tetrahedral interstitial.
‡ Substitutional.

species.[4] The lattice location of group II species is characterized primarily by a tetrahedral interstitial component.[94] The effects of a high concentration of donors on the location of these species has not been investigated and the results are limited to rather high concentrations of group II atoms.

The behavior of group I elements is of interest because early work had suggested that doping by these elements was due to their occupation of interstitial positions.[95] Channeling effect experiments indicate that they are not on substitutional sites.[96] However, there was no indication of a preference for the tetrahedral interstitial position. Since only a small fraction of the atoms implanted are electrically active, it is not possible to correlate this electrical activity with a particular interstitial position.[97]

The lattice location measurements summarized in Table 2 were carried out before the importance of beam effects and flux peaking (see Section 1) were realized. Consequently, the results on the fraction of the implanted species located off substitutional positions may not be reliable. For example, the early results which indicated that only ~50% of the arsenic implanted in silicon was substitutionally located have been shown to be incorrect because of beam effects.[44] Such effects may also be responsible for the low (~40%) substitutional component observed in selenium-implanted silicon.[94] With the understanding of flux peaking effects has come the realization that detailed angular scans through several symmetry axes, and perhaps also through symmetry planes, are necessary to reach reliable conclusions concerning the lattice position of dopants located off substitutional sites (Section 1). This was not done in the early work and we have therefore not quoted results on the fraction of implanted atoms in interstitial vs. substitutional positions. When such detailed measurements have been made (ytterbium in silicon,[20] for example) it has been possible to suggest a fairly precise interstitial location for the dopant. Flux peaking effects have also been seen for zirconium, hafnium, thallium, and mercury implanted into silicon.[98]

3.5. Electrical Effects in Ion-Implanted Layers in Silicon

Preceding sections have indicated that there is now a fair description of the range and disorder distribution in implanted silicon, the identity of defects is somewhat in hand, and the annealing of the defects follows a pattern similar to that in other irradiation studies. The descriptions are complex but manageable. One would expect that the electrical nature of the implanted layer would exhibit far greater complexities. However, for the conventional group III and V dopants such as boron and phosphorus the picture is comparatively simple.[4] If one anneals to sufficiently high temperatures, so that most of the radiation damage effects are annealed away, all the implanted dopant ions become electrically active. An example of this is shown in Figure 18, which gives the number of holes per cm² vs. the dose of implanted boron ions.[99] There is a 1:1 relation over four orders of magnitude between the number of holes and the number of boron ions per cm². This behavior holds for cases where the concentration of implanted species is less than the equilibrium solubility at the anneal temperature.

Perhaps the most thoroughly investigated aspect of ion implantation in silicon is the electrical behavior of the implanted layer. The electrical characteristics are of course crucial in the application of implantation to device fabrication. Hall effect measurements have

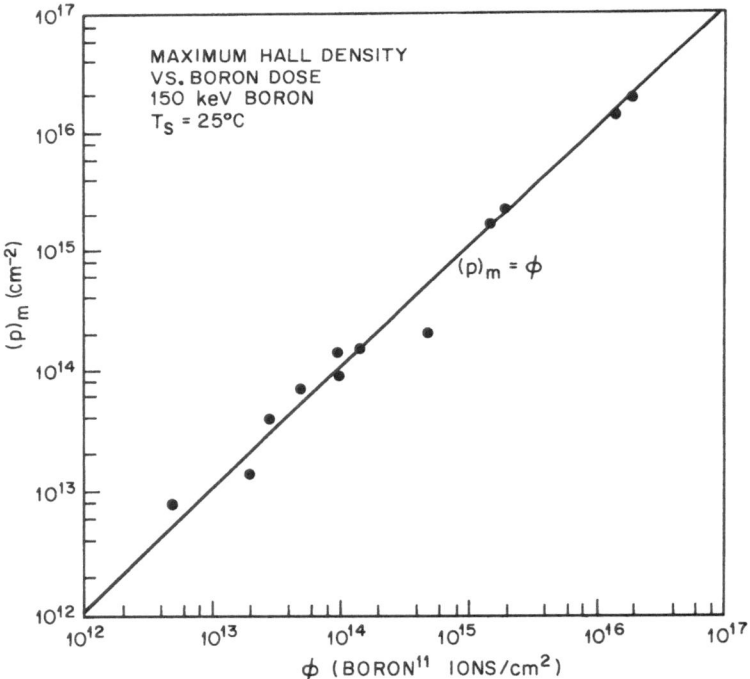

Fig. 18. Relation between the number of electrically active centers and the number of implanted ions in boron-implanted silicon (adapted from MacRae[99] by Stephen[57]).

been utilized, layer removal techniques have been developed to determine the carrier concentration profile, conventional techniques such as sheet resistivity and angle lap have been employed, the junction characteristics have been evaluated, and minority carrier lifetimes have been measured.

Often the first step in evaluating an implanted layer is to make Hall effect measurements using a *p–n* junction to provide electrical isolation between the substrate and the implanted layer. It was early recognized that these Hall effect measurements provide weighted averages of the number of carriers per cm^2 and the mobility. This arises because of the variation of dopant concentration and hence carrier mobility throughout the implanted layer. These mobility variations can be accounted for by using layer removal techniques and differential Hall measurements.[4] In the following sections we will utilize results of these Hall effect measurements to discuss annealing characteristics, solubility effects, and energy levels of implanted layers.

3.5.1. Anneal Characteristics

Figure 19 shows the results of a study of annealing conditions for obtaining a high degree of electrical activity for phosphorus implanted into Si substrates at room temperature.[100] The ratio of the number of electrons/cm^2 to the number of implanted phosphorus atoms/cm^2 is plotted vs. the phosphorus dose. The pronounced minima in the curves occur at a dose just below that required to form a continuous amorphous layer overlapping the ion distribution. At higher doses the rise in the relative number of electrons reflects the epitaxial reordering of the amorphous layer on the underlying single-crystal material. In this epitaxial regrowth process the dopants are incorporated into substitutional lattice sites and nearly all the compensating defects are removed.

For almost all dopant species investigated the formation and annealing of an amorphous layer plays the dominant role in the anneal characteristics. Figure 20 shows this behavior for a series of dopant species implanted into silicon to concentrations such that an amorphous layer is formed.[101] Upon anneal the number of carriers per cm^2 increases abruptly in the temperature range where the amorphous layer reorders. At high anneal temperatures there is a decrease in the number of carriers, indicating that precipitation or outdiffusion effects had occurred. The relation between the number of carriers/cm^2 and

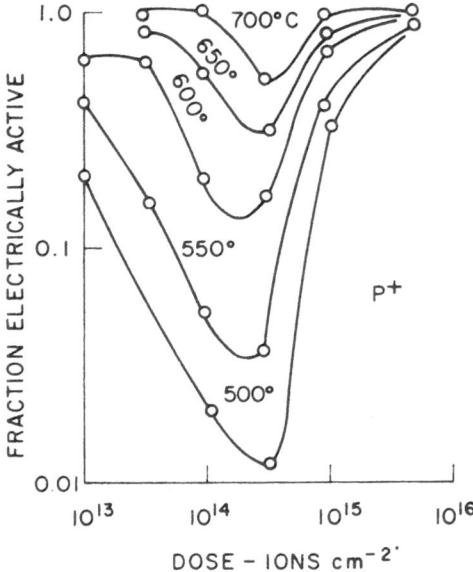

Fig. 19. Isochronal annealing of silicon implanted with 280-keV phosphorus ions at room temperature. The ratio of the number of carriers to the number of implanted ions is plotted as a function of the ion dose for 30-min anneals at the temperatures indicated (from Crowder[100]).

the number of implanted dopant atoms is discussed in the following section.

Not all anneal characteristics exhibit the simple behavior shown in Figures 19 and 20. The classic exception is the behavior of boron implanted in silicon. In this case the number of holes/cm² exhibits negative anneal behavior.[4] The number of holes decreases over a limited range of anneal temperatures (500–600°C) and then increases at higher temperatures. The combination of channeling effect and Hall measurements indicated that this negative anneal behavior was due to the motion of boron off substitutional sites.[102,103] At higher anneal temperatures nearly all the boron reoccupies substitutional sites.

We emphasize that temperatures of 800–900°C are required to remove compensating defects in heavily disordered regions where an amorphous layer is not formed. This was first observed in evaluation of hot-substrate implants. However, the effect can have a major

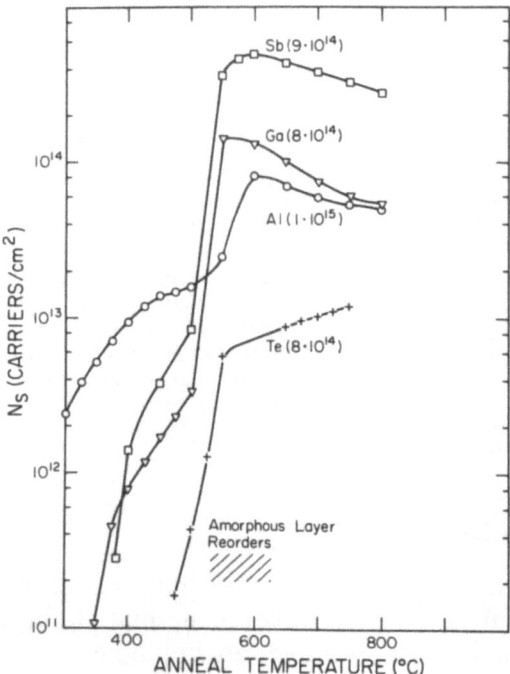

Fig. 20. Anneal characteristics for silicon samples implanted with the ions shown to doses such that an amorphous layer was formed. The abrupt rise in the curves reflects the reordering of the amorphous layer (from Johansson and Mayer[101]).

influence in the depth distribution of the carrier concentration. Crowder[54,104] has measured the carrier concentration profile of two wafers implanted with 3×10^{14} 280-keV ^{31}P ions/cm^2 and annealed at 600°C. At the original position of the amorphous–crystalline interface (prior to anneal) there was a transition region after anneal in which the carrier concentrations were anomalously low. This indicates the presence of compensating defects. Higher anneal temperatures are required for their removal.

The lifetime of minority carriers is very sensitive to defects. Figure 21 shows isochronal annealing of minority carrier lifetime in silicon irradiated with 1.5 MeV silicon ions.[105] With increasing dose higher anneal temperatures are required. It was reported that no lifetime recovery was observed until the anneal temperature approached 900°C for a dose of 1.25×10^{15} Si ions/cm^2.

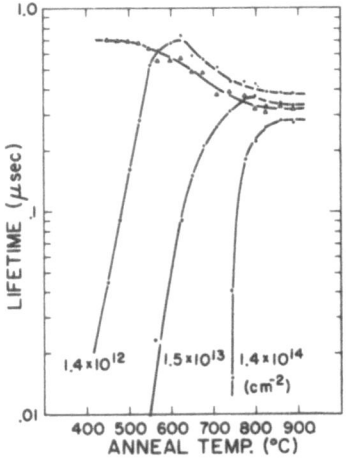

Fig. 21. The isochronal annealing of minority carrier lifetime in 0.5 ohm-cm *p*-type silicon irradiated with 1.5-MeV silicon ions to the indicated doses. The lifetime in unirradiated diodes that were otherwise similarly treated is shown for comparison (from Davies and Roosild[105]).

3.5.2. Solubility Effects

Ion implantation affords the possibility of introducing dopant species to concentrations well above the thermal equilibrium solubility values in the solid. For example, an implanted dose of $10^{15}/cm^2$ (somewhat less than a monolayer) distributed over a 1000 Å corresponds to a concentration of 10^{20} atoms/cm^3. This value is significantly above the equilibrium solubility for many impurities in silicon. Under normal annealing conditions at temperatures of 850–900°C the dopant atoms are sufficiently mobile so that concentrations greater than solubility cannot be maintained. The excess atoms either precipitate or diffuse out so that their concentration approaches solubility values at the anneal temperature. The solubility limitation has been noted for group III dopants in Si, for example.[4]

There are two methods by which it has been possible to exceed the solubility limits. Both involve utilizing implantation so that the major part of the lattice disorder is annealed at a temperature well below that at which an appreciable diffusion of implanted atoms occurs. One method that was first utilized is to implant at temperatures between 300 and 450°C. At these temperatures the lattice disorder around the track of the ion anneals sufficiently fast so that there is not a build-up of appreciable disorder. In this case substitutional concentrations well above the equilibrium solubility have been achieved.[4]

The other approach involves forming an amorphous layer and then annealing only to temperatures such that the amorphous layer has just reordered. Antimony-implanted layers exhibit such behavior (see Figure 20). This technique of forming an amorphous layer and then annealing at low temperatures to produce reordering and high substitutional concentrations does not always work. For such species as zinc, cadmium, and mercury which have high diffusion coefficients in silicon the implanted species is observed to outdiffuse in nearly the same temperature range as the reordering of the amorphous layer occurs.[68,94]

3.5.3. Ionization Level Measurements

One of the attractive features of ion implantation is that it affords a unique opportunity to study the ionization level (or activation energy) of unusual dopant species. The concept was that it would be possible to implant nearly any dopant species and then anneal to temperatures of 850–900°C to remove compensating defects. Hall effect measurements would then be made as a function of temperature to obtain the ionization energy. Because the carrier concentration and hence carrier mobility vary with depth, it is necessary to make such measurements in combination with layer removal. Such differential Hall measurements have been used with success for indium-implanted layers.

The measurement procedure is not easy to carry out and there are many opportunities for error. Figure 22 (upper) shows differential Hall measurements of carriers vs. reciprocal temperature for phosphorous implants in silicon annealed to 850°C.[107] For the low dose $(3 \times 10^{12} \text{ P/cm}^2)$ implant the measured activation energy (0.039 eV) corresponds to published values; while for the higher dose $(3 \times 10^{13}/\text{cm}^2)$ implant the activation energy has decreased due to the high dopant concentration. This shift in the ionization level can also be seen in the differential Hall measurements on tellurium-implanted samples (Figure 22, lower). Only the lowest dose implant (where the carrier concentration at 300°K was less than $10^{17}/\text{cm}^3$) exhibited behavior close that predicted from prior measurements on bulk doped samples $(E_A = 0.15 \text{ eV})$.

To obtain values for the activation energy, the carrier concentration must be kept below $10^{17}/\text{cm}^3$. If low-energy implantations are used, where the average penetration depth is the order of 1000 Å, the number of dopants per cm^2 will be of the order of 10^{12}. For this surface concentration it is very difficult to make sure that surface effects are not playing a dominant role. One solution to this problem

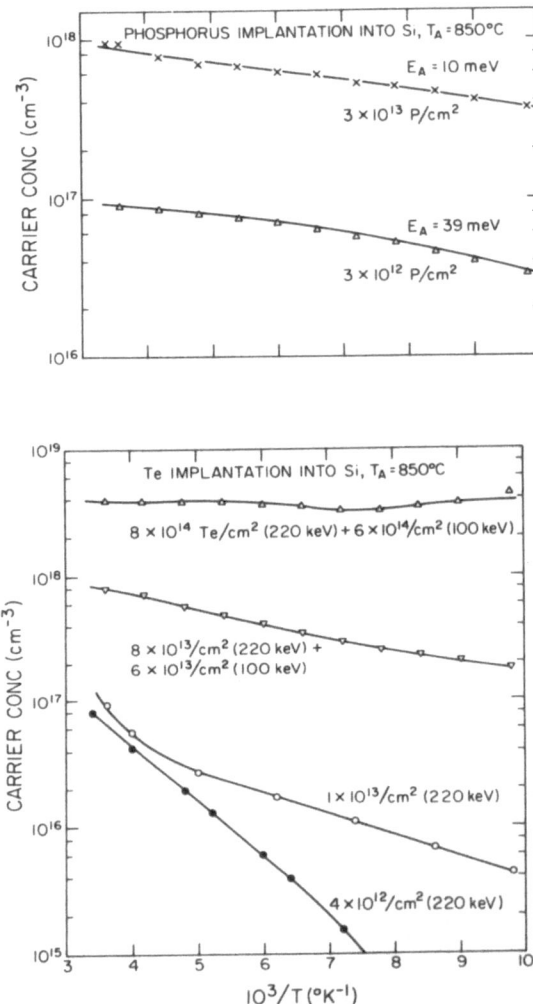

Fig. 22. Differential Hall measurement of carrier concentration vs. temperature for phosphorus (upper) and tellurium (lower) implantations into silicon (from Lee[107]).

is to use a series of low-dose implants at different energies to achieve a nearly flat implantation profile to depths of 5000 Å or greater. This procedure requires ion energies of 300 keV or more, depending on the dopant species.

4. Compound Semiconductors

4.1. Introduction

The application of ion implantation to silicon has been very successful in that controlled electrical behavior and high-quality *p–n* junctions can be obtained. There is a general feeling for the behavior of implanted layers for a wide variety of implantation and anneal parameters. In device technology, implantation has been able to compete successfully with a standard and well-established diffusion technology.[57] The natural extension of the work in silicon is to consider implantation processes in compound semiconductors. In this case diffusion technology is at a rather primitive stage as compared to silicon processing, so that implantation processes would not have to compete with a well-established technology. A more fundamental consideration is to determine whether implantation will be successful in introducing electrically active donors or acceptors into these materials.

It is necessary to adopt some criteria to decide whether implantation doping is successful in a compound semiconductor. We utilize our experience with silicon for guidelines. The following criteria may be somewhat arbitrary, but we suggest four points: implanted species on substitutional sites, minimum number of lattice defects, one-to-one correspondence between the number per cm^2 of carriers and of implanted dopants for low-dose implants, and adequate encapsulation. Compared to silicon requirements, only the last point introduces an additional constraint for compound semiconductors.

Conventional dopants are usually substitutional; therefore the first requirement for implantation doping is that the dopants be substitutionally located following implantation and annealing. This requirement can be studied by channeling effect measurements. Substitutional dopants are a necessary but not sufficient requirement for good electrical activity. In order to achieve carrier concentrations equivalent to those obtained by conventional processing, the compensating effects of the defects must be removed by postimplantation annealing. Other properties of the implanted material, such as photoluminescence efficiency or minority carrier lifetime, may be more sensitive to defects than is the carrier concentration. This may place even more stringent requirements on the annealing of damage for some applications. Finally, compounds often dissociate when annealed to the temperature required to remove the postimplantation damage.[108,109] The simplest method of avoiding this dissociation is to encapsulate the implanted samples. The material used for encapsulation must adhere to the sample during annealing, prevent the loss of the constituents of the compound and the implanted species, and be easily removable following the anneal.

Associated with these four criteria of substitutional lattice location, electrical activity, minimum defect concentration, and encapsulation are a host of other problems. The electrical properties of the substrate material may change during the anneal cycle, and outdiffusion of the implanted dopant can occur. Both phenomena can cause misinterpretation of the anneal results. Also, diffusion of the implantation-produced defects can result in a compensated region below the implanted layer.[110] This often leads to undesirable device performance. We should also note that deviations from stoichiometry and changes in the native defect concentration may also result from the implantation step.[110a]

In the following subsections we will discuss the range of implanted ions and describe the production and annealing of lattice disorder in III–V compound semiconductors. Then we will review the electrical characteristics of donor- and acceptor-implanted GaAs. In the last subsection we consider the results on implantation in GaP to achieve photoluminescence. The work on implantation in II–VI compound semiconductors is too fragmentary to give an overall picture or even to determine if the implantation process meets any of the four criteria listed above. The proceedings of the last two implantation conferences[7,9] and a recent review article[110b] can be referred to for recent work in this field.

4.2. Ranges

Measurements of ranges of implanted ions in compounds have not been as extensive as in silicon. One of the difficulties is that layer removal techniques are not as well established in these materials. The techniques which appear to have the greatest promise are chemical etching[111] and mechanical polishing.[112]

With the exception of beryllium, most of the ions of interest for doping GaAs have relatively high mass, so that the electronic stopping coefficient need not be accurately known in order to predict their projected range. Therefore the calculations of the projected range based on the LSS theory should be fairly reliable. In profiling experiments with krypton implanted into GaAs it was observed[112] that the random component of the krypton distribution agreed well with the projected range calculated from LSS theory. Backscattering measurements[113,114] on GaAs samples implanted with 200-keV tellurium (see the solid curve in Figure 23) have yielded values for the depth and width of the tellurium distribution which were also in agreement with the LSS theory.

Whitton *et al.*[112,115,116] have made a number of studies of channeling in GaAs and GaP. They have observed the effects of orientation, temperature, and lattice disorder on the profile of the channeled ions. The data for 400-keV krypton in GaAs show a definite channeling tail when the sample has been carefully aligned so that no symmetry axis or plane was parallel to the beam.[115] Since the critical angles for channeling of dopant ions in GaAs are fairly large (about 4–5°), it is reasonable to expect some channeling of implanted ions even when the beam is not aligned with a channeling direction. About the same channeling tail was found for a high-dose sulfur implant performed at 150°C (where most of the disorder anneals during implantation) as was observed for a low-dose implant at room temperature.[116] We shall see later in considering electrical characteristics that it is often desirable to implant at an elevated temperature in GaAs so that the buildup of disorder is minimized. Under these conditions a significant channeling tail may be present for aligned implants.

The possibility that diffusion effects may also change the distribution of implanted ions should be considered. For example, the deep profile of sulfur implanted in GaAs has been attributed[111] to diffusion of the sulfur. Outdiffusion effects can occur. An example of outdiffusion is shown in Figure 23 (dashed curve) for tellurium implanted in GaAs at room temperature to doses such that an amorphous layer was formed.[114] After annealing to 750°C the distribution had

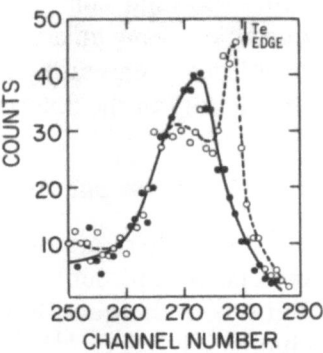

Fig. 23. Depth distribution of tellurium implanted into GaAs as determined by backscattering measurements. The arrow indicates the position of tellurium on the surface. The solid curve was obtained following implantation at room temperature and the dashed curve after annealing at 750°C with an SiO$_2$ layer covering the surface (from Harris *et al.*[114]).

shifted toward the surface. As in the case of implants in silicon, the amount of outdiffusion seems to depend on the total amount of disorder which was built up during implantation. The buildup of disorder can be reduced by implantation at higher temperatures. For example, no evidence for outdiffusion was found in GaAs samples that were implanted at elevated temperatures (up to 350°C) and annealed to temperatures as high as 900°C.[117] Diffusion effects should depend on the ion species. At present there is not enough evidence to give general guidelines.

4.3. Radiation Effects and Lattice Disorder

In evaluating ion-induced radiation damage in silicon, we had considerable prior information on the nature of the microscopic defects as provided by previous studies of fast neutron, energetic electron, and gamma radiation damage. The situation in GaAs and indeed in the III–V compounds in general is much different. There is no information on the microscopic nature of the defects from such measurements as electron paramagnetic resonance and optical absorption. However, there has been considerable work done in evaluating radiation damage effects from the standpoint of determining the orientation dependence of damage production, threshold displacement energy, and the annealing of electrically active defects.[118] These measurements do not give us any guidance about microscopic processes involved in the production and annealing of lattice disorder in compound semiconductors.

We have shown that the presence of disorder-induced defects plays a role in determining the electrical behavior of the implanted layer. However, even in the case of implants in silicon the defects responsible for the high-temperature annealing have not been identified. Although this is not a satisfactory solution, one can ignore questions about microscopic defects, in that high electrical activities have been obtained by using high anneal temperatures. The issue of the identity of the compensating defects can thereby be bypassed.

4.4. Production and Anneal of Lattice Disorder

Since there are no measurements of the identity of microscopic defects, we consider here results on the total amount of lattice disorder. Various techniques have been utilized to measure this disorder: channeling effect,[119–121] near-band-edge optical absorption,[122]

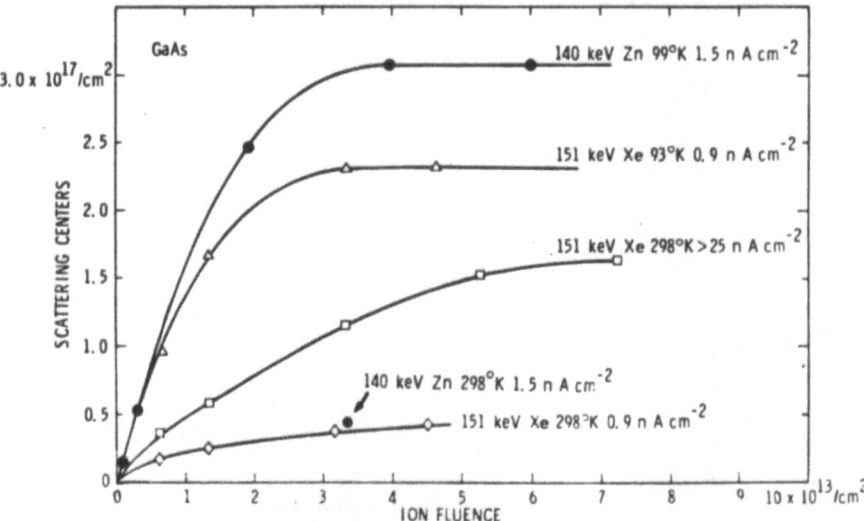

Fig. 24. Channeling effect measurement of lattice disorder in GaAs vs. implantation fluence (dose). All analyses were performed below 100°K (from Weisenberger et al.[128]).

optical reflection,[123-125] Raman scattering,[126] and "Coates–Kikuchi" patterns.[125,127] The results of all these investigations are in general agreement for the specific cases studied. Consequently in this review we will stress most heavily the results that have been obtained by channeling effect measurements. Detailed series of investigations have been made in a number of laboratories.

Figure 24 shows the dependence of lattice disorder in GaAs on implantation temperature and dose rate for xenon and zinc ions.[128] One can see a factor of ten difference between the number of scattering centers for zinc implanted at 99°K and that for an implant at 298°K. This points out the influence of annealing during implantation at room temperature. Similar annealing effects can be deduced from the xenon implantation at 298°K, where one notices that there is more disorder for the high-dose-rate implantation than for the low-dose-rate implantation. This dose rate dependence of disorder is a typical feature of implantation behavior when annealing is taking place.[27] Similar buildup and saturation of disorder have been observed in other compound semiconductors.[129] As in the case of silicon, an amorphous layer can be formed at high implant doses. Such layers have been studied by Raman spectroscopy in GaAs, GaP, InSb, and GaSb.[88]

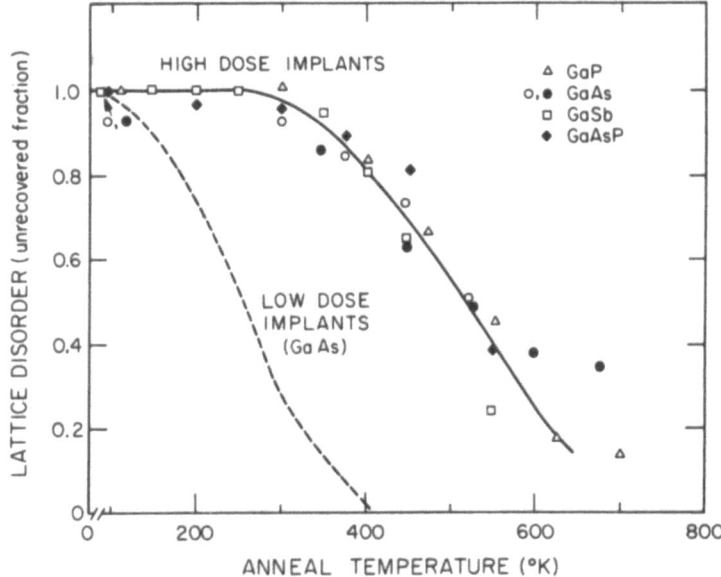

Fig. 25. Channeling effect measurements of the anneal characteristics of lattice disorder in several III–V compounds. The dashed curve represents the anneal of a low-dose implant in GaAs. The solid curve is the annealing results when an amorphous layer has been formed (from Picraux[130]).

Anneal characteristics of the lattice disorder are shown in Figure 25 for low-temperature implantation.[130] For low dose implants the dashed curve indicates that by temperatures of 130°C most of the disorder has vanished. Similar results were noted for GaP, GaSb, and $GaAs_xP_{1-x}$.

Figure 25 also shows the annealing results where an amorphous layer has been created by high-dose implantation. In this case the anneal of disorder for all four materials follows nearly the same behavior with a rather broad annealing peak, having a midpoint at about 530°K. One should note from these curves that the anneal of this amorphous layer exhibits a much broader range than that observed[4] in silicon or in germanium, where regrowth occurs in a relatively narrow temperature range. Channeling spectra indicated a narrowing of the disorder peaks from the low-energy side with increase in anneal temperature.[130] This suggests that the lattice reordering occurs epitaxially at the interface between the amorphous layer and the substrate as was observed in the case of silicon.[131]

As we will see, there is some importance in evaluating the properties of hot-substrate implantations in GaAs and GaP. As one might

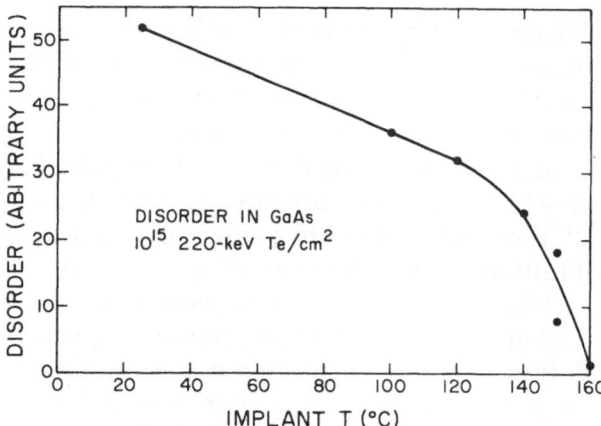

Fig. 26. Channeling effect measurement of lattice disorder
in GaAs implanted with 10^{15} 220-keV tellurium ions/cm^2
at the temperatures shown (from Eisen *et al*[117]).

anticipate from the anneal characteristics, the amount of disorder
decreases rapidly as the implantation temperature is increased above
room temperature.[116,117,120,129] Figure 26 shows the amount of disorder vs. implantation temperature for an implantation at 1 μA/cm^2
of 10^{15} tellurium ions/cm^2 in GaAs.[117] An amorphous layer was
formed by this tellurium dose and dose rate at temperatures up to
140°C. Around 150°C there was a large decrease in the amount of
disorder and at temperatures of 160°C the disorder has almost
vanished. Similar results have been noted by Whitton and Bellavance[116] for sulfur implantation in gallium arsenide.

4.5. Compound Semiconductors—Lattice Location

There have not been extensive channeling effect measurements
of the lattice location of implanted species in compound semiconductors. One of the reasons is that high-mass substrates such as GaAs or
GaP limit the application of the backscattering technique to implanted
species of mass equal to or greater than that of cadmium. It is only
recently that efforts have been made to employ ion-induced X rays
in lattice location experiments.[132]

In III–V semiconductors the lattice location of dopant species
has been studied in the cases of tellurium[114] in GaAs and bismuth
in GaP.[133] These cases show that the implanted dopant occupies
substitutional sites under conditions such that gross amounts of
disorder are not present. For bismuth in GaP, detailed measurements

169

of the angular dependence of the channeling effect indicate that the bismuth is located on the phosphorus sublattice. It is difficult to extrapolate these results to other cases. In silicon it was found that dopant species were not always substitutionally located.

Infrared localized vibrational mode absorption has been employed to study the location of implanted aluminum and phosphorus in GaAs.[134] These impurities are known to give rise to extremely localized vibrational modes (extending over less than one lattice constant) which are infrared-active and appear as absorption bands at frequencies higher than the maximum phonon frequency of the pure lattice. From the integrated absorption it was estimated that nearly all of the aluminum is located on gallium sites and the phosphorus is on arsenic sites after annealing at 900°C. The site symmetry was found to be approximately tetrahedral. This work is being extended to silicon[135] and nitrogen[136] implanted in GaAs.

4.6. GaAs—Electrical Measurements

There is considerable difficulty entailed in making electrical measurements in ion-implanted GaAs. There are three general problems: substrate, contacts, and layer removal. As a semiconductor, GaAs has not been purified to the same extent as silicon. Consequently, there are variations in the properties of substrate materials. The best-quality GaAs is produced by epitaxy. Material with carrier concentrations less than 10^{15} per cm^2 can be obtained, but it is often difficult to collect an adequate supply of good epitaxial material. Semi-insulating GaAs (chromium-doped) has been used in Hall effect studies of implanted layers. This material has the drawback of containing not only large concentrations of chromium ($\gtrsim 10^{17}/cm^3$) but comparable amounts of oxygen, carbon, and nitrogen.[111] There is always the possibility that the behavior of the implanted layer might be dominated by impurity effects or variations in the concentration of chromium.

Contacts to GaAs can also present a severe problem, especially for *n*-type material. Evaporated metal contacts form Schottky barriers and alloyed contacts tend to penetrate deeply into the GaAs.[137,138] Considerable care must be taken to form Ohmic contacts without shorting out the junction region. Semiinsulating material is often used in this context, since the contacts can extend through the implanted layer into the substrate without deleterious effects.

To obtain a profile of the carrier concentration in the implanted layer, it is necessary to remove controlled thicknesses of material without introducing damage. The anodic oxidation technique used in silicon cannot be applied by GaAs and therefore chemical etching procedures are used. This etching technique is not as well established and consequently it is necessary to take strong precautions to make sure that the implanted layer etches uniformly and reproducibly.[111,117]

As a result of these factors, there have not been as many results on the evaluation of implanted layers in GaAs. In spite of considerable effort in at least three laboratories, there are gaps in the available data. It is still possible, however, to make a reasonably cohesive picture of progress in GaAs.

4.6.1. Anneal Characteristics

Some of the initial efforts in the evaluation of implanted layers in GaAs were to determine if good electrical activity could be obtained without encapsulation of the implanted layer.[139,140] Figure 27 shows data on *n*-type GaAs implanted with 10^{15} zinc atoms/cm^2 at 1 MeV and annealed at 600°C with no protective covering on the surface.[140] After a 30-min anneal the number of holes per cm^2 was about 1% of the number of implanted zinc atoms. The number of holes decreased after 45 min of annealing and after 1 hr had decreased by two orders of magnitude. This points out the need for protecting the surface of the sample during anneal treatment. It has been noted, for example, that there is a pronounced release of arsenic from implanted surfaces at temperatures as low as 300°C.[109]

Anneal temperatures of 800–900°C are generally required to obtain the maximum number of carriers per cm^2 in implanted layers. Figure 28 shows the anneal characteristics of samples implanted at room temperature with 45-keV magnesium ions.[141] After implantation the samples were coated with a 2000-Å-thick sputtered layer of SiO$_2$ prior to annealing. The Hall measurements indicated that for samples implanted with doses less than 10^{14} magnesium ions/cm^2 nearly 100% electrical activity was observed after annealing at 800°C. For higher-dose implantations there was no appreciable increase in the maximum number of carriers. The only exception to the requirement for anneal temperatures of 800°C or higher was found for the case of beryllium implanted in GaAs.[141] In this case an anneal temperature of 600°C was adequate to obtain greater than 80%

171

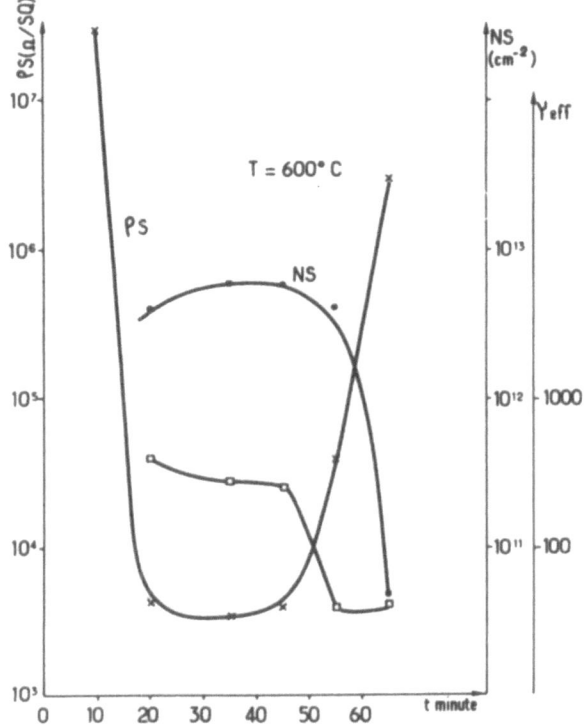

Fig. 27. Dependence of sheet resistance, surface carrier concentration, and mobility on isothermal anneal temperature for an *n*-type GaAs sample implanted at room temperature with 10^{15} 1.0-MeV zinc ions/cm^2 and annealed to 600°C with no surface protection (from Favennec[140]).

electrical activity of the implanted atoms. This lower anneal temperature requirement probably results from the fact that less disorder is produced by the relatively light beryllium ions than by heavier ions such as cadmium or zinc.

4.6.2. *p*-Type Dopants

The common *p*-type dopants in GaAs are the group II elements such as zinc and cadmium, which, when occupying gallium sites, act as acceptors. As mentioned previously, lattice location measurements have not been made for these dopants and we can only infer the lattice site positions from the results of electrical measurements. For implantation doses of less than 10^{14} ions/cm^2 there is nearly one-to-one

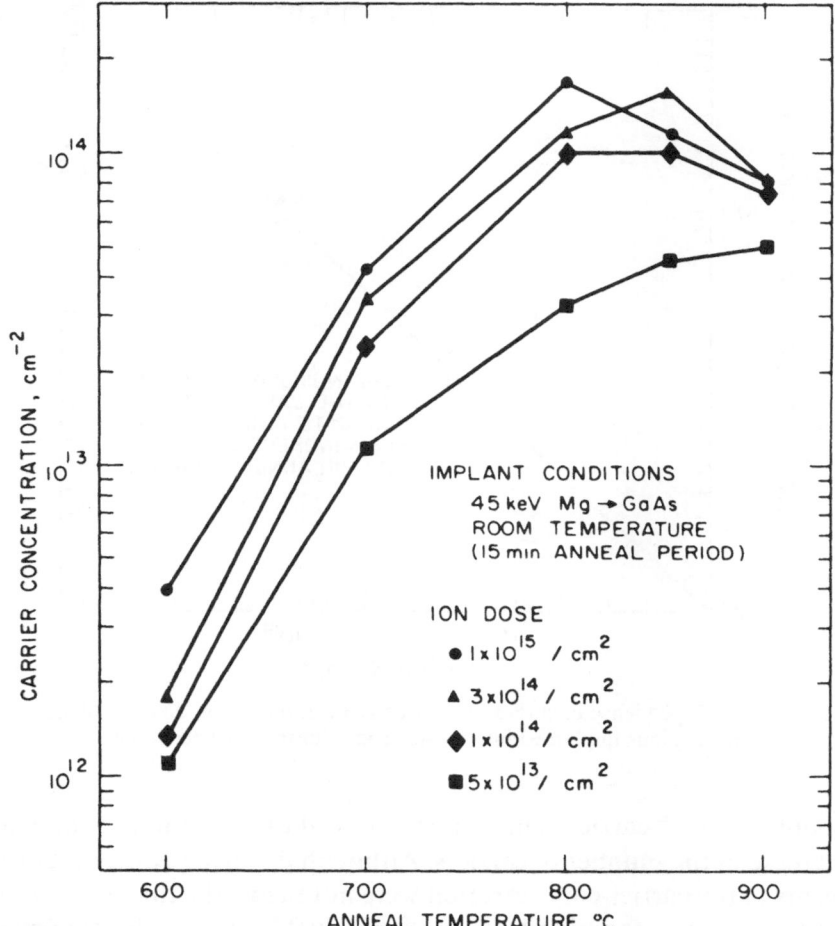

Fig. 28. Electrical properties of magnesium-implanted layers in GaAs (from Hunsperger *et al.*[141]).

correspondence between the number of implanted ions and the number of holes for Be,[141] Mg,[141] Zn,[142-145] and Cd.[142] These results imply a high substitutional component for the implanted dopants.

The dependence of surface carrier concentration on ion dose for 60-keV cadmium ions implanted at room temperature is shown in Figure 29.[142] For samples annealed at temperatures of 800–900°C there is a one-to-one correspondence between the number of carriers and the ion dose up to approximately 10^{14} ions/cm². For larger doses the value of the surface carrier concentration tends to saturate at

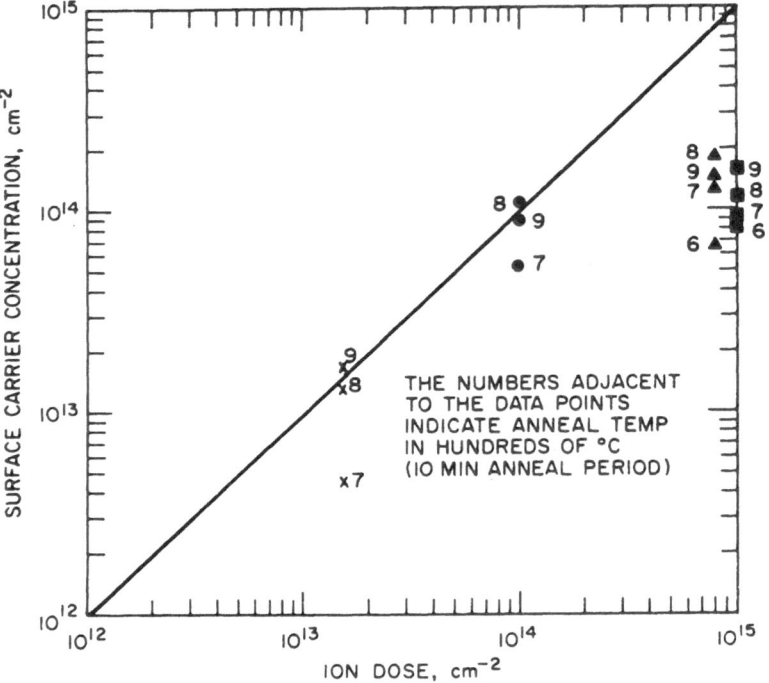

Fig. 29. Dependence of surface carrier concentration on ion dose for 60-keV cadmium ions implanted into GaAs (from Hunsperger and Marsh[142]).

about 2×10^{14} carriers/cm². Longer anneal times did not result in an increase in the number of carriers. Although detailed profile measurements of the carrier concentration were not made, the thickness of the *p*-type layer was found to be approximately 0.1 μm. For this thickness 2×10^{14} carriers/cm² corresponds to an average carrier concentration of about 2×10^{19}/cm², which is the reported lower limit for the solid solubility of cadmium in GaAs.[146] Similarly, for magnesium a maximum carrier concentration of about 2×10^{19}/cm³ was found, a value which has been reported for the maximum magnesium concentration obtained in epitaxially grown layers.[147] The situation for zinc implantations is not as clear. As shown in Figure 30, the number of carriers/cm² for 20-keV zinc implants at room temperature saturates at 10^{14}/cm².[143] Similar results were found by Hunsperger and Marsh.[142] If it is assumed that the zinc distribution follows the LSS range predictions, the value of 10^{14} carrier/cm² would correspond to carrier concentrations greater than 10^{20} per cm². This would be close to the reported solubility of zinc in GaAs of 3×10^{20}/cm² at

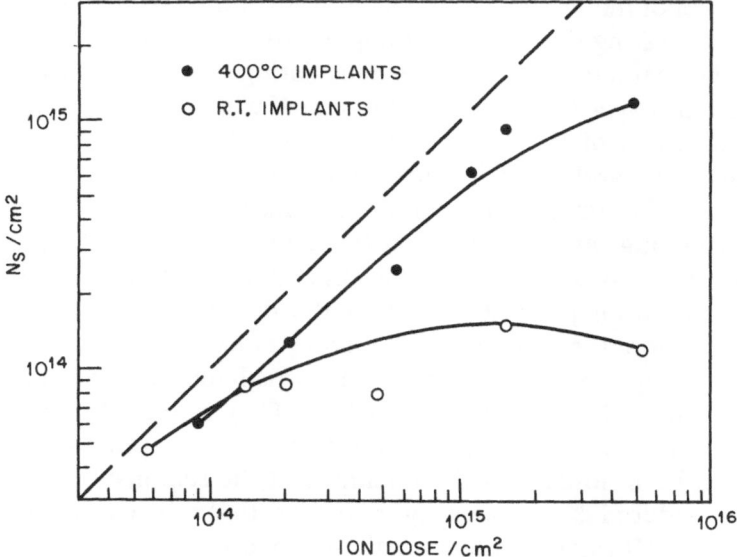

Fig. 30. Surface carrier concentration as a function of the dose of zinc ions implanted into GaAs. The samples were annealed at 900°C for 10 min (from Fujimoto *et al.*[143]).

900°C.[148] However, for 400°C implants the carrier concentration saturates at about $10^{15}/cm^2$, as was also noted by Hunsperger and Marsh.[142] The increase in the number of carriers/cm² for hot-substrate implants may be due to diffusion of the zinc atoms to greater depth during implantation.

The mobility values in the implanted layers correspond to bulk values from Sze and Irvin[149] (for implantation anneals up to 800–900°C). This indicates again that the majority of the disorder in the implanted layer has been annealed. However, there is some residual disorder, as deduced from Coates–Kikuchi patterns even after annealing at 900°C.[150]

4.6.3. Formation of the Semiinsulating Layer

It was early recognized that particle bombardment produces high-resistivity layers in GaAs.[151] This effect has been used to advantage for isolation of junction devices in GaAs by proton bombardment.[152] However, for zinc and cadmium implantation in *n*-type substrates, localized defects remain even after high-temperature annealing and cause formation of a semiinsulating layer.[110,153] The

175

formation of the semiinsulating layer is attributed to a deep diffusion of compensating defects during implantation or subsequent anneals.

The characteristics of the semiinsulating layers were investigated in detail in *n*-type GaAs samples implanted with zinc or cadmium.[153] The thickness of the implanted layer was found to depend upon implantation and anneal temperatures and substrate doping concentration. For implantation at room temperature the thickness of the semiinsulating layer decreases with increasing anneal temperature. Temperatures of 900°C were sufficient to nearly eliminate the layer.

It has been proposed[153] that the defects responsible for the compensation are complexes involving arsenic vacancies and the substrate dopants. Further support for this hypothesis was provided by the work of Itoh and Kushiro.[154] They observed that arsenic implantations serve to reduce the thickness of the intrinsic layer. It has also been noted that the thickness of the semiinsulating region can be reduced by implanting zinc ions through a thin dielectric coating.[153] It was suggested that this technique limits arsenic vacancy generation during implantation.

4.6.4. *n*-Type Dopants

The conventional *n*-type dopants in GaAs are the group VI elements such as sulfur and tellurium, which act as donors when on arsenic sites. The initial results obtained on implanted layers in GaAs for the *n*-type dopants were disappointing in that very low electrical activities resulted.[111,142,155] Because of the success with *p*-type dopants, various permutations of the implantation parameters were attempted to achieve better electrical activity in *n*-type layers. The most recent results for tellurium implantation indicate that good electrical activity can be attained by the proper choice of implantation temperature and encapsulating layer.[114-117]

The most complete early investigation was that of Sansbury and Gibbons,[111] who studied the properties of ion–implanted silicon, sulfur, and carbon in GaAs. The samples were implanted at room temperature up to doses of 10^{16} ions/cm^2. In most cases the surface was protected after implantation by a layer of silicon dioxide approximately 2000 Å thick deposited at 300°C. The doping efficiency was quite low, with 11 % efficiency obtained for silicon implants and 6 % or less for sulfur implantations. Hunsperger and Marsh also reported low doping efficiency (approximately 1 %) for samples annealed at temperatures as high as 800 or 900°C.[142] For the sulfur implants the

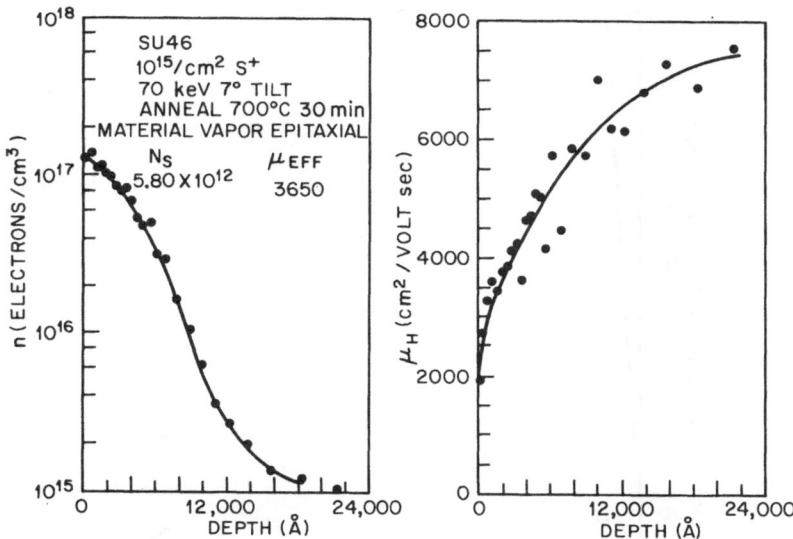

Fig. 31. Electron concentration and mobility profile for sulfur implanted into epitaxial GaAs (from Sansbury and Gibbons[111]).

dopant profiles, as shown in Figure 31, exhibited deep diffusion characteristics.[111] The profile extends an order of magnitude further into the sample than predicted from the LSS theory. This depth was attributed to radiation damage enhanced diffusion of the sulfur. It should be noted that the mobility values shown on the right-hand side of Figure 31 are comparable to the mobility values for bulk doped material. Low doping efficiency was also found for germanium-implanted[156] and tellurium-implanted[157] GaAs.

It was first thought that the formation of an amorphous layer would lead to high electrical activity as in the case of implants in silicon. Various attempts in this direction by implanting at low temperature[158,159] were not successful. There was no increase in electrical activity. In fact it was found that the production of an amorphous layer in GaAs led to lower electrical activity and mobility than if the implantation was carried out under conditions which did not produce amorphous layers.

A series of studies was then carried out by Eisen and co-workers[114,117] to investigate the influence of implantation temperature and surface protection on the properties of tellurium-implanted layers in GaAs. By use of photoluminescence measurements it was found that a substantial reduction in the amount of lattice disorder could be obtained by implantations at 150°C or higher.

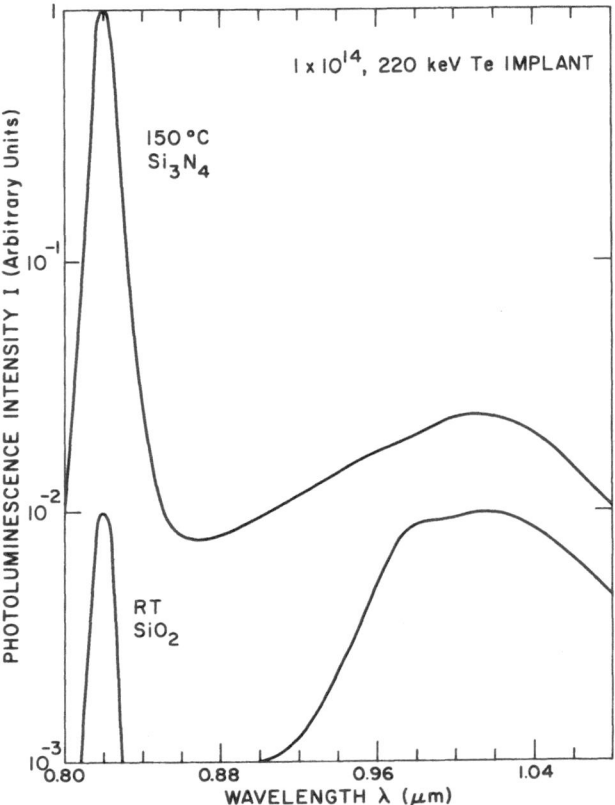

Fig. 32. Photoluminescence spectra for two samples of epitaxial GaAs implanted with 10^{14} tellurium ions/cm^2. The implant temperatures and dielectric layers used during annealing are labeled on the figure (from Harris *et al.*[114]).

Figure 32 shows the photoluminescence spectrum obtained at 77° with the use of illumination from a helium–neon laser.[114] The peak at 8200 Å is due to band-to-band recombination and is indicative of the amount of lattice disorder. Samples implanted at 150°C and annealed at 750°C had a band gap intensity about the same as that for an unimplanted sample, a value approximately 100 times that for room-temperature implants annealed at the same temperature.

The broad peak observed at 1 μm is attributed to a gallium vacancy–tellurium complex.[160] It was found that the intensity of this peak is much higher relative to the band gap peak for samples annealed with silicon dioxide on the surface than with silicon nitride. This result suggests that gallium vacancies are present in the implanted layer. Previous results on diffusion of gallium in SiO_2 suggest that the presence of gallium vacancies in the implanted layer is due to outdiffusion of the gallium through the oxide layer during anneal.[161] Gallium vacancy–tellurium complexes are thought to be acceptors[160] and hence may compensate the implanted tellurium.

In our work[114] Hall effect measurements indicated very low electrical activity for room-temperature implants, irrespective of the dielectric coating used. In view of this and the photoluminescence results emphasis was shifted to hot-substrate implantations and nitride coatings. In this case the silicon nitride was deposited by reactive sputtering. The major difficulty was found to be nonadherence of the silicon nitride layers. Figure 33 is a scanning electron microscope photo of a tellurium-implanted sample which was coated with silicon nitride and annealed to 750°C for 15 min.[117] In the tellurium-implanted region (upper right-hand portion) the silicon nitride has bubbled and in one case has ruptured. In the nonimplanted area there

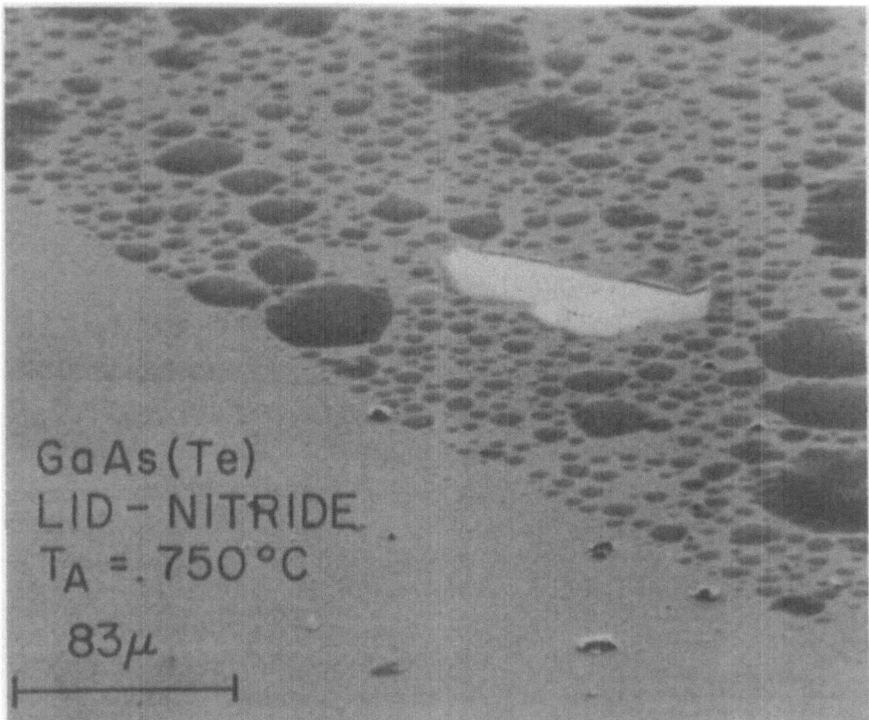

Fig. 33. Scanning electron micrograph of a tellurium-implanted GaAs sample which was coated with silicon nitride and annealed at 750°C (from Eisen *et al.*[117]).

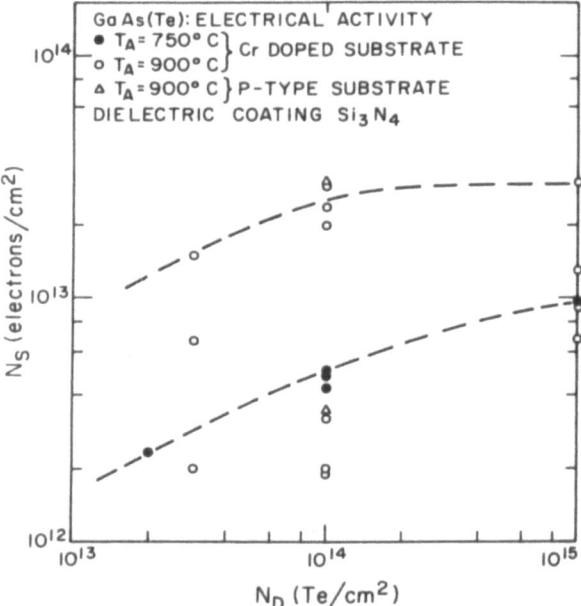

Fig. 34. Hall effect measurements of the surface electron concentration of GaAs samples implanted with 220-keV tellurium ions at 350°C and annealed under the conditions shown in the figure (from Eisen *et al.*[117]).

was no significant bubbling of the nitride. In spite of these difficulties, the electrical measurements were very encouraging. The results of measurements on 20 samples implanted with tellurium at 350°C and annealed at 750 or 900°C are shown in Figure 34 for samples coated with silicon nitride. The lower dashed line represents the 750°C anneal results and the upper dashed line represents the 900°C anneal results. The higher anneal temperature gave more electrical activity (the low points were obtained on samples with poor nitride adhesion). In fact, in one sample implanted with a dose of 3×10^{13} Te ions/cm^2 and annealed to 900°C, 50% electrical activity was obtained. Measurements of the carrier density as a function of depth for a sample implanted with 10^{14} Te ions/cm^2 through a 300-Å-thick nitride layer showed a carrier concentration peak of 8×10^{18} electrons/cm^3.[117] This value is approximately the maximum electron concentration which has been attained by doping GaAs with tellurium during growth.[162]

It should be noted that Schottky barrier capacitance voltage measurements on these hot-substrate-implanted samples coated with a nitride layer did not reveal the presence of an intrinsic layer after anneal.[114] The electron concentration merged smoothly with the epitaxial layer background doping. Again this suggests that the particular dielectric coating used may influence the growth of an intrinsic layer.

These results indicate that by proper choice of the implantation temperature and dielectric coating, electron concentrations close to those observed for bulk doped samples can be obtained. The implantation of sulfur and silicon in gallium arsenide should be reexamined in light of these results.

4.7. Photoluminescence in GaP

Photoluminescence measurements provide another method of testing the efficiency of ion implantation. We consider here the results of a series of experiments on implantation of bismuth into GaP.[133,163,164] Under thermal diffusion processing bismuth acts as an isoelectronic trap in GaP since it replaces a host atom (phosphorus) which is in the same column of the periodic table.[165] The system exhibits an intense, low-temperature luminescence spectrum with sharp lines. Hence high-resolution optical studies offer the possibility of providing information on the lattice location and microscopic environment of the implanted ion. For example, the observation of normally forbidden lines in the spectrum would indicate that the bismuth atoms were situated in regions of severe disorder.

Bismuth offers another attractive feature in that it is a high-mass element. Consequently, its lattice location in GaP can be determined directly by channeling effect measurements.

4.7.1. Room-Temperature Implants

The investigations of Merz and co-workers[133,163,164] followed essentially the same pattern as those employed in the evaluation of electrically active dopants in GaAs. The production and annealing of lattice disorder were determined from channeling measurements. The percentage of bismuth on substitutional sites was then compared with the photoluminescence efficiency.

The amount of lattice disorder for 100-keV bismuth implants initially increased linearly with dose and saturated at about 10^{13} atoms/cm^2.[164] To prevent decomposition of the implanted layer during anneal, it was necessary to coat the sample on one side with a

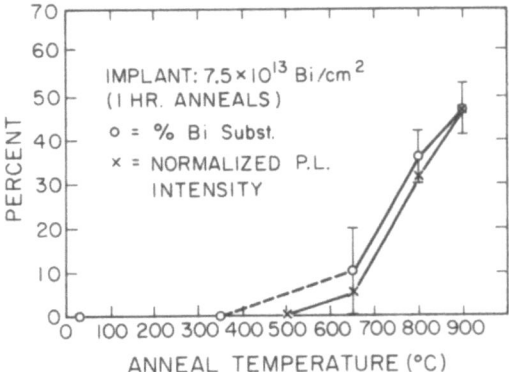

Fig. 35. Channeling effect and photoluminescence measurements for GaP samples
implanted with bismuth at room temperature and
annealed to the temperature shown. The channeling effect results (○) show the fraction of the
bismuth located on substitutional lattice sites
(from Feldman *et al.*[164]).

silicon oxide layer. For samples implanted with $\lesssim 2 \times 10^{13}$ Bi
ions/cm^2, disorder annealing occurred at temperatures of $\sim 450°$C
and for doses of $> 7.5 \times 10^{13}$ ions/cm^2 at temperatures of approximately 750°C.

For the low-dose implants significantly higher annealing temperatures were required to achieve substantial luminescence ($T \gtrsim$
700°C; see Figure 36) than were required to reduce the lattice disorder.
Figure 35 shows the substitutional fraction of the bismuth, as determined by channeling, compared to the photoluminescence intensity.[164] The relative photoluminescence intensity follows the fraction
of bismuth on substitutional sites. However, the intensity was approximately a factor of ten less than that expected from the measured
substitutional concentration of bismuth atoms. This was taken as an
indication of the presence of appreciable damage in the form of lattice
distortion or competing damage centers even after annealing at
900°C.

4.7.2. Hot-Substrate Implantations

In order to reduce the amount of disorder, implantations were
carried out at 450°C.[133] Channeling effect measurements made
after these implantations showed that the lattice disorder was greatly
reduced and that the bismuth was predominantly on substitutional

sites. This is similar to the results found in tellurium-implanted GaAs. In addition, for GaP it was possible to make detailed angular scans (as discussed in Section 1) to show that $\sim 75\%$ of the bismuth atoms were on phosphorus sites (isoelectronic replacement).

However, photoluminescence anneal curves indicated only a modest improvement in efficiency over those found in room-temperature implanted samples. This is shown in Figure 36, which plots the dependence of luminescence efficiency on anneal temperature for samples implanted at room temperature and at 450°C.[133] In both cases the intensity increases sharply between 700 and 800°C and the intensity is greater for the low-dose implant.

Feldman *et al.*[164] point out that the luminescence can be quenched by either severe local distortion or strain or by competing

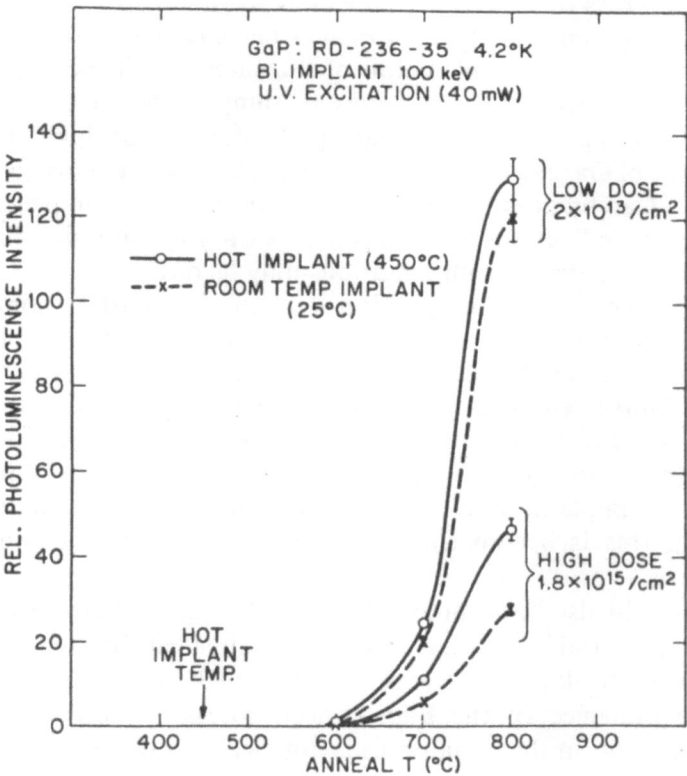

Fig. 36. Relative photoluminescence intensity of the bismuth iso-electronic trap in GaP as a function of anneal temperature for high- and low-dose samples implanted at room temperature and 450°C (from Merz *et al.*[133]).

centers. The existence of strain was evident from the identification of forbidden lines in the spectrum. However, not all the annealing behavior of the luminescence was attributed to strain. We suggest that the oxide coating may play a dominant role as was found in the case of tellurium implants in GaAs. We feel that improved luminescence efficiency might be found if silicon nitride coatings were used.

4.8. Summary

In the past three years there have been a sufficient number of studies on the behavior of implanted layers in GaAs so that a picture is emerging. However, what can be said about other materials? First of all, the problems of obtaining good electrical activity will be difficult. In many cases it is even hard to obtain reproducible substrates. Although gross disorder may anneal at relatively low temperatures, substantially higher temperatures will be required to remove compensating defects. Encapsulation of the samples will be required at these high temperatures to prevent sample dissociation. Silicon dioxide coatings may not be adequate due to the high diffusion coefficient of Ga, Zn, and Cd; silicon nitride or other coatings will be required. Careful attention must be paid to implantation parameters. The present indications are that it is necessary to minimize disorder; therefore hot-substrate implantations may be desired.

We believe that the situation is open in regard to compound semiconductors other than GaAs as far as electrical properties are concerned. To our knowledge there have not been any studies that were complete enough to answer the simple criteria outlined in Section 4.1. Junctions have been obtained by implanting sulfur and zinc in InSb and InAs[166] and antimony in PbTe[167] and PbS.[167a] Studies of implantation into SiC have been rather detailed;[168] however, this is a somewhat unusual case of a compound semiconductor.

It should also be pointed out that GaAs can be characterized as a covalent material.[170] It has not been established that implantation can be used to dope more ionic material such as ZnS or ZnO. The possible influence of the high concentrations of intrinsic defects usually present in these materials is not clear at present.[171]

5. High-Dose Implantation and Compound Formation

In the previous sections we considered the case where the concentration of implanted atoms is less than 1 at. %. The dose of the implanted species seldom exceeded a few monolayers; i.e., it was less than 10^{16} ions/cm^2. Under these conditions sputtering effects did not play a major role. When we consider higher dose implantations, with doses greater than 10^{17} ions/cm^2, sputtering effects can be dominant.

Sputtering, the erosion of solid surfaces by energetic particles, has attracted the interest of experimentalists and theoreticians over the past 100 years. Extensive reviews have been published,[172,173] and the recent theoretical treatment of Sigmund[174,175] agrees well with experimental results. The subject, however, is complex and we will only indicate a few features which are germane to the later discussion.

Sputtering is basically a collision phenomenon, with collision cascades often dominant over single-collision events. The sputtering yield is defined as the number of sputtered atoms per incident ion and is a function of the nature of the incident ion, ion energy, and substrate. It can vary from much less than unity to values of ten or greater. For a given ion–substrate combination the sputtering yield is a maximum at about the energy at which the nuclear stopping is maximum, and it decreases at higher energies.

For low-dose implantations sputtering cannot remove enough of the substrate material to greatly affect the distribution of the implanted species. As the implanted dose is increased sputtering may result in significant loss of the implanted atoms. When the sputtering yield is greater than unity the concentration of the implanted species saturates at a value which depends on the value of the sputtering yield and the range of the implanted ions.

One can attain appreciable concentrations of the implanted species if the sputtering yield is less than unity. For systems in which this is not the case there are three alternatives: (1) Implant at a low energy where the sputtering yield is negligible. These energies are less than about 50 eV, so that the penetration of the implanted ions is only a few angstroms. Consequently, high substrate temperatures are required so that the implanted atoms can diffuse into the material. The difference between this approach and normal diffusion techniques is that an energetic ion can overcome a barrier at the surface. (2) Implant at a sufficiently high energy that the desired concentration can be reached before too much of the substrate has been eroded. It may be difficult to attain such energies for heavy ions with commercial implantation systems. (3) Maintain the substrate at high enough temperature that the implanted species can diffuse deeper into the substrate than the depth to which erosion will be produced by sputtering.

To our knowledge, the first approach has only been used in the oxidation of lead by low-energy O_2^+ bombardment.[176] Evaporated layers of lead were bombarded with oxygen ions at energies less than 50 eV and at temperatures between 100 and 200°C. PbO layers several hundred angstroms thick were formed. Since the range of the oxygen ions is only a few angstroms, oxygen diffusion is the rate-limiting mechanism in the oxidation process. The activation energy of 1 eV found here agreed with earlier results from thermal oxidation experiments.

Most attempts to achieve high concentrations of implanted species in systems where the sputtering coefficient is greater than unity have relied upon high energies or high temperatures or combinations thereof. We will review here some of the applications of these techniques to implantations in silicon, metals, and thin metal films. The recent reports of inhibition of corrosion and formation of superconducting layers open exciting new areas of research. The work in these areas has not yet been published[9,176a] and we can only give a brief summary of the findings.

5.1. Silicon

One of the most dramatic effects in high-dose implantations in Si is the metallic conduction behavior observed for metal ion implantation.[177,178] The concentration of metal ions is sufficiently great that binary systems of the type $Si_{1-x}M_x$ have been obtained. This was most easily accomplished for aluminum in silicon, the system which has been investigated in the most detail, since the sputtering coefficient is less than unity.

Figure 37 shows values for the log of the sheet conductivity vs. dose for aluminum implants at different energies.[178] The dashed horizontal line corresponds to the transition to metallic conduction. With increased implantation energy larger doses of metal ions are required to reach the concentration level at which the transition to metallic conduction occurs.

The exact mechanism leading to metallic conduction has not yet been pinned down. The dashed-line fit for 4-keV implants is based on a model for statistically arranged hydrogenlike atoms with phonon-assisted electron hopping between atoms. However, other models could be involved; for example, nucleation processes could lead to the formation of metallic aggregate centers. The conduction mechanism would then correspond to thermionic emission processes between

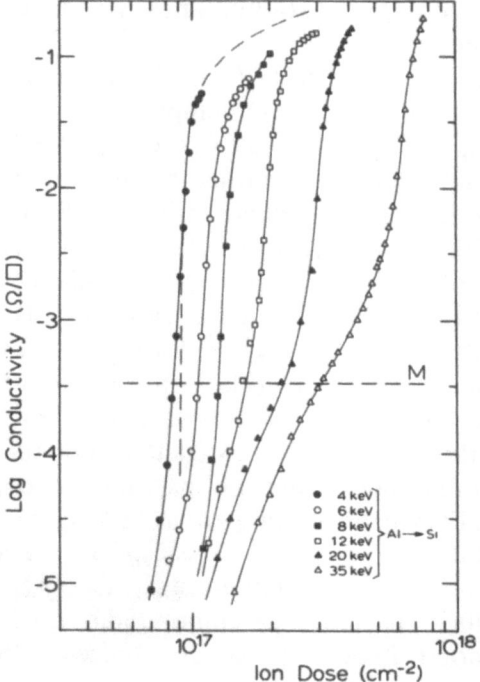

Fig. 37. Conductivity of the Si–Al system vs. ion dose for different implantation energies. The horizontal dashed line denoted by *M* separates the areas of metallic (above) and nonmetallic (below) conduction (from Kräutle and Kalbitzer[178]).

metallic islands. Metallic conduction is lost when the samples are heated to temperatures between 400 and 600°C. Scanning electron probe micrographs indicate that metal precipitates are formed.

The formation of continuous SiO_2 films by oxygen-ion bombardment has been reported.[179–181] Wantanabe and Tooi[179] used 60-keV oxygen ions at doses up to 10^{18} ions/cm² and temperatures up to 300°C. The oxide thickness depended on ion energy only and was not increased by increased implantation doses. Effects of radiation damage were evidenced, as was also suggested in the work of Pavlov and Shitova.[180] Freeman *et al.*[181] have made a detailed study of the properties of oxide and nitride layers formed by implanting large doses ($\sim 2 \times 10^{18}$ ions/cm²) of 40-keV ions. They concluded that this technique cannot be regarded in any way as an alternative to the thermal methods of preparation since it produces considerable damage in the silicon.

Attempts have been made to form SiC by implanting 200-keV carbon ions into silicon at room temperature.[182-183] Following annealing at temperatures of $\gtrsim 850°C$ crystalline SiC was observed to form. The compound was identified by its characteristic phonon absorption band. Under these implantation conditions (peak carbon concentration of 10 ± 3 at. %) a layer of SiC was not formed, but rather microcrystallites of SiC surrounded by bulk silicon. Similar results were found for Si_3N_4 for anneal temperatures of 1000°C. In more recent work Akimenchenko *et al.*[184] formed SiC by implantation of 4×10^{17} silicon ions/cm² at 35 keV into diamond at room temperature. The synthesized SiC was identified by optical absorption and electroreflection spectra. The spectra indicated that implantation of silicon ions in diamond produced SiC with a strongly disordered crystal structure. During heat treatment at 1000°C the α-SiC changed to the cubic β-SiC modification. No electrical evaluation was reported; however, weak photoluminescence was observed at 77°K.

These results indicated that compounds could be formed in silicon and diamond. From a practical standpoint the results were somewhat disappointing for the oxide and nitride layers. Further evaluation of the characteristics of the SiC layers is required.

5.2. Implantation in Metal Films

Ion implantation into thin films can produce changes in structure and composition. The resistance of thin films can be increased by orders of magnitude and their adherence to glass substrates increased significantly, and even electromigration properties can be influenced. The subject has a long history[185] and has recently been reviewed by Stroud.[186]

Since thin-film structures can have thicknesses less than the range of the implanted ions, atoms near the substrate–film interface can be transferred across the boundary. This leads to an enhancement of adhesion that can be quite significant; an increase in adhesion of two orders of magnitude was found for aluminum films on glass substrates after argon bombardment.[187] More recent studies of the same system have indicated that 2–3 aluminum atoms per argon ion can be forward-sputtered into the substrate. They point out that this recoil implantation process[189] could lead to the introduction of significant amounts of oxygen into silicon, for example, if phosphorus

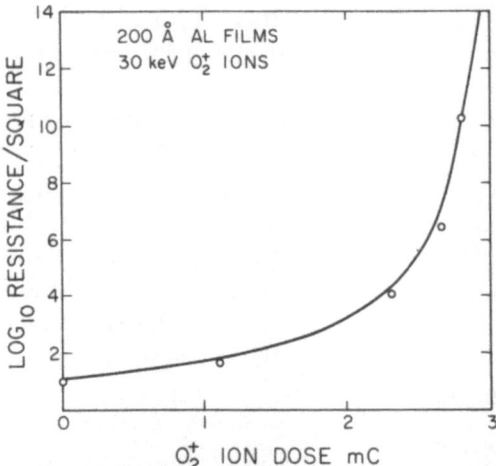

Fig. 38. Resistance of aluminum films bombarded by O_2^+ ions as a function of ion dose (from Stroud *et al.*[191]).

ions were implanted through a thin oxide layer covering a silicon substrate.

The electrical properties of bombarded thin films can be changed due to disorder and damage effects.[186] However, a more interesting effect lies in the possibility of changing the film composition.[186,190] An example of this is the increase in resistance of Al films bombarded by O_2^+ ions shown in Figure 38.[191] Here the resistance of a 200-Å Al film is plotted vs. the dose of 30-keV O_2^+ ions. Transmission electron microscope and diffraction studies did not indicate the presence of Al_2O_3, but transmission infrared spectroscopy showed the presence of aluminates and silicates in the implanted film. It was concluded that oxygen-ion implantation leaves much of the aluminum polycrystalline but converts some of it to a completely amorphous oxide phase.

Typical beam doses required to produce significant changes in the properties of thin films are greater than $10^{16}/cm^2$. In terms of implantation systems $1\,\mu A/cm^2$ for 1 hr is required for a dose of 2.25×10^{16} ions/cm^2. Consequently, although there are potential applications, commercial success depends on whether simple and cheap ion beam generators capable of large currents can be produced.

5.3. Implantation in Metals

The same implantation system considerations apply to implantation in metals where applications involving oxidation, sliding friction, or other changes in surface properties are possible. This has stimulated the Harwell group to design and test a large-volume implantation system.

At the Third International Ion Implantation Conference a series of papers were given on the influence of ion implantation on the properties of metal surfaces.[192-197] It was found that oxidation could be either reduced or enhanced by implantation of different species. The magnitude of the effect depended upon the ion species, ion dose, and substrate material. Opposite effects for the same implanted ion were observed in titanium and stainless steel. Frictional force measurements indicated that changes in the mechanical properties could be induced by ion implantation. These results suggest that exciting new applications of implantation may be forthcoming. The superconducting transition temperature of metals may also be altered by implantation.[198]

One of the basic considerations in implantation in metals is whether or not compounds can be formed. There has been one report by Arminen *et al.*[199] on the formation of Cu–Al compounds. Polycrystalline aluminum was implanted with 30- and 60-keV copper ions. The properties of the implanted layer were evaluated by nuclear back-scattering and transmission electron microscopy. Large concentrations of copper were obtained, with a maximum concentration corresponding to about that of Cu_2Al. Transmission microscopy indicated that precipitates had formed. Evidence was found that the erosion of the substrate by sputtering was inhibited by the presence of Al_2O_3 layer on the surface.

6. Bombardment Effects

The doping effects of implanted ions were emphasized in previous sections. There are a number of interesting and important effects associated with the disorder produced by ion bombardment. We will only treat these topics in cursory fashion because they do not involve the chemical or electrical nature of the implanted species.

Proton bombardment of GaAs, for example, is known to produce a very high-resistivity layer in either *n*- or *p*-type material.[151] This has been used by several workers in making electronic[152] and optical[200] devices in GaAs. The disadvantage of using a damage effect in device production is that the thermal stability is likely to be lower than for chemical doping. For example, the high resistivity produced by proton bombardment of GaAs is not stable above about 300°C.[152]

One particular application of proton bombardment of GaAs and GaP has been to make optical waveguides by utilizing the changes in index of refraction of the material produced by the proton bombardment.[200-202] Bombardment of glasses by protons and other light ions also produces changes in the index of refraction. Again, optical waveguides have been made in this way.[203]

Bombardment of magnetic garnet materials with light ions (protons, helium ions, and neon ions have been studied) has produced changes in the magnetic anisotropy of these materials.[204,205] It was originally thought that these changes in the material were due to chemical effects, but more recent studies have indicated they are due to stress induced by the damage created during bombardment. These effects may be of technological importance in that they provide a simple means of suppressing the formation of "hard bubbles" in these materials, making it possible to produce higher-speed bubble memories. It is also possible to use bombardment to form localized channels in which the bubbles may be guided.

The formation of voids in metals and alloys as a result of high-dose fast-neutron irradiation is an important problem in the development of fast breeder reactors. Ion bombardment has been extensively used to simulate high-dose neutron irradiations. This field has been the subject of a separate conference and will not be discussed further here.[206]

Ion bombardment of semiconductors and insulators results in the formation of amorphous layers in many cases. One of the more interesting aspects of the work of Kelly and co-workers is the change in the composition of oxides under bombardment.[207-208] They have observed, for example, that a transition from crystalline TiO_2 to an amorphous oxide to crystalline Ti_2O_3 occurs for progressively increasing bombardment doses of rare gas ions.[208] They suggest that this may be a general phenomenon whenever the oxide becomes amorphous as a result of ion bombardment and reduction is chemically possible. They have observed the following changes: MoO_3 to MoO_2, TiO_2 to Ti_2O_3, and WO_2 to $W_{18}O_{49}$. The thickness of the layer of reduced oxide is greater than the range of the bombarding ions. Kelly suggests that this is caused by diffusion of oxygen vacancies into the material. He suggests four criteria: (1) lower oxides exist, (2) the binding energy for oxygen is small, (3) an amorphous phase can be produced by bombardment, and (4) the oxygen vacancy has a high diffusion coefficient. In some cases these effects can be observed rather easily. For example, the increase in conductivity of MoO_3 and V_2O_5

under bombardment is caused by the formation of the lower oxides, MoO_2 and V_2O_3, which have metallic conduction behavior at room temperature.

Acknowledgment

This work was partially supported by the Defense Research Projects Agency of the Department of Defense and was monitored by Air Force Cambridge Research Laboratories under Contract No. F19628-74-C-0038.

References

1. J. W. Mayer and O. J. Marsh, in *Applied Solid State Science* (R. Wolfe, ed.), Vol. 1, pp. 239–342, Academic Press, New York (1969).
2. G. Dearnaley, Ion bombardment and implantation, *Rep. Progr. Phys.* **32**, 405–491 (1969).
3. L. N. Large, Ion Implantation: A new method of doping semiconductors, *Contemp. Phys.* **10**, 277–298, 505–531 (1969).
4. J. W. Mayer, L. Eriksson, and J. A. Davies, *Ion Implantation in Semiconductors*, Academic Press, New York (1970).
5. F. H. Eisen and L. T. Chadderton (eds.), *Ion Implantation* (Proc. First Int. Ion Implantation Conf.), Gordon and Breach, New York (1971).
6. *Proceedings of the European Conference on Ion Implantation, Reading, England, September* 1970, Peter Peregrinus Ltd., Stevenage, England (1970).
7. I. Ruge and J. Graul (eds.), *Ion Implantation in Semiconductors* (Proc. Second Int. Ion Implantation Conf.), Springer-Verlag, Berlin (1971).
8. S. Namba (ed.), *Ion Implantation in Semiconductors* (Proc. U.S.–Japan Seminar), Japan Society for the Promotion of Science (1972).
9. B. L. Crowder (ed.), *Ion Implantation in Semiconductors and Other Materials* (Proc. Third Int. Ion Implantation Conf.), Plenum Press, New York (1973).
10. R. S. Nelson, *The Observation of Atomic Collisions in Crystalline Solids*, John Wiley, New York (1968).
11. D. V. Morgan (ed.), *Channeling in Solids*, John Wiley, New York (1973).
12. J. Koch and K. O. Nielsen (eds.), *Electromagnetic Isotope Separators and Their Applications* (Proc. Int. Conf. Aarhus, Denmark, June 1965), North-Holland, Amsterdam (1965); also in *Nuc. Inst. Meth.* **38**, 1–325 (1965).
13. Proc. Int. Conf. Atomic Collision and Penetration Studies with Energetic (KeV) Ion Beams, Chalk River, Nuclear Laboratories, Canada, September 1967, *Can. J. Phys.* **46**, 449–782 (1968).
14. D. W. Palmer, N. W. Thompson, and P. D. Thownsend (eds.), *Atomic Collision Phenomena in Solids* (Proc. Int. Conf., Brighton, England, September 1970), North-Holland, Amsterdam (1970).
15. S. Andersen, K. Bjorkqvist, B. Domeij, and N. J. E. Johansson (eds.), *Atomic Collisions in Solids IV* (Proc. Int. Conf., Gausdal, Norway, September 1971), Gordon and Breach, New York (1972).
15a. D. S. Gemmell, Channeling and related effects in the motion of charged particles through crystals, *Rev. Mod. Phys.* **46**, 129–228 (1974).
16. J. Lindhard, Influence of crystal lattice on motion of energetic charged particles, *Kgl. Danske Vid. Selsk., Mat. Fys. Medd.* **34**(14), 1–64 (1955).

16a. J. A. Davies, J. Denhartog, L. Eriksson, and J. W. Mayer, Ion implantation of silicon. I. Atom location and lattice disorder by means of 1.0 MeV helium ion scattering, *Can. J. Phys.* **45**, 4053–4073 (1967).

17. W. K. Chu, J. W. Mayer, M.-A. Nicolet, T. M. Buck, G. Amsel, and F. Eisen, Principles and applications of ion beam techniques for the analysis of solids and thin films, *Thin Solid Films* **17**, 1–41 (1973).

18. S. T. Picraux, J. A. Davies, L. Eriksson, N. G. E. Johansson, and J. W. Mayer, Channeling studies in diamond type lattices, *Phys. Rev.* **180**, 873–882 (1969).

19. J. A. Davies, Chapter 13 in Ref. 11.

20. J. Andersen, O. Andreasen, J. A. Davies, and E. Uggerhøj, The use of channeling-effect techniques to locate interstitial foreign atoms in silicon, *Rad. Eff.* **7**, 25–34 (1971); also in Ref. 5, pp. 315–324.

21. D. Van Vliet, On the spatial distribution of channeled ions, *Rad. Eff.* **10**, 137–156 (1971).

22. F. H. Eisen and E. Uggerhøj, Experimental investigation of the energy and depth dependence of flux peaking, *Rad. Eff.* **12**, 233–240 (1972); also in Ref. 15, pp. 377–384.

23. L. C. Feldman, E. N. Kaufman, J. M. Poate, and W. M. Augustyniak, Lattice location of impurities implanted into metals: Carbon in iron, in Ref. 9.

24. R. B. Alexander and J. M. Poate, An exacting test of the channeling technique for atom location: Br implanted into Fe, in Ref. 9.

25. D. Sigurd and K. Björkqvist, Channeling studies of ion implanted silicon, *Rad. Eff.* **17**, 209–220 (1973).

26. S. T. Picraux, W. L. Brown, and W. M. Gibson, Lattice location by channeling angular distributions, *Phys. Rev. B* **6**, 1382–1394 (1972).

27. F. H. Eisen and B. Welch, Flux and fluence dependence of disorder produced during implantation of ^{11}B in silicon, *Rad. Eff.* **7**, 143–148 (1971); also in Ref. 5, pp. 459–464.

28. E. Bøgh, Defect studies in crystals by means of channeling, *Can. J. Phys.* **46**, 653–662 (1968).

29. L. C. Feldman and J. W. Rodgers, Depth profiles of the lattice disorder resulting from ion bombardment of silicon single crystals, *J. Appl. Phys.* **41**, 3776–3782 (1970).

30. J. E. Westmoreland, J. W. Mayer, F. H. Eisen, and B. Welch, Analysis of disorder distributions in boron implanted silicon, *Rad. Eff.* **6**, 161–174 (1970); also in Ref. 5, pp. 31–44.

31. F. H. Eisen, Chapter 14 in Ref. 11.

32. F. H. Eisen, G. J. Clark, J. Bøttiger, and J. M. Poate, Stopping power of energetic helium ions transmitted through thin silicon crystals in channeling and random directions, *Rad. Eff.* **13**, 93–100 (1972); also in Ref. 15, pp. 181–188.

33. J. Bøttiger and F. H. Eisen, Conversion from an energy scale to a depth scale in channeling experiments, *Thin Solid Films* **19**, 239–246 (1973).

34. G. H. Kinchin and R. S. Pease, The displacement of atoms in solids by radiation, *Rep. Prog. Phys.* **18**, 2–50 (1955).

35. P. Sigmund, On the number of atoms displaced by implanted ions or energetic recoil atoms, *Appl. Phys. Lett.* **14**, 114–117 (1969).

36. D. A. Marsden, G. R. Bellavance, J. A. Davies, M. Martini, and P. Sigmund, The energy dependence of lattice disorder in ion-implanted silicon, *Phys. Stat. Sol.* **35**, 269–275 (1969).

37. J. Bøttiger and J. A. Davies, Chalk River Nuclear Laboratories, Private communication.
38. K. B. Winterbon, Chalk River Nuclear Laboratories, Private communication.
39. L. T. Chadderton and F. H. Eisen, On the annealing of damage produced by boron implantation of silicon single crystals, *Rad. Eff.* **7**, 129–138 (1971); also in Ref. 5, pp. 445–454.
40. J. J. Grob, A. Ghitescu, and P. Siffert, Lattice disorder produced in GaAs by cadmium implantation, in Ref. 9.
41. E. Bøgh, P. Høgild, and I. Stensgaard, Spatial distribution of defects in ion bombarded silicon and germanium, *Rad. Eff.* **7**, 115–122 (1971); also in Ref. 5, pp. 431–438.
42. J. K. Hirvonen, W. L. Brown, and P. M. Glotin, Structural differences in light and heavy ion disorder in Si studied by single and double alignment channeling techniques, in Ref. 7, pp. 8–16.
43. G. Fladda, P. Mazzoldi, E. Rimini, D. Sigurd, and L. Eriksson, Evidence of a replacement reaction between ion implanted substitutional Tl dopants and interstition Si atoms, *Rad. Eff.* **1**, 249–256 (1969).
44. J. Haskell, E. Rimini, and J. W. Mayer, Channeling measurements in As-doped Si, *J. Appl. Phys.* **43**, 3425–3431 (1972).
45. F. H. Eisen, J. D. Haskell, E. Rimini, and J. W. Mayer, The effect of ion implantation on the lattice location of As in arsenic-doped Si, in Ref. 9.
46. J. Lindhard, M. Scharff, and H. Schiøtt, Range concepts and heavy ion ranges, *Kgl. Danske Vid. Selskab., Mat. Fys. Medd.* **33** (14), 1–39 (1963).
47. F. H. Eisen, Channeling of medium-mass ions through silicon, *Can. J. Phys.* **46**, 561–572 (1968).
48. H. E. Schiøtt, Approximations and interpolation rules for ranges and range stragglings, *Rad. Eff.* **6**, 107–114 (1970); also in Ref. 5, pp. 197–203.
49. K. B. Winterbon, Heavy-ion range profiles and associated damage distributions, *Rad. Eff.* **13**, 215–226 (1972); also in Ref. 15, pp. 429–440.
50. W. S. Johnson and J. F. Gibbons, *Projected Range Statistics in Semiconductors*, Stanford University Book Store (1969).
51. J. F. Ziegler, B. L. Crowder, G. W. Cole, J. J. E. Baglin, and B. J. Masters, Boron atom distributions in ion-implanted silicon by the (n, ^{4}He) nuclear reaction, *Appl. Phys. Lett.* **21**, 16–17 (1972).
52. R. P. Gittins, D. V. Morgan, and G. Dearnaley, The application of the ion microprobe analyzer to the measurement of the distribution of boron ions implanted into silicon crystals, *J. Phys. D* **5**, 1654–1663 (1972).
53. T. E. Seidel, Distribution of boron implanted silicon, in Ref. 7, pp. 47–57.
54. B. L. Crowder, The influence of the amorphous phase on ion distributions and annealing behavior of group III and group V ions implanted into silicon, *J. Electrochem Soc.* **118**, 943–952 (1971).
55. J. F. Gibbons, Ion implantation in semiconductors part I: Range distribution theory and experiments, *Proc. IEEE* **55**, 295–319 (1968).
56. K. B. Winterbon, P. Sigmund, and J. B. Sanders, Spatial distribution of energy deposited by atomic particles in elastic collisions, *Kgl. Danske Vid. Selsk., Mat. Fys. Medd.* **37** (14), 1–73 (1970).
57. J. Stephen, Ion implantation in semiconductor device technology, *Radio Electron. Eng.* **42**, 265–283 (1972).

58. J. L. Whitton, in Ref. 11.
59. V. G. K. Reddi and A. Y. C. Yu, Ion implantation for silicon device fabrication, *Solid State Technology* **1972** (October), 35–41.
60. G. Dearnaley, J. H. Freeman, E. A. Gard, and M. A. Wilkins, Implantation profiles of ^{32}P channeled into silicon crystals, *Can. J. Phys.* **46**, 587–595 (1968).
61. R. A. Moline, Ion-implanted phosphorus in silicon: Profiles using C–V analysis, *J. Appl. Phys.* **42**, 3553–3558 (1971).
62. V. K. G. Reddi and J. D. Sansbury, Channeling of phosphorus ions in silicon, *Appl. Phys. Lett.* **20**, 30–31 (1972).
63. J. A. Galaktionova, V. M. Gusev, V. G. Naumenko, and V. V. Titov, Conductivity and carrier density distributions in silicon doped by ion bombardment, *Soviet Phys.—Semiconductors* **2**, 656–659 (1968).
64. V. K. G. Reddi and J. D. Sansbury, Channeling and dechanneling of ion implanted phosphorus in silicon, *J. Appl. Phys.* **44**, 2951–2963 (1973).
65. R. A. Moline and G. W. Reutlinger, Phosphorus channeled into silicon: profiles and electrical activity, in Ref. 7, pp. 58–69.
66. P. Blood, G. Dearnaley, and M. A. Wilkins, The depth distribution of phosphorus ions implanted into Si crystals, in Ref. 9.
67. R. L. Minear, D. G. Nelson, and J. F. Gibbons, Enhanced diffusion in Si and Ge by light ion implantation, *J. Appl. Phys.* **43**, 3468–3480 (1972).
68. O. Meyer and J. W. Mayer, Enhanced diffusion and out-diffusion in ion-implanted silicon, *J. Appl. Phys.* **41**, 4166–4174 (1970).
69. S. Namba, K. Masuda, K. Gamo, and M. Iwaki, Enhanced diffusion in ion implanted silicon, in Ref. 8, pp. 49–58.
70. D. K. Brice, Spatial distribution of energy deposited into atomic processes in ion-implanted silicon, *Rad. Eff.* **6**, 77–88 (1970); also in Ref. 5, pp. 101–112.
71. K. L. Brower, F. L. Vook, and J. A. Borders, Depth distribution of EPR centers in 400 keV O$^+$ ion-implanted silicon, *Appl. Phys. Lett.* **16**, 108–111 (1970).
72. H. J. Stein, F. L. Vook, and J. A. Borders, Depth distribution of divancies in 400 keV O$^+$ ion-implanted silicon, *Appl. Phys. Lett.* **16**, 106–108 (1970).
73. J. W. Mayer, Ion implantation-lattice disorder, *Rad. Eff.* **8**, 269–278 (1971).
74. D. K. Brice, Ion implantation depth distributions: Energy deposition into atomic processes and ion locations, *Appl. Phys. Lett.* **16**, 103–106 (1970).
75. E. Bøgh, Private communication.
76. F. H. Eisen, B. Welch, J. E. Westmoreland, and J. W. Mayer, Lattice disorder produced in silicon by Boron ion implantation, in Ref. 14, pp. 111–127.
77. E. P. EerNisse, Investigation of ion implantation damage with stress measurements, in Ref. 7, pp. 17–22.
78. S. I. Tan and B. S. Berry, Internal friction study of point defects in boron implanted silicon, in Ref. 9.
79. F. L. Vook and H. J. Stein, Relation of neutron to ion damage annealing in Si and Ge, *Rad. Eff.* **2**, 23–30 (1969).
80. S. T. Picraux, W. H. Weisenberg, and F. L. Vook, Low temperature channeling measurements of ion implantation lattice disorder in single crystal silicon, *Rad. Eff.* **7**, 101–108 (1971); also in Ref. 5, pp. 411–418.
81. J. Bøttiger, J. A. Davies, D. V. Morgan, J. L. Whitton, and K. B. Winterbon, Damage profiles in ion-implanted semiconductors at low (25°K) temperatures, in Ref. 9.

82. S. T. Picraux and F. L. Vook, Ionization, thermal, and flux dependences of implantation disorder in silicon, *Rad. Eff.* **11**, 179–192 (1971).

83. F. L. Vook and H. J. Stein, Evidence for vacancy motion in low temperature ion-implanted silicon, *Rad. Eff.* **6**, 11–18 (1970); also in Ref. 5, pp. 9–16.

84. K. L. Brower, Electron paramagnetic resonance of the neutral ($S = 1$) one-vacancy-oxygen center in irradiated silicon, *Phys. Rev. B* **4**, 1968–1982 (1971).

85. K. L. Brower and W. Beezhold, Electron paramagnetic resonance of the lattice damage in oxygen-implanted silicon, *J. Appl. Phys.* **43**, 3499–3506 (1972).

86. T. Matsumori, T. Kobayashi, H. Maekawa, and T. Izumi, Strain induced effects on EPR centers in silicon generated by ion implantation, in Ref. 9.

87. H. J. Stein, F. L. Vook, D. K. Brice, J. A. Borders, and S. T. Picraux, Infrared studies of crystallinity in ion implanted silicon, *Rad. Eff.* **6**, 19–26 (1967); also in Ref. 5, pp. 17–24.

88. J. E. Smith, Jr., M. H. Brodsky, B. L. Crowder, and M. I. Nathan, Raman scattering in amorphous Si, Ge, and III–V semiconductors, *J. Non-Cryst. Solids* **8–10**, 179–184 (1972).

89. M. Tamura, T. Ikeda, and N. Yoshihiro, Electron microscopic study of crystal defects in boron- and phosphorus-implanted silicon, *J. Japan Soc. Appl. Phys.* (Suppl.) **40**, 9–15 (1971).

90. M. D. Matthews, Electrical and electron microscopy observations on defects in ion implanted silicon, *Rad. Eff.* **11**, 167–178 (1971).

91. B. J. Masters, J. M. Fairfield, and B. L. Crowder, Radiochemical determination of range profiles in silicon, *Rad. Eff.* **6**, 57–62 (1970); also in Ref. 5, pp. 81–86.

92. A. U. MacRae, Recent advances in ion implanted junction-device technology, in Ref. 7, pp. 329–334.

93. J. A. Borders and K. L. Brower, EPR of substitutional group V impurities in silicon, *Rad. Eff.* **6**, 135–140 (1970).

94. J. Gyulai, O. Meyer, R. D. Pashley, and J. W. Mayer, Lattice location and dopant behavior of group II and VI elements implanted in silicon, *Rad. Eff.* **7**, 17–24 (1971); also in Ref. 5, pp. 297–304.

95. J. O. McCaldin, in *Progress in Solid State Chemistry* (H. Reiss, ed.), Vol. 2, pp. 9–25, Pergamon Press, New York (1965).

96. O. Meyer and J. W. Mayer, Analysis of Rb and Cs implantations in silicon by channeling and Hall effect measurements, *Solid-State Electron.* **13**, 1357–1362 (1970).

97. H. R. Bilger, M.-A. Nicolet, and O. Meyer, Electrical properties of alkali implants in silicon, in Ref. 6, pp. 70–71.

98. B. Domeij, G. Fladda, and N. G. E. Johansson, Anomalously high yields of elastically scattered ^{12}C Ions from Zr, Hf, Tl, and Hg atoms implanted into silicon, *Rad. Eff.* **6**, 155–160 (1970); also in Ref. 5, pp. 425–430.

99. A. U. MacRae, Device fabrication by ion implantation, *Rad. Eff.* **7**, 59–64 (1971); also in Ref. 5, pp. 363–368.

100. B. L. Crowder, Applications of ion implantation for new device concepts, *J. Vac. Sci. Tech.* **8**, S71–S78 (1971).

101. N. G. E. Johansson and J. W. Mayer, Hall effect measurements of Sb and Ga implanted silicon; anneal behavior and comparison with other species, *Solid-State Electron.* **13**, 123–130 (1970).

102. J. C. North and W. M. Gibson, Channeling study of boron implanted silicon, *Appl. Phys. Lett.* **16**, 126–129 (1970).

103. G. Fladda, K. Björkgvist, L. Eriksson, and D. Sigurd, The lattice location of borons ions implanted into silicon, *Appl. Phys. Lett.* **16**, 313–315 (1970).

104. B. L. Crowder, The role of damage in the annealing characteristics of ion-implanted Si, *J. Electrochem. Soc.* **117**, 671–674 (1970).

105. D. E. Davies and S. A. Roosild, Irradiation defects and the electrical quality of ion implanted silicon, *Solid State Electron.* **14**, 975–983 (1971).

106. R. D. Pashley, Electrical properties of indium implanted silicon, *Rad. Eff.* **11**, 1–8 (1971).

107. F. Lee, California Institute of Technology, Private communication.

108. E. Munoz, W. L. Snyder, and J. L. Moll, Effect of arsenic pressure on heat treatment of liquid epitaxial GaAs, *Appl. Phys. Lett.* **16**, 262–265 (1970).

109. S. T. Picraux, Increased release of As during Annealing of Sb implanted GaAs, in Ref. 9.

110. R. G. Hunsperger, O. J. Marsh, and C. A. Mead, The presence of deep levels in ion implanted junctions, *Appl. Phys. Lett.* **13**, 295–297 (1968).

110a. R. Heckingbottom and T. Ambridge, Ion implantation in compound semiconductors—an approach based on solid state theory, *Rad. Eff.* **17**, 31–36 (1973).

110b. P. L. Degan, Ion implantation doping of compound semiconductors, *Phys. Stat. Sol.(a)* **16**, 9–42 (1973).

111. J. D. Sansbury and J. F. Gibbons, Properties of ion implanted silicon, sulfur, and carbon in gallium arsenide, *Rad. Eff.* **6**, 269–276 (1970); also Ref. 5, pp. 253–260.

112. J. L. Whitton, G. Carter, J. H. Freeman, and G. A. Gard, The implantation profiles of 10, 20, and 40 keV ^{85}Kr in gallium arsenide, *J. Mat. Sci.* **4**, 208–217 (1969).

113. D. Sigurd, California Institute of Technology, Private communication.

114. J. S. Harris, F. H. Eisen, B. Welch, J. D. Haskell, R. D. Pashley, and J. W. Mayer, Influence of implantation temperature and surface protection on tellurium implantation in GaAs, *Appl. Phys. Lett.* **21**, 601–603 (1972).

115. J. L. Whitton and G. Carter, The implantation profiles of energetic heavy ions in GaAs, GaP, and Ge, in Ref. 14, pp. 615–632.

116. J. L. Whitton and G. R. Bellavance, Ion implantation of sulfur into GaAs, GaP and Ge monocrystals, *Rad. Eff.* **9**, 127–132 (1971).

117. F. H. Eisen, J. S. Harris, B. Welch, R. D. Pashley, D. Sigurd, and J. W. Mayer, Properties of Tellurium implanted GaAs, in Ref. 9.

118. F. H. Eisen, III–V compound review, *Rad. Eff.* **9**, 235–242 (1971).

119. J. E. Westmoreland, O. J. Marsh, and R. G. Hunsperger, Lattice disorder produced in GaAs by 60 keV Cd ions and 70 keV Zn ions, *Rad. Eff.* **5**, 245–250 (1970).

120. G. Carter, W. A. Grant, J. D. Haskell, and G. A. Stephens, Radiation damage by implanted ions in GaAs and GaP, *Rad. Eff.* **6**, 277–284 (1970); also in Ref. 5, pp. 261–268.

121. J. S. Harris and F. H. Eisen, The annealing of damage in ion implanted GaAs, *Rad. Eff.* **7**, 123–128 (1971); also in Ref. 5, pp. 439–444.

122. J. A. Borders, Near band edge optical absorption produced by ion implantation in GaAs, *Appl. Phys. Lett.* **18**, 16–18 (1971).

123. G. A. Shifrin and R. G. Hunsperger, Effect of ion-implantation damage on the optical reflection spectrum of gallium arsenide, *Appl. Phys. Lett.* **17**, 274–276 (1970).

124. D. D. Sell and A. U. MacRae, Optical detection of surface damage in GaAs induced by argon ion implantation, *J. Appl. Phys.* **41**, 4929–4932 (1970).

125. R. G. Hunsperger, E. D. Wolf, G. A. Shifrin, O. J. Marsh, and D. M. Jamba, Measurement of lattice damage caused by ion-implantation doping of semiconductors, *Rad. Eff.* **9**, 133–138 (1971).

126. R. S. Peercy, Raman scattering of ion implanted GaAs, *Appl. Phys. Lett.* **18**, 574–576 (1971).

127. E. D. Wolf and R. G. Hunsperger, Measurement of ion implantation lattice damage in (111) GaAs using the scanning electron microscope, *Appl. Phys. Lett.* **16**, 526–529 (1970).

128. W. H. Weisenberger, S. T. Picraux, and F. L. Vook, Low temperature channeling measurements of ion implantation lattice disorder in GaAs, *Rad. Eff.* **9**, 121–126 (1971).

129. J. D. Haskell, W. A. Grant, G. A. Stephens, and J. L. Whitton, The influence of various parameters on radiation damage in GaP; in Ref. 7, pp. 193–198.

130. S. T. Picraux, Disorder annealing in III–V semiconductors after ion implantation at low temperature, *Rad. Eff.* **17**, 261–267 (1973).

131. J. W. Mayer, L. Eriksson, S. T. Picraux, and J. A. Davies, Ion implantation of silicon and germanium at room temperature. Analysis by means of 1.0-MeV helium ion scattering, *Can. J. Phys.* **46**, 663–673 (1967).

132. J. F. Chemin, I. V. Mitchell, and F. W. Saris, Lattice location of low-Z impurities in medium-Z targets using ion-induced X-rays, in Ref. 9.

133. J. L. Merz, D. L. Mingay, W. M. Augustyniak, and L. C. Feldman, Implantation of Bi into GaP. III. Hot-implant behavior, in Ref. 7, pp. 182–191.

134. L. H. Skolnik, W. G. Spitzer, A. Kahan, and R. G. Hunsperger, Infrared localized-vibrational-mode absorption of ion-implanted aluminum and phosphorus in gallium arsenide, *J. Appl. Phys.* **42**, 5223–5229 (1971).

135. L. H. Skolnik, W. G. Spitzer, A. Kahan, F. Euler, and R. G. Hunsperger, Localized vibrational mode adsorption of ion implanted silicon in GaAs, *J. Appl. Phys.* **43**, 2146–2150 (1972).

136. A. H. Kachare, W. G. Spitzer, A. Kahan, F. K. Euler, and T. A. Whatley, Ion-implanted nitrogen in gallium arsenide, *J. Appl. Phys.* **44**, 4393–4399 (1973).

137. J. S. Harris, Y. Nannichi, G. L. Pearson and G. F. Day, Ohmic contacts to solution-grown gallium arsenide, *J. Appl. Phys.* **40**, 4575–4581 (1969).

138. J. Gyulai, J. W. Mayer, V. Rodriquez, A. Y. C. Yu, and H. J. Gopen, Alloying behavior of Au and Au–Ge on GaAs, *J. Appl. Phys.* **42**, 3578–3585 (1971).

139. R. G. Hunsperger and O. J. Marsh, Electrical properties of zinc and cadmium ion implanted layers in gallium arsenide, *J. Electrochem. Soc.* **116**, 488–492 (1969).

140. P. N. Favennec, Implantation of zinc into GaAs at 1 MeV, in Ref. 7, pp. 174–181.

141. R. G. Hunsperger, R. G. Wilson, and D. M. Jamba, Mg and Be ion implanted GaAs, *J. Appl. Phys.* **43**, 1318–1320 (1972).

142. R. G. Hunsperger and O. J. Marsh, Electrical properties of Cd, Zn, and S ion implanted layers in GaAs, *Rad. Eff.* **6**, 257–262 (1970); also in Ref. 5, pp. 247–252.

143. M. Fujimoto, Y. Kawasaki, and T. Imai, Electrical properties of zinc ion implanted layers in gallium arsenide, in Ref. 8, pp. 165–170.

144. M. A. Littlejohn, J. A. Hauser, and K. L. Monteith, The electrical properties of 60 keV zinc ions implanted in semi-insulating gallium arsenide, *Rad. Eff.* **10**, 185–190 (1971).

145. V. M. Zelevinskaya, G. A. Cochurin, N. B. Pridachin and L. S. Smirnov, Doping of gallium arsenide by the implantation of zinc ions, *Soviet Phys.—Semiconductors* **4**, 1529–1532 (1971).

146. B. Goldstein, Diffusion of cadmium and zinc in gallium arsenide, *Phys. Rev.* **118**, 1024–1027 (1960).

147. H. Kressel and F. Z. Hawrylo, Ionization energy of Mg and Be acceptors in GaAs, *J. Appl. Phys.* **41**, 1865 (1970).

148. L. L. Chang and G. L. Pearson, The solubilities and distribution coefficients of Zn in GaAs and GaP, *J. Phys. Chem. Solids* **25**, 23–30 (1964).

149. S. Sze and J. Irvin, Resistivity mobility and impurity levels in GaAs, Ge, and Si at 300°K, *Solid State Electron.* **11**, 599–602 (1968).

150. R. G. Hunsperger and E. D. Wolf, Anneal behavior of Cd Ion implanted GaAs, *J. Electrochem Soc.* **118**, 1847–1851 (1971).

151. K. Wohlleben and W. Beck, Change of concentration and mobility of charge carriers in GaAs due to irradiation with protons, *Z. Naturforsch.* **21a**, 1057–1071 (1966).

152. A. G. Foyt, W. T. Lindley, C. M. Wolfe, and J. P. Donnelly, Isolation of junction devices in GaAs using proton bombardment, *Solid State Electron.* **12**, 209–214 (1969).

153. R. G. Hunsperger and O. J. Marsh, Anneal behavior of defects in ion-implanted GaAs diodes, *Met. Trans.* **1**, 603–607 (1970).

154. T. Itoh and Y. Kushiro, The effects of arsenic ion implantation in GaAs, in Ref. 7, pp. 168–173.

155. V. M. Zelevinskaya and G. A. Kachurin, Annealing of gallium arsenide doped by implantation of Group VI Ions, *Soviet Phys.—Semiconductors* **5**, 1455–1456 (1972).

156. V. M. Zelevinskaya and G. A. Kachurin, Behavior of germanium implanted in GaAs by ion bombardment, *Soviet Phys.—Semiconductors* **5**, 1011–1013 (1971).

157. J. W. Mayer, O. J. Marsh, R. Mankaious, and R. Bower, Zn and Te implantations into GaAs, *J. Appl. Phys.* **38**, 1975–1976 (1967).

158. J. S. Harris, The effects of dose rate and implantation temperature on lattice damage and electrical activity in ion implanted GaAs, in Ref. 7, pp. 157–167.

159. R. G. Hunsperger, Hughes Research Laboratories, Private communication.

160. E. W. Williams, A photoluminescence study of acceptor centers in gallium arsenide, *Brit. J. Appl. Phys.* **18**, 253–262 (1967).

161. J. Gyulai, J. W. Mayer, I. V. Mitchell, and V. Rodriguez, Outdiffusion through silicon oxide and silicon nitride layers on gallium arsenide, *Appl. Phys. Lett.* **17**, 332–334 (1970).

162. I. V. Mitchell, J. W. Mayer, J. K. Kung, and W. G. Spitzer, Investigation of Te-doped GaAs annealing effects by optical and channeling-effect measurements, *J. Appl. Phys.* **42**, 3982–3987 (1971).

163. J. L. Merz, L. C. Feldman, and E. A. Sadowski, Ion implantation of Bi into GaP: I. Photoluminescence, *Rad. Eff.* **6**, 285–292 (1970); in Ref. 5, pp. 269–276.

164. L. C. Feldman, W. M. Agustyniak, and J. L. Merz, Implantation of Bi into GaP: II. Channeling studies, *Rad. Eff.* **6**, 293–300 (1970); also in Ref. 5, pp. 277–284.

165. J. D. Cuthbert and D. G. Thomas, Optical properties of tellurium as an isoelectronic trap in cadmium sulfide, *J. Appl. Phys.* **39**, 1573–1580 (1968).

166. P. J. McNally, Ion implantation in InAs and InSb, *Rad. Eff.* **6**, 149–154 (1970); also in Ref. 5, pp. 405–410.

167. J. P. Donnelly, T. C. Harman, A. G. Foyt, and W. T. Lindley, *p–n* junction photodiodes in PbTe prepared by Sb^+ ion implantation, *Appl. Phys. Lett.* **20**, 279–281 (1972).

167a. J. P. Donnelly, T. C. Harmon, A. G. Foyt, and W. T. Lindley, PbS photodiodes fabricated by Sb^+ ion implantation, *Solid State Elec.* **16**, 529–534 (1973).

168. R. R. Hart, H. L. Dunlap, O. J. Marsh, Disorder produced in SiC by ion bombardment, *Rad. Eff.* **9**, 261–266 (1971).

169. R. R. Hart, H. L. Dunlap, and O. J. Marsh, Backscattering analysis and electrical behavior of SiC implanted with 40 keV indium, in Ref. 7, pp. 134–140.

170. S. Kurtin, T. C. McGill, and C. A. Mead, Fundamental transition in the electronic nature of solids, *Phys. Rev. Lett.* **22**, 1433–1436 (1969).

171. D. G. Thomas (ed.), *II–VI Semiconducting Compounds*, W. A. Benjamin, New York (1967).

172. G. Carter and J. S. Colligon, *Ion Bombardment of Solids*, American Elsevier, New York (1968).

173. R. J. MacDonald, The ejection of atomic particles from ion bombarded solids, *Adv. Phys.* **80**, 457–524 (1970).

174. P. Sigmund, Theory of sputtering. I. Sputtering yield of amorphous and polycrystalline targets, *Phys. Rev.* **184**, 383–416 (1969).

175. P. Sigmund, Collision theory of displacement damage, ion ranges, and sputtering, *Revue Roum. Phys.* **17**, 823–870 (1972).

176. J. Geerk and O. Meyer, Oxidation of lead by low energy $O_2{}^+$ bombardment, *Surface Sci.* **32**, 222–230 (1972).

176a. Conference on Applications of Ion Beams to Metals, Albuquerque, New Mexico, October, 1973.

177. H. Kräutle and S. Kalbitzer, Conductive properties of the ion implanted binary system $Si_{1-x}Al_x$, in Ref. 7, pp. 499–506.

178. H. Kräutle and S. Kalbitzer, Ion implanted silicon–metal systems, in Ref. 9.

179. M. Watanabe and A. Tooi, Formation of SiO_2 films by oxygen-ion bombardment, *Japan J. Appl. Phys.* **5**, 737–738 (1966).

180. P. V. Pavlov and E. V. Shitova, The structure of oxide films obtained by oxygen ion bombardment of a silicon surface, *Soviet Phys.—Doklady* **12**, 11–13 (1967).

181. J. H. Freeman, G. A. Gard, D. J. Mazey, J. H. Stephen, and F. B. Whiting, Formation of insulating layers by the use of reactive ion beams, in Ref. 6, pp. 74–80.

182. J. A. Borders, S. T. Picraux, and W. Beezhold, Formation of SiC in silicon by ion implantation, *Appl. Phys. Lett.* **18**, 509–511 (1971).

183. J. A. Borders and W. Beezhold, Infrared studies of SiC, Si_3N_4, and SiO_2 Formation in ion-implanted silicon, in Ref. 7, pp. 241–247.

184. I. P. Akimenchenko, V. S. Vavilov, V. V. Galkin, V. S. Ivanov, V. V. Krasnopevtsev, and Yu. V. Milyutin, Synthesis of silicon carbide by implantation of Si ions in diamond, *Soviet Phys.—Semiconductors* **6**, 1039–1041 (1972).

185. J. J. Trillat, *Ionic Bombardment, Theory and Application*, Gordon and Breach, New York (1962).

186. P. T. Stroud, Ion bombardment and implantation and their application to thin films, *Thin Solid Films* **11**, 1–26 (1972).

187. L. E. Collins, J. G. Perkins, and P. T. Stroud, Effect of ion bombardment on the adhesion of aluminum films on glass, *Thin Solid Films* **4**, 41–45 (1969).

188. J. G. Perkins and P. T. Stroud, Transmission sputtering and recoil implantation from thin metal films under ion bombardment, *Nuc. Inst. Meth.* **102**, 109–115 (1972).

189. R. S. Nelson, The theory of recoil implantation, *Rad. Eff.* **2**, 47–50 (1969).

190. J. G. Perkins, Conductive properties and microstructure of metal SiO cermet thin films produced by recoil atom implantation, *J. Non-Crystal. Solids* **3**, 349–364 (1972).

191. P. T. Stroud, H. M. Lindsay, and J. G. Perkins, Some preliminary studies of the structure of ion bombarded thin films, *Vacuum* **23**, 125–130 (1973).

192. G. Dearnaley, P. D. Goode, W. S. Miller, and J. F. Turner, The influence of ion implantation upon the high temperature oxidation of titanium and stainless steel, in Ref. 9.

193. J. E. Antill, M. Bennett, G. Dearnaley, F. Fern, P. D. Goode, and J. F. Turner, The effects of yttrium ion implantation upon the oxidation and oxide adhesion of an austenitic stainless steel, in Ref. 9.

194. N. E. W. Hartley, G. Dearnaley, and J. F. Turner, frictional changes induced by the ion implantation of steel, in Ref. 9.

195. J. F. Turner, W. Temple, and G. Dearnaley, Possible radiation enhanced diffusion of nickel ions in titanium, in Ref. 9.

196. A. D. Street, W. A. Grant, and G. Carter, Surface passivation of metals by ion bombardment, in Ref. 9.

197. G. Linker, M. Gettings, and O. Meyer, Ion implantation and radiation damage in vanadium, in Ref. 9.

198. O. Meyer, Nuclear Research Center, Karlsruhe, Private communication.

199. E. Arminen, A. Fontell, and A. V. Lindroos, Anomalously high collection of copper ions implanted in aluminum, *Phys. Stat. Sol. (a)* **4**, 663–673 (1971).

200. E. Garmire, H. Stoll, A. Yariv, and R. G. Hunsperger, Optical waveguiding in proton-implanted GaAs, *Appl. Phys. Lett.* **21**, 87–88 (1972).

201. S. Somekh, E. Garmire, A. Yariv, H. L. Garvin, and R. G. Hunsperger, Channel optical waveguide directional couplers, *Appl. Phys. Lett.* **22**, 46–47 (1973).

202. M. K. Barnoski, R. G. Hunsperger, R. Wilson, and G. Tangonan, Proton implanted GaP optical waveguide, *J. Appl. Phys.* **44**, 1925–1926 (1973).

203. W. L. Brown, New implantation areas, in Ref. 7, pp. 430–438.

204. R. Wolfe, J. C. North, R. L. Barns, M. Robinson, and H. J. Levinstein, Modification of magnetic anisotropy in garnets by ion implantation, *App. Phys. Lett.* 19, 298–300 (1971).

205. J. C. North and R. Wolfe, Ion implantation effects in magnetic bubble garnets, in Ref. 9.

206. J. W. Corbett and L. C. Ianniello (eds.), *Proc. 1971 Int. Conf. on Radiation Induced Voids in Metals*, AEC Symposium Series, No. 26.

207. H. M. Naguib and R. Kelly, On the increase in the electrical conductivity of MoO_3 and V_2O_5 Following ion bombardment, *J. Phys. Chem.* **33**, 1751–1759 (1972).

208. T. Parker and R. Kelly, Electrical and structural changes in ion-bombardment TiO_2, in Ref. 9.

Semiconductor Surfaces

S. Roy Morrison
Stanford Research Institute, Menlo Park, California

1. Introduction

The most interesting characteristic of a semiconductor, and the one that dominates most of its behavior, is its limited supply of current carriers. There may be a few carriers available in the conduction band —extra electrons over and above those tied up in valence bonds. There may be a few carriers available in the valence band—missing valence electrons termed holes. Or, particularly for nonequilibrium conditions, there can be both conduction band electrons and valence band holes. The limited supply of current carriers dominates the bulk properties of a semiconductor in familiar ways, leading to unique optical and electronic effects in the solid. The limited supply of carriers dominates the surface properties of a semiconductor also, leading to unique optical, electronic, and chemical effects arising at the semiconductor/gas or semiconductor/liquid interface.

In this chapter we discuss the interaction of these electrons and holes with groups at the semiconductor surface. Such interaction is of interest whether the chemical identity of the surface group is known or unknown. If unknown, one classifies[1,2] the reacting surface group simply in terms of its electronic energy level, its "surface state," and determines its influence on the electrical and optical properties of the semiconductor. The importance of carrier/surface group interaction arises in solid-state physics because such groups can and do dominate the electronic behavior of semiconductors unless the experiment is carefully designed to avoid surface problems. If the chemical identity of the surface group is known, the behavior becomes even more interesting, for both electrical and chemical influence can be recognized. The importance of carrier/surface group interactions in surface chemistry is equally great as in surface physics, although at present the chemical behavior is understood in less detail. For example, the role of semiconductors as practical redox catalysts is well known, but the theoretical analysis describing their behavior[3–7] has not been adequate. Most chemists agree, however, that ionized adsorbates, arising by interaction with holes or electrons from the semiconductor, are important intermediate species in many catalytic redox reactions. In other areas as widespread as corrosion (involving electron transfer to adsorbed oxygen ions on the semiconducting oxide), thermionic or photoelectron emission (involving electron transfer from adsorbed cesium ions or their equivalent), or pigmented polymers (subject to degradation through oxidation of organic binders by holes) problems involving a semiconductor surface and electron and/or hole transfer at the surface appear regularly.

The fundamental reason for the importance of such carrier/ surface group reactions is easily demonstrated. Holes from a semiconductor arriving at the surface are strong oxidizing agents, conduction band electrons strong reducing agents. Thus hole and electron transfer implies active redox processes. Where the surface physicist refers to hole capture by a surface state, the surface chemist can view it as oxidation of an adsorbate (or other surface group) by a hole from the semiconductor. Conversely, electron capture is reduction of a sorbate. The picture becomes most clear in discussions of photoeffects, where band-gap radiation produces both holes and electrons which migrate to the surface and provide a variety of interesting redox reactions (photocatalysis).

Although the qualitative features describing hole and electron capture are simplest to understand in chemical terminology, the more quantitative features of the effects, not only the chemical effects but the electrical effects used to monitor them, are far easier to describe in solid-state terminology. Thus much of this chapter will be denoted to describing the complex chemical/physical interactions at the surface in terms of surface states—surface electronic energy levels which can exchange electrons and/or holes with the conduction or valence band of the semiconductor.

A careful distinction must be made among three types of interaction when discussing redox reactions on the semiconductor surface: (a) formation of a new phase, (b) adsorption with local interaction only, and (c) ionosorption, where there is electron exchange between the surface group and the semiconductor bands, removing or injecting carriers. An example of (a) is oxide growth on germanium. The oxygen removes valence electrons from the germanium, but it also removes the germanium atom from the semiconductor, so, excepting a small shrinkage of the crystal dimensions, this strong chemical reaction has no direct influence on the electrical properties. It is certainly not rewarding to try to express such a chemical reaction in terms of a surface state model. An example of (b) might be the adsorption of oxygen on germanium at a temperature too low for atom rearrangement. Actually it is observed[8] that with a reasonably low oxygen pressure at room temperature oxygen only adsorbs on clean germanium to about 0.6 monolayer. Here, as will be discussed in more detail later, the interaction appears to be only a local bonding, with no electron or hole exchange with the semiconductor. Perhaps the bonding may be a partly covalent bond:

$$\begin{array}{ccccc} & O & & O & \\ & \diagup \diagdown & & \diagup \diagdown & \\ -Ge & \!\!-\!\!\!-\!\!\!-\!\! & Ge\!-\!Ge & \!\!-\!\!\!-\!\!\!-\!\! & Ge- \\ | & & | \quad | & & | \end{array}$$

The energies of the electrons in these bonds are apparently too low to permit electron exchange with the conduction band. Again it is not rewarding to analyze the process in terms of surface states—the usual chemical bonding representations describe the reaction much better. An example of (c) is the adsorption of oxygen on germanium in the form of O_2^-, where an acceptor energy level can be formed lower than the germanium conduction band and capture of free electrons from the conduction band can occur. This latter case, ionosorption, is an example of the case of interest here, where the surface state description is useful. The key factor is the electron transfer—it is for the description of this event that the surface state model is needed.

In Section 2 we review the models which permit a qualitative understanding of the origin and expected location of the various types of surface states, preparatory to the later discussions of equilibrium statistics (Section 3) and of the kinetics of the carrier exchange and associated redox reactions (Section 5).

2. Surface States at the Semiconductor Surface

2.1. The Surface State Description of Redox Reactions

A description of surface phenomena in terms of electronic energy levels (the band model) is preferred when one is interested in semiconductors in order to provide a useful connection to the description of bulk semiconductor properties. First, the model is convenient to describe the origin and behavior of the holes and electrons involved in the surface reactions. Second, if electronic energy levels are used to describe the surface groups of interest, then simple measurements of the electrical properties of the semiconductor can be interpreted and can provide the appropriate parameters needed in the theory. Therefore reduction of oxidizing agents by conduction band electrons will be discussed in terms of electron transfer between a semiconductor band and the unoccupied localized electronic energy level associated with a surface group. This localized energy level is called a surface state. Again, oxidation of reducing agents by holes from the valence band will be described in terms of hole transfer to the occupied electronic energy level of the reducing agent. The level is an occupied surface state. We define a surface state as any localized electronic energy level at the semiconductor surface capable of exchanging free electrons or holes with the semiconductor. Such energy levels arise from free major sources. Two are obvious—impurities on the surface or gases adsorbing from the gas phase. The third source, less obvious, is the aperiodicity associated with the termination of the solid, and leads to the presence of surface states even on a clean semiconductor surface.

An example of a surface state associated with impurity ions on a surface is copper on germanium. Copper ions provide an energy level near the middle of the germanium band gap[9,10] and are extremely active in hole and electron capture. They arise during the pretreatment of the crystal; in the washing process copper precipitates onto germanium from solutions as dilute[11] as 5×10^{-8} M. Surface states due to adsorbed gases on semiconductors have always been of interest. Oxygen, water, ammonia, and many other gases tend to cause electron injection into or extraction from the semiconductor, forming ions on the surface. There is less practical interest in surface states on a "clean" surface, but great interest from a fundamental point of view, for this is one case where results of surface measurements can be relied on to be reproducible. If the solid is covalent (germanium, silicon), surface states on the clean surface are expected because there are unsatisfied valence electrons at the discontinuity which will behave differently from the normal electrons, providing unique energy levels localized at the surface. If the solid is ionic (zinc oxide), such states are expected because the electrostatic potential at a surface cation or anion site is different from that at a bulk cation or anion site, so the energy levels association with these surface ions is unique, giving rise to localized energy levels—surface states.

2.2. Quantum-Mechanical Description of Surface States

It is beyond the scope of this chapter to present the detailed mathematics of the quantum mechanical descriptions of surface states; for such analyses we recommend the review by Davison and Levine.[12] The objective in this section is to present a description of the methods used and the results obtained, providing thereby a basis to understand (a) the expected location of surface states and (b) the accuracy to be anticipated from such calculations.

Both surface states on a clean surface and surface states due to adsorbates can be described by quantum analysis. The characteristics can be calculated by several approximations.[12] We will discuss the most common and most interesting in solid-state chemistry, namely the tight binding approximation, and only mention the crystal potential or Konig–Penney model and the Mathieu potential method. We will describe only the linear chain approximation, omitting the complex two- or three-dimensional calculations.

In the tight binding model[12–16] one assumes the atoms in the crystal are far enough apart that the electronic wave functions can be described as only slight perturbations of the atomic orbitals. Consider N atoms in a linear chain. If there were no perturbations from neighboring atoms, one would have N energy levels all at the same energy. A small perturbation, however, is included, but so small that direct interaction is permitted only between nearest neighbors—an exchange interaction. This interaction splits the N energy levels into a band. In the mathematical analysis the two key parameters are α, the Coulomb integral describing the energy associated with the electron on a given lattice site, and β, the resonance integral, describing the interaction energy between nearest neighbors. It is convenient to consider a crystal with alternating A and B atoms, making the system if desired analogous to a ZnO crystal if A and B are different and to a Ge crystal if A and B are the same. The list of parameters then increases to include both α_A and α_B, the Coulomb attraction integrals of A and B, respectively.

The end atom on the chain only has one neighbor, so is different. The Columb integral will be different (and will be described by α_A' or α_B') and, if it is a foreign adsorbate, the resonance integral will be different, and will be described by β^*. Putting these values into the

Schrödinger equation yields in addition to the bands of levels complex solutions representing wave functions localized at the end atom and decaying in amplitude as one moves into the crystal.

The model that develops is roughly as follows. If the parameters α_A and α_B differ, there are two bands of energy levels, one associated with each of the α's, the higher of which can be considered the conduction band. With the ZnO analogy, this would be equivalent to one band of allowed states with the electrons localized on Zn^{2+} ions (conduction band, normally empty) and the other band for electrons on O^{2-} ions (valence band, normally full). Surface states arise if the surface Zn^{2+} has sufficiently higher electron affinity than the bulk Zn^{2+}. That is, if

$$\alpha_B{}' > \alpha_B \qquad (1a)$$

or

$$\alpha'_{Zn^{2+}} > \alpha_{Zn^{2+}} \qquad (1b)$$

then there are surface state levels somewhat below the conduction band which can capture electrons (acceptor states). If the surface O^{2-} has a lower electron affinity than the bulk, that is, if

$$\alpha_A > \alpha_A{}' \qquad (2a)$$

or

$$\alpha_{O^{2-}} > \alpha'_{O^{2-}} \qquad (2b)$$

then there are surface state levels above the valence band which can give up electrons (donor states). Such surface states identified with a different α value at the surface are called Tamm states. If the crystal is monatomic (A is the same species as B), there is only one band and one solves to determine if there are localized surface states outside this band. It turns out they are present if

$$(\alpha' - \alpha)^2/\beta^2 > 1 \qquad (3)$$

that is, if there is a large difference in magnitude between the Coulomb integral for the surface atom and that for a bulk atom, sufficient to overcome the homogenizing influence of the exchange integral β, a Tamm surface state appears out of the band.

In the case of an adsorbed species the general principles are similar but the mathematics become more complex, and the existence condition for surface states becomes more complex.

In the crystal potential, or Kronig–Penney, model, Bloch wave functions are used instead of atomic wave functions. Again it is possible to define conditions for the existence of Tamm states. One added possibility which can be accounted for with this approach is that the bands can be so broad that they cross. Then another type of surface state forms—the Shockley state.[17] Koutecky[18,19] interprets these Shockley states as free valences on the surface. He notes that they only appear when the exchange energy is very high relative to the Coulomb integral. The electrons associated with these free valences have, so to speak, no neighbors to interact with and for these therefore the effect of this large exchange energy is missing, and they have an energy level out of the band region.

From the above models one can extract a qualitative picture of what to expect in the way of surface states. Tamm states or states due to chemisorbed species are expected if the electron affinity of the surface species differs substantially from that of the bulk species. Such a substantial difference is expected for the surface ions of ionic crystals, since the electrostatic potential at a surface ion differs from that of a bulk ion, and in this case one expects surface cations to provide acceptor levels near the conduction band and surface anions to provide donor levels near the valence band. Tamm states are also expected when foreign species are on the surface, but in this case the location of the states and their occupancy (before electron transfer) depend on the chemical identity of the species or the complex formed. Shockley states are expected in some cases, particularly for covalent crystals.

If one wishes to calculate quantitatively from such quantum mechanical arguments the surface state energy levels to be expected, the exercise is disappointing, for the results are extremely sensitive to the details of the surface potential well. One calculation[12,20] that illustrates this clearly is the Mathieu problem, a particularly simple variation of voltage with distance in the crystal. Here the potential in the "lattice" is made a cosine function, so the mathematics can be solved analytically. The surface is introduced as a potential barrier at a distance x_0 from one of the potential maxima, as shown in Figure 1. The effective shape of the surface potential well is varied by varying x_0 from zero to π. It is found that the energy and even the existence of surface states in a given forbidden gap are sensitive functions of the value assumed for x_0. The "mathematical" difficulty—the sensitivity of the results to the details of the potential well chosen—parallels very closely the difficulty from a chemical point of view in describing the adsorption site, so for the near future a quantum mechanical picture can be expected to yield only qualitative

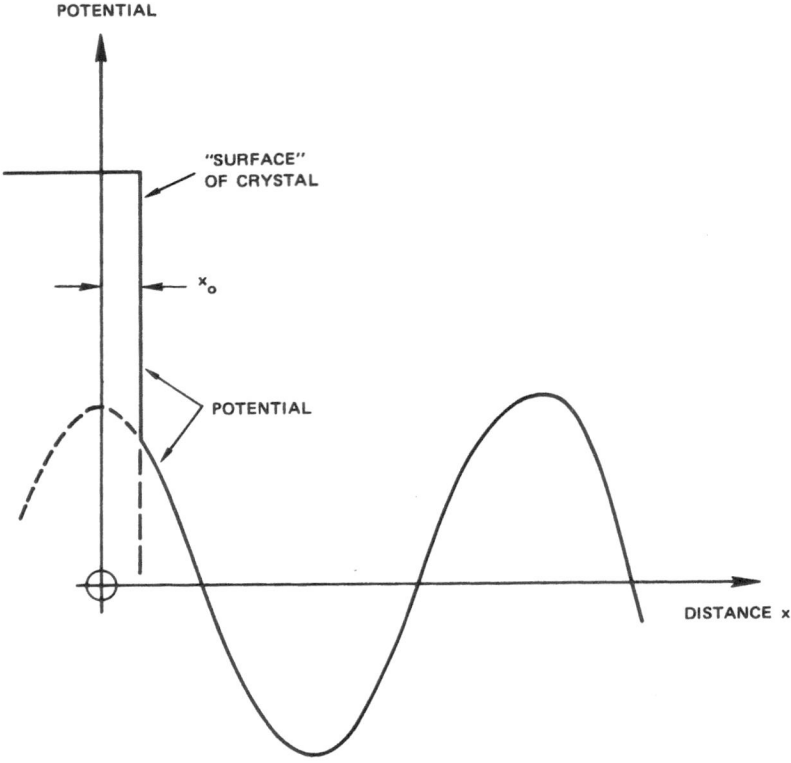

Fig. 1. Potential vs. distance from the crystal surface, as used in the Mathieu problem.

features.[21] Even for surface states on the clean surface the problems are very serious, for not only must reasonable potential variations α and β be assumed for the calculation, but also the geometry is poorly defined. Reconstruction of the surface structure (the shift of surface atoms from their expected lattice positions) occurs, and experimentally it was found by Henzler[22] that different pretreatment of the sample can cause different reconstruction and this leads to different densities of surface states.

2.3. Semiclassical Description of Surface States

With the varied forms of adsorption sites presented by solids to surface species, there are many possible local chemical interactions between surface groups and the solid. Any simple theory cannot hope to encompass all the possible complexities. The models available

are limited to very simple surface species and sorbate/sorbent interactions. Each author has attempted to estimate the change in electronic
energy level of an adsorbate as it moves to the solid surface from an
environment where its energy levels are known. For example, Green
and Lee[23] start with the electron affinity or ionization potential of
the molecule in the gas phase and calculate the energy level of the
adsorbed ion from this, the work function of the solid, and expressions
for adsorbate/solid interactions with the solid treated as a uniform
dielectric. Mark and his co-workers[24–26] develop the Madelung
model, again starting with the electron affinity of the gaseous species,
and analyze the electrostatic effect as the molecule approaches the
surface, treating the solid as an array of point charges. Another
method is to start with a picture of the adsorbate in solution,[27] with
knowledge of its properties obtained from electrochemistry, and
estimate the changes as the species moves to the surface. In this
section we attempt briefly to review these approximations in order to
develop a general understanding of the predictability of surface state
energies. In addition some of the chemical properties of the surface
groups will be explored in order to suggest how these properties are
reflected in a surface state model.

The analysis of clean surfaces by Levine and Mark[24] and the
analysis of adsorbates by Mark[25] provide an interesting bridge
between the quantum mechanical models discussed in the preceding
section and the semiclassical models. The model is confined to ionic
crystals and assumes the primary gas/solid interaction is due to the
electrostatic forces between point charges on lattice sites and the
charge on the surface species under study. In essence it is the Madelung
theory, with a Madelung potential calculated at a surface site—either
the potential at the site of the surface lattice ion when calculating
surface state energies on a clean surface, or at the site of the adsorbed
ion for surface states due to adsorbates. This potential is the potential
due to both the positive and negative ions, summed over all the ions in
the lattice, except the one at the site under study.

The calculations show an important difference depending on
whether the site is within or above the top layer of lattice ions. The
results for sites in the top layer of lattice ions show similarity to the
quantum mechanical theories above—as it should, of course, since both
cases describe the end ion of a chain (or crystal) of lattice ions. To
illustrate the behavior, consider Figure 2, showing schematically the
electron affinity of an oxygen and a zinc ion (the electron energy levels
in the neighborhood of a ZnO lattice). A gas-phase Zn^{2+} ion has a

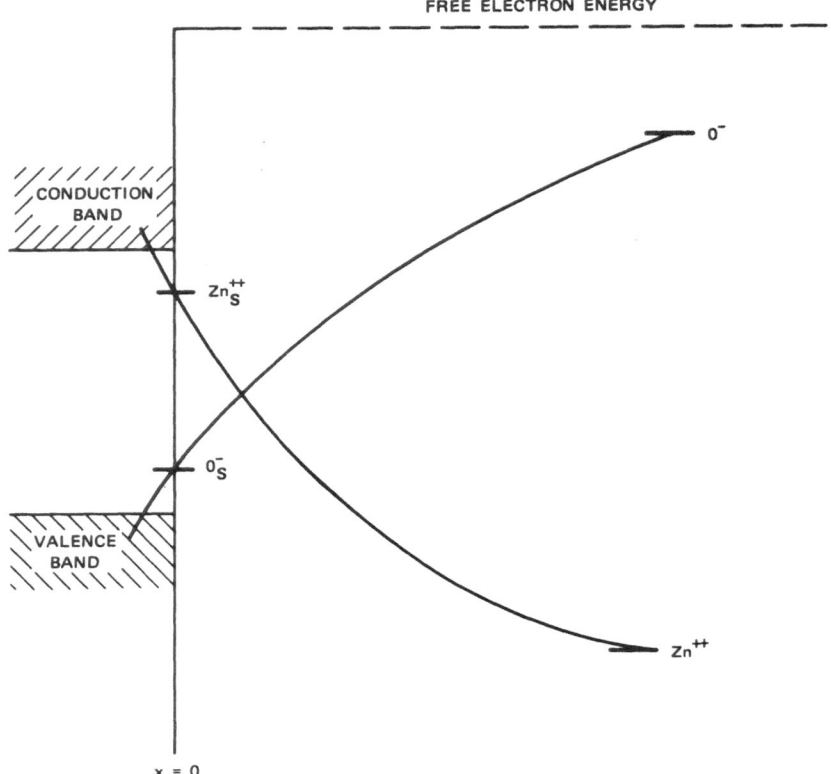

FREE ELECTRON ENERGY

Fig. 2. The electron affinity of zinc and oxygen ions as they approach their
lattice positions in the ZnO crystal.

very high electron affinity, and thus a very low energy level. As it nears
its lattice site in the ZnO surface plane $x = 0$ it obtains O^{2-} neighbors,
and these, together with the other ions in the lattice, provides a
potential at the adsorption site repulsive to electrons, so the electron
affinity of the adsorbing Zn^{2+} is less, and the energy level is higher.
If the ion is in the surface plane, as indicated in the figure by the
subscript s, it will have several O^{2-} ions as nearest neighbors, and the
repulsive potential will be greater, so the energy level will be even
higher. Finally, if the Zn^{2+} ion is within the crystal, at a bulk lattice
site, it has a full quota of O^{2-} nearest neighbors, the repulsive potential
is maximum, and the energy level is at its highest. If we equate the
energy level of the bulk Zn^{2+} ion with the energy of the conduction
band edge (neglecting band broadening for this simple argument),
then it is seen that the energy level of the surface ions Zn_s^{2+} is lower,
providing acceptor states somewhat below the conduction band. A

Zn^{2+} cation beyond the last surface layer corresponding to an adsorbed ion would have an even lower energy level. Mark shows that if only the Madelung potential is taken into consideration, the energy level of an ion on an adsorption site differs only slightly from the energy level of the gas-phase ion.

In summary, the Madelung model predicts a spectrum of energy levels for surface states on an adsorbate-free surface analogous to those derived quantum mechanically, with acceptor states just below the conduction band and donor states (by analogous arguments to the above) just above the valence band. These surface states are associated with the surface layer of the lattice. If adsorbed species are present, new surface states are obtained the energy level of which is the energy level of the isolated species, slightly modified by the Madelung potential as it nears the ionic lattice.

This general model for the clean surface, with acceptor surface states just below the conduction band and donor surface states just above the valence band, appears to describe qualitatively the observations on many ionic compound semiconductors. For example, Swank[28] examined the band structure for a series of cleaved ionic crystals, using photoelectron emission, contact potential, and surface photovoltage. He found that electron capture occurred at the surface of *n*-type crystals and hole capture on *p*-type crystals. This is in qualitative agreement with the Madelung or quantum mechanical predictions, where with an *n*-type crystal electron capture is expected, as in Figure 3(a), whereas with a *p*-type crystal (Figure 3b) the donor levels are suitably located to capture holes (give electrons to unoccupied levels in the valence band). Figure 3 indicates the band bending associated with the charged surface states, which was the parameter Swank actually monitored; the double layer analysis justifying this picture of band bending is discussed in Section 3.

The Madelung approach, like the quantum mechanical analysis, is extremely valuable from the point of view of obtaining an insight into the origin of surface states (on ionic crystals) but is of little use in predicting values for surface state energy levels. It suffers from two factors: First, it does not account for chemical interaction at the surface, including local chemical bonding and local reconstruction (shifts of surface atoms in the surface plane from their expected sites), and second, the value obtained is extremely sensitive to the ionicity of the compound. It is possible that if one surface state were measured to calculate the effective ionicity, one could then make some predictions for other surface states, but such has not yet been done.

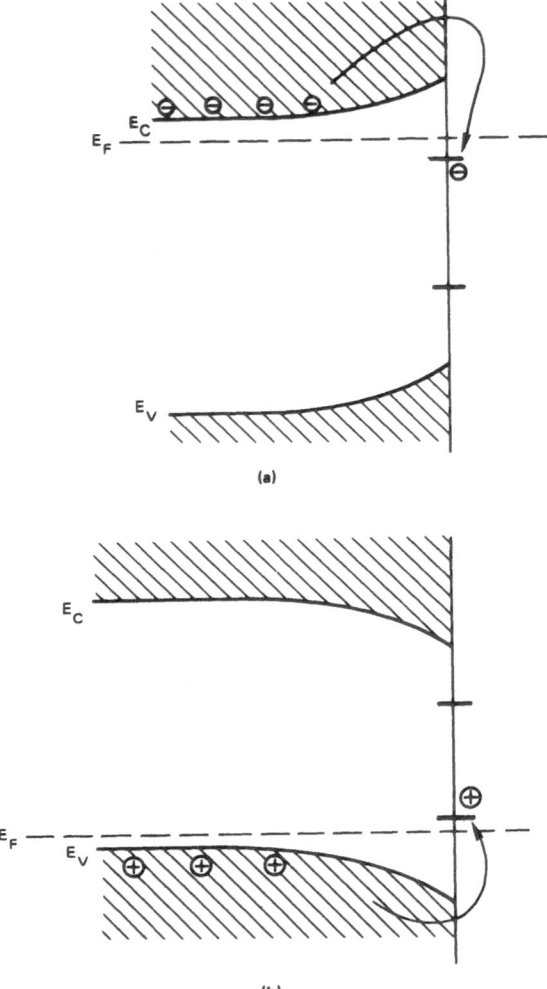

Fig. 3. Carrier capture by surface states on cleaved ionic crystals. (a) Electrons are captured on the acceptor surface states if the crystal is *n* type. (b) Holes are captured on the donor states if the crystal is *p* type.

Green and Lee[23] suggest a similar approach, the major difference being that instead of the Madelung theory to calculate the electrostatic energy upon adsorption, they use more general models, based on image forces and on induced dipoles in the lattice atoms and the adsorbate. This broadens the applicability of the approach to include nonionic semiconductors.

It is clear that these crude approximations do not describe the complete picture. For example, Green and Lee conclude that the only gas-phase species with a favorable heat of adsorption is oxygen, when adsorbed as O_2^-. Other species do become chemisorbed, of course, so clearly these analyses of adsorption are not detailed enough, and must be extended to include more complex chemical processes, such as dissociation and local adsorbate/surface bonding. The extension of such analysis to include such chemical considerations has not been considered rewarding enough as yet.

The other approach is to choose adsorbates where little local interaction is expected. For some cases nonvolatile, deposited ions can be chosen which will lead to much simpler interaction than gas-phase adsorbates. The analysis of ions deposited from solution has been treated by Morrison.[27] To avoid chemical complications, stable complexes, with the ligand groups forming a shell to prevent local sorbate/sorbent interactions, were preferred. Because of difficulties in obtaining the necessary parameters appropriate to the Madelung approach, a surface state estimation starting from the measured electrochemical properties of the species in solution (the standard redox potential was used.

The electrochemical approach is a development of a concept of Beck and Gerischer,[29] who suggested that for electron transfer from a semiconductor to an ion in solution one could expect the energy level associated with the ion to be related to the standard redox potential associated with the ion. That is, if the electron transfer process involves reduction of the ion X^{2+} to X^+ or X, the unoccupied energy level should be related to the redox potential of the couples X^{2+}/X^+ or X^{2+}/X, respectively. Morrison showed that such a relation should be valid only for one-equivalent couples, and that the relation was particularly simple for the case when the chemical environment was similar for the two ions in the couple. That is, the simple approximation applies when there is no large change in the chemical form (no dimerization, for example) or in the solvation spheres after a change in the oxidation state. In this case the energy level in solution (which is equivalent to a surface state energy level) should obey the expression

$$E = E^0 + \text{const} \tag{4}$$

where E^0 is the standard redox potential for the one-equivalent couple. The constant, essentially locating the bands of the semiconductor relative to the redox potential scale, is determined by experiment.

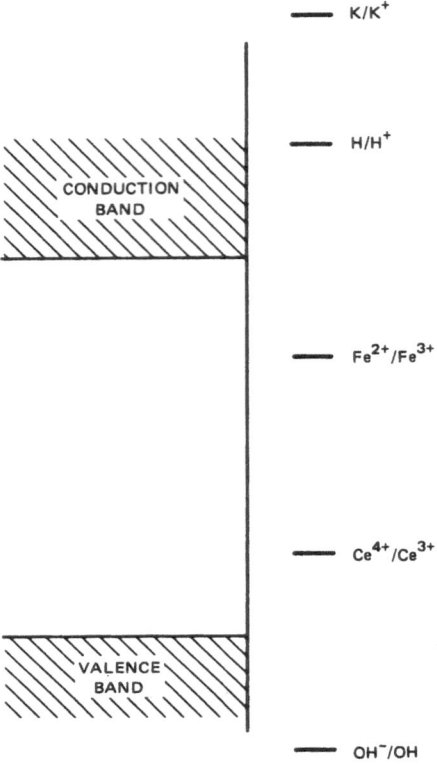

Fig. 4. Expected relative energy levels
of ions near a semiconductor surface.

The qualitative features of Eq. (4) are shown in Figure 4. Here, as in the rest of this chapter, where the chemistry of surface states is to be discussed, we will use the notation that surface state A^-/A^{2-} is the surface state associated with the redox couple A^-/A^{2-} where the redox couple can exchange electrons and holes with the semiconductor according to

$$A^- + e \rightleftarrows A^{2-} \tag{5}$$

or

$$A^- \rightleftarrows A^{2-} + p \tag{6}$$

Thus the surface state A^-/A^{2-} is the chemical species A^- if the surface state is unoccupied by an electron and is the species A^{2-} if the surface state is occupied by an electron.

As indicated then in Figure 4, a strong reducing agent such as potassium, with a high E^0, has a high electronic energy level, tending to inject its electrons into the semiconductor. In the simplest model the same energy level, now unoccupied, obtains for K^+, and its high energy means electrons from the solid are not easily captured. (K^+ is not easily reduced.) Other redox couples are shown, with lower energy levels corresponding to a lower redox potential.

In actual practice, of course, there will often be local interaction between the solid and the adsorbate (for example, the formation of a covalent or coordination bond between the sorbate ion and a surface group on the sorbent). Such interaction will affect the energy levels, causing large divergence from the simple picture of Figure 4. For example, if the semiconductor is germanium, the formation of a covalent Ge—H bond, stabilizing the electrons, would substantially lower the energy level of hydrogen atoms.

The application of these concepts to the gas–solid interface is a more demanding extrapolation.[30] The requirement is the same, namely that there be no large shift in chemical environment, now going from the ion in aqueous solution to the ion adsorbed or deposited on the surface. However, it is less likely that the requirement can be approached in this case. In solution the ion is solvated, having interacted with protons or hydroxyl ions and having a sphere of polarized water molecules associated with it. On the surface the ion is adsorbed, and some local adsorption interaction with the solid surface can usually be expected.

For a few particular simple cases the extrapolation of the electrochemical approach to the analysis of the surface states associated with adsorbed ions may be a reasonable approximation. This is the case where the ion has a strongly bound shell of ligand groups, which provide a constant environment for the ion relatively unaffected by such details as whether the complex is in solution or adsorbed on the surface. Thus there is evidence that iron if complexed by cyanide [the $Fe(CN)_6^{-4}/Fe(CN)_6^{-3}$ couple] shows the same energy level on ZnO whether in liquid or deposited on the solid. (There are apparently temporal fluctuations in the energy level with the ion in the liquid phase, as discussed in the next section.) Another example is hexachloroiridate, the couple $IrCl_6^{-2}/IrCl_6^{-3}$. There is evidence that the conduction band of ZnO is near the iron cyanide level, and the iridium hexachloride level is about 0.8 eV below, in agreement with the prediction of the simple model (E^0 for hexachloroiridate is about 0.7 V below E^0 for iron cyanide). A study of such species can be expected to provide valuable information about the behavior of simple surface states.

If the assumption is not valid that the chemical environment of the species is unchanging upon change of phase (upon adsorption from the gas phase in the Madelung approach, or upon adsorption from the liquid phase in the electrochemical approach), then the adjustment of surface states due to chemical interaction must be included in the calculation, as discussed in the next section.

2.4. The Chemical Behavior of Surface States

As described in the preceding section, to describe the chemical properties of the adsorbate and the local chemical interaction between the surface species and the solid can require a major change in the surface state picture. Such "chemical" effects will not only occur with adsorbates but can occur on the clean surface, as shown by the observation of reconstruction of surface atoms (a shift in the atomic positions of the atoms in the surface plane because, having no neighbors normal to the plane, they interact more strongly with each other). The purpose of this section is to examine the modification of the simple surface state models required to include modest chemical interaction and transformation steps.

Three types of interaction will be discussed. The first is solid/adsorbate interaction (or in the case of the clean surface those interactions leading to reconstruction), and how this influences the surface state model of the adsorbate. The second is the interaction between the surface state and an adjacent polar medium, either a polar liquid phase, a polar coadsorbed species, or a polar semiconductor sorbent. The third is the two-equivalent behavior of some chemical species, and how such behavior can be described using surface states.

The first interaction for discussion is the influence on surface states of local chemical bonding between the surface state group and the solid. We will not attempt a description of the types of bonds possible, since this would involve a review of bonding theories of chemistry. In principle, traditional methods (molecular orbital theories, crystal field theories, ionicity theories) of the estimation of single bond energies[31,32] must be extended to describe the possible multiple interactions at the surface.[33] In Refs. 31 and 33 such problems are discussed, emphasizing strictly covalent bonding. If a strong local sorbate/sorbent interaction occurs, there will be two effects on the surface state picture. First, the overall complex must now be considered the surface state, which may have an electronic energy level almost unrelated to the energy levels of the isolated adsorbate. Second, the bonding will usually be different depending on whether the surface state is occupied or unoccupied by an electron, and if the bonding is different, there will be a shift of ion position and a shift in the surface state energy.

This shift in ion position is a key factor, since it relates to the Franck–Condon principle. In the analysis to follow we will consider the surface group to shift from the configuration S_1 as the favorable configuration (or site) when unoccupied by an electron to the configuration S_2 as the favorable configuration when the group is occupied by an electron. Such a Franck–Condon configurational shift can be expected either with Tamm states (capture of an electron on a Tamm state may cause "reconstruction" on a microscopic scale), or can be expected with adsorbed ions (an H atom will interact with the surface quite differently than an H^+ ion).

If we designate the surface species as A^- when unoccupied and as A^{2-} when occupied, electron exchange can be written in chemical notation:

$$e + S_1 \cdot A^- \rightleftarrows S_1 \cdot A^{2-} \tag{7}$$

$$S_1 \cdot A^{2-} \rightleftarrows S_2 \cdot A^{2-} \tag{8}$$

or

$$S_2 \cdot A^{2-} \rightleftarrows S_2 \cdot A^- + e \tag{9}$$

$$S_2 \cdot A^- \rightleftarrows S_1 \cdot A^- \tag{10}$$

The valences indicated are for convenience in illustration, and do not affect the generality of the analysis. Here we denote $S_1 \cdot A^-$ as the most stable complex of A^- on the surface. Thus S_1 can be visualized as a particularly attractive surface configuration with the surface state unoccupied, or a particularly attractive surface site for A^-. The symbol S_2 is the corresponding configuration with the surface state occupied, or "site" for A^{2-}, and we assume it is not the same as S_1. Thus Eq. (7) represents electron transfer to the unoccupied surface state in its dominant form. By the Franck–Condon principle the electron transfer is rapid and the ion reorientation slow, so Eq. (8) is the chemical transformation representing the reconstruction of the surface now that the surface group is reduced. At equilibrium Eq. (7) can be represented by a Fermi distribution:

$$[S_1 \cdot A^{2-}]/[S_1 \cdot A^-] = \exp\{-(E_{t1} - E_F)/kT\} \tag{11}$$

where E_{t1} is the surface state when the surface group is associated with the site S_1, and the square brackets represent concentrations. Equation (8) at equilibrium can be described by the free energy ΔG^{2-} of the reaction:

$$[S_2 \cdot A^{2-}]/[S_1 \cdot A^{2-}] = \exp(-\Delta G^{2-}/kT) \qquad (12)$$

where ΔG^{2-} is negative, since the S_2 configuration is by definition the low-energy configuration when the surface group is occupied.

Similarly the electron transfer reaction Eq. (9) must be described at equilibrium by the Fermi distribution

$$[S_2 \cdot A^{2-}]/[S_2 \cdot A^-] = \exp\{-(E_{t2} - E_F)kT\} \qquad (13)$$

and the chemical transformation Eq. (8) by its free energy

$$[S_1 \cdot A^-]/[S_2 \cdot A^-] = \exp(-\Delta G^-/kT) \qquad (14)$$

where E_{t2} is the surface state when the surface group is associated with the site S_2 and ΔG^- is the free energy (also negative) for switching sites with an unoccupied surface state.

The solution of these equations shows that

$$E_F = \tfrac{1}{2}(E_{t2} + E_{t1}) + \tfrac{1}{2}(\Delta G^{2-} - \Delta G^-)$$
$$+ kT \ln\{[S_2 \cdot A^{2-}]/[S_1 \cdot A^-]\} \qquad (15)$$

and

$$E_{t1} - E_{t2} = -(\Delta G^- + \Delta G^{2-}) \qquad (16)$$

Effectively, from Eq. (15) the system can be described at equilibrium by one surface state E_{eff}:

$$E_{eff} = \tfrac{1}{2}(E_{t2} + E_{t1}) + \tfrac{1}{2}(\Delta G^{2-} - \Delta G^-) \qquad (17)$$

For kinetic considerations (capture cross section, etc.), however, it must be recognized that the effect of chemical interactions is to separate the surface state into two levels, one dominantly unoccupied, one dominantly occupied, with the unoccupied level above the occupied by an amount determined by the chemical transformation energies, as in Eq. (16). The energy levels are thus as shown in Figure 5.

The second form of chemical interaction which is expected to be common is the effect of a polar medium or polar absorbate adjacent to the surface state. The effect can be described generally as follows. Either before or after electron transfer to the surface state the medium will change its polarization to reflect the new charge distribution. Thus, for example, electron transfer to a surface state will be followed by dipole reorientation in the dielectric medium, reflecting the arrival of

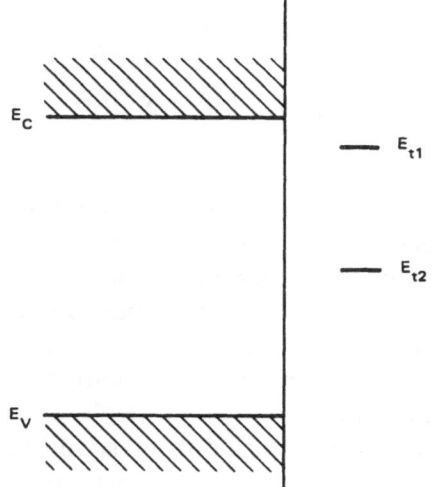

Fig. 5. Splitting of the surface state energy levels by Franck–Condon effects. E_{t1} is the energy of the unoccupied surface state with the ions in their lowest energy configuratio.. E_{t2} is the energy of the occupied surface state with the ions in their lowest energy configuration for this valence state.

the electron. If we let the symbol S_1 represent the dipole configuration (polarization) of minimum energy for the surface state unoccupied, and S_2 the dipole configuration of minimum energy when the surface state is occupied, Eqs. (7)–(17) apply, where now the ΔG's represent the change in free energy associated with a change the polarization between S_1 and S_2 with the surface state occupied and unoccupied, respectively. Thus one result of the presence of such an adjacent polar medium is to split the energy levels of the occupied and unoccupied states, as with other chemical processes.

Another result of the local sorbate/sorbent interaction which is particularly important where there is a polar medium relates to the kinetics of electron transfer to and from the surface state. Normally in solid-state electronics one considers only phonon interaction when examining the rates of electron exchange from one energy level to another. With the possibility, as on the surface, of configurational changes, one has to consider the possibility that the energy level itself shifts due to thermal fluctuations of the ion position. Then the kinetically active process in surface-state capture may not be associated with phonons in the solid. For example, the kinetically most rapid route for electron capture, in lieu of Eqs. (7) and (8), may be

$$S_1 \cdot A^- \to S_3 \cdot A^- \tag{18}$$

$$S_3 \cdot A^- + e \to S_s \cdot A^{2-} \tag{19}$$

$$S_3 \cdot A^{2-} \to S_2 \cdot A^{2-} \tag{20}$$

where S_3 represents a special ion configuration (an "activated intermediate") such that the electronic levels are equal [there is no energy change during reaction (19)]. Such behavior can occur when any of the local chemical effects are important, but it can be shown that when a highly polar medium is present the "configuration coordinate" route has a much lower activation energy than the phonon interaction route.

Analysis of electron transfer from solids to ions in aqueous solution has been made by Marcus.[34] Fluctuations in the orientation of water dipoles (fluctuations in the polarization of water treated as a dielectric medium) were shown to produce large temporal fluctuations in the electronic energy levels for electrons on ions. The model has been expanded by Gerischer[35] in several studies, where the temporal fluctuations have been treated as a surface state "band." There is recent evidence,[36,37] as will be discussed in Sections 5.3 and 5.4, confirming the wide fluctuations of energy level in aqueous solution.

Such temporal fluctuations in surface state energy levels will also be expected[38] due to coadsorbed water from the gas phase, or when the surface states are on a highly polar semiconductor such as TiO_2. The mathematics describing the behavior is much more complex, unfortunately. However, much early work on the influence of water on semiconductors, where electrons were found to be injected upon adsorption of water, must be reexamined. The influence of adsorbed water may often be to cause temporal fluctuations in the surface states of, and thus permit electron injection from, adsorbed oxygen, O_2^- or O^-,[38] rather than to provide direct electron injection.

A third chemical behavior which must be described by an adequate surface state representation is the multiequivalent behavior of some chemical species. Many chemical species are unstable in certain oxidation states, and can be termed "two-equivalent." For example, Mg is stable as Mg or Mg^{2+} but not as Mg^+. However, hole capture by Mg or electron capture by Mg^{2+} will result in the formation of the unstable radical ion, and a further charge transfer process must occur to form a stable species. Unless it is in a very unusual environment, Mg^+ will either capture an electron to become zero-valent, or give up an electron to become Mg^{2+}. This instability must be described by the surface state model in order to describe two-equivalent foreign species. Not that two-equivalent behavior is necessarily restricted to foreign adsorbates; even Tamm-type surface states may show this behavior, as will be discussed for the case of germanium.

The detection of the two-equivalent behavior of surface states has been made primarily at the semiconductor/electrolyte interface, where the reduction or oxidation of the radical appears as added electrolytic current. For example, with an *n*-type ZnO electrode there is negligible anodic current in the dark, since electrons cannot be injected into the ZnO conduction band at the surface. Upon illumination under the anodic potential an anodic current due to the photo-produced holes is measured. The holes are produced in the bulk ZnO and move to the surface, registering an anodic current. The hole current to the surface is maintained constant[39] by maintaining the light intensity constant. The addition of arsenic (III) to the electrolyte causes the current to increase substantially, and this is due to the following reactions:

$$p + As^{3+} \rightarrow As^{4+} \tag{21}$$

$$As^{4+} \rightarrow As^{5+} + e \tag{22}$$

Here hole capture by the As^{3+} leads to the unstable radical, which, there being no available electrons to capture, releases an electron into the conduction band of the solid.

Such behavior can be described[40] using a surface state model, but since two electrons are involved per surface group, two surface states must be included. The two surface states required for the description in the arsenic example are As^{3+}/As^{4+} and As^{4+}/As^{5+}, as shown in Fig. 6. If the surface state is in the chemical form of As^{3+}, the lower surface state is occupied (and the upper surface state does not exist yet since neither As^{4+} nor As^{5+} is present). Capture of a photoproduced hole according to Eq. (21) empties the surface state As^{3+}/As^{4+}, and since the ion is in the form of As^{4+}, the upper surface state is in existence in its occupied form. From Figure 6, where it is emphasized that the ordinate in the band diagram is the energy of an electron, it is obviously an energetically favorable process to remove the As^{4+} either by electron capture on the lower state (returning the species to As^{3+}) or by electron injection (or hole capture) from the upper surface state (oxidizing the species further to As^{5+}). If the radical is oxidized to As^{5+}, the upper state is unoccupied and the lower state is nonexistent.

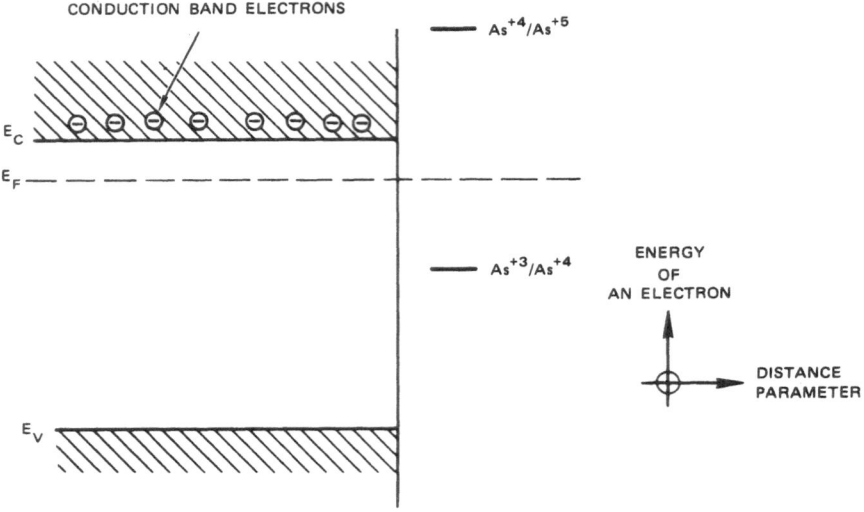

Fig. 6. Two surface state energy levels required to describe a two-equivalent chemical species, as illustrated by arsenic. The energy required to remove the first electron from As^{3+} is more than the energy required to remove a second electron.

It is clear from Figure 6 that for two-equivalent species the description must always include both energy levels. If one energy level of the surface group is active (can capture or inject an electron or hole), then the other must be, too. This situation arises because it requires less energy to transfer the second electron than the first. Thus if by any mechanism (optical, thermal, electrical) electron exchange with one level is stimulated, then the possibility of electron exchange with the other is unavoidable.

Some of the consequences of this complex behavior are described in Ref. 40. The most interesting is the minority carrier trapping to be expected. After reactions such as Eqs. (21) and (22), capturing holes, the reverse reaction can be highly improbable, because of the difficulty of capturing an electron on the upper energy level shown in Figure 6.

Such two-equivalent behavior is not necessarily restricted to foreign adsorbates. For example, Brattain and Garrett[41] found that as holes are captured at the germanium surface, oxidizing the germanium from Ge^0 to Ge^{4+}, electron injection is observed. Again the effect is noted during electrochemical measurements. This can be interpreted in a manner analogous to the preceding discussion.[30]

Further observations and theory of the behavior of two-equivalent surface states are discussed in Section 5.2.

2.5. Adsorbed Gases as Surface States

Gases adsorbed from the ambient atmosphere provide an important source of surface states, and significant differences exist between the analysis of such surface states and the analysis of non-volatile surface states. The two most important differences relate to the density of states and to the inevitable complexity of the chemical behavior. The importance from a physical point of view of surface states due to adsorption has been recognized since the physics of semiconductor surfaces was first studied. In many of the early studies[1,2] a systematic cycle of gaseous ambient, between dry ozone as an oxidizing ambient and wet nitrogen as a reducing ambient, was commonly used to vary the surface properties. The cycle of gases was sometimes known as the "Brattain–Bardeen cycle."[42] The gaseous ambient (and adsorption) was found to dominate the surface properties, but the behavior was too complex for quantitative analysis, and the analysis of the studies of the "slow states" due to adsorption[43–45] has always lagged the analysis of "fast states," as the surface states were termed which are found at the solid/solid interface (such as surface states between Ge and a GeO_2 layer) where adsorption/desorption is not a problem.

The equilibrium adsorption of a polyatomic molecule to form a surface state may be represented by

$$nS_1 + X_n(g) \rightleftarrows nS_1 \cdot X \tag{23}$$

where Eq. (23) is the adsorption of the neutral molecule X_n, with dissociation of the molecule into atoms if appropriate, followed by local sorbate/sorbate interaction at the surface leading to the surface configuration $S_1 \cdot X$. Then, with $S_1 \cdot X$ identified with $S_1 \cdot A^-$ of Eq. (7), the analysis of Eqs. (7)–(17) applies.

There is an important difference between the expression for the Fermi energy E_F, Eq. (15), in the case of equilibrium adsorption and the expression in the case of equilibrium occupation of nonvolatile surface states. In the case of equilibrium adsorption of the neutral molecule the concentration $[S_1 \cdot X]$ becomes essentially independent of electron transfer processes. The value of $[S_1 \cdot X]$ is determined by an appropriate adsorption isotherm or isobar describing adsorption according to Eq. (23). Then $[S_1 \cdot X]$ is a constant of the expression (15) as opposed to the case of nonvolatile surface states, where the total surface state density is the constant.

If Henry's law adsorption can be assumed (coverage proportional to pressure) and the surface state is one-equivalent, then the solution of Eqs. (15) and (23) can be manageable; however, often the chemistry is not so simple.

As an illustration to show the possible chemical complexity of adsorption on semiconductors, the adsorption of oxygen can be considered. With $O_2(g)$ a gas-phase molecule and $O_2(s)$ a sorbed molecule, the adsorption steps can be written:

$$O_2(g) \rightleftarrows O_2(s) \tag{24}$$

$$e + O_2(s) \rightleftarrows O_2^- \tag{25}$$

$$e + O_2^- \rightleftarrows [O_2^{2-}]^* \tag{26}$$

$$[O_2^{2-}]^* \rightleftarrows 2O^- \tag{27}$$

$$e + O^- \rightleftarrows O^{2-} \tag{28}$$

where the steps are simplified in that possible local interaction with surface groups is ignored. Equation (24) is described at equilibrium by an adsorption isotherm, Eq. (25) by a surface state. The form $[O_2^{2-}]^*$ in Eq. (26) is intended to represent the kinetically most active "activated intermediate" for the reduction/dissociation of O_2^- and the reverse transition, the oxidation/dimerization of O^-. Then Eqs. (26) and (28) at equilibrium are represented by surface states (equilibrium defined by the Fermi energy), while Eq. (27) is defined by the free energy of a chemical transformation. Clearly the analysis of adsorption can be very complex, and when this analysis is further complicated by the fact that the Fermi energy is a function of the total surface charge (through the double layer, as discussed in the next section) it is understandable that most analyses of adsorption on semiconductors have required oversimplification.

Some recent inroads have been made in describing adsorbed oxygen on semiconductors through various methods of separating the O_2^-, O^- and O^{2-} forms. For example, the form O_2^- has an easily identifiable triplet in its electron spin resonance. In other studies Chon and Pajares[46] have shown by combining Hall and adsorption measurements on ZnO that at oxygen pressures the order of microns, below about 200°C adsorption occurs as O_2^- and above about 200°C adsorption as O^- predominates. Such knowledge is of use for studies[47] of the various forms independently as surface states.

3. The Double Layer at the Semiconductor Surface

3.1. The Role of the Surface Double Layer in Physical and Chemical Processes

With surface states present there normally will be an excess of charge at a semiconductor surface. There can be an excess of electrons or a deficiency of electrons. Since the system as a whole must be electrically neutral, the surface state charge must be compensated, leading to an electrical double layer. If the semiconductor/liquid interface is under consideration, the compensating charge can be either in the liquid or the semiconductor, but if the semiconductor/gas interface is under consideration, the charge must be wholly compensated in the semiconductor. We will concern ourselves primarily with charge compensation in the semiconductor. In Section 3.2 the mathematical analysis will be developed, but first general considerations will be explored, relating the double layer to chemical and physical processes at the surface.

Consider a semiconductor upon which is deposited an electron-accepting surface state, so the surface becomes negative. Electrons are captured from the semiconductor; if the semiconductor is *n* type, the electrons normally come from the conduction band, leaving behind the positive donor ions to form the compensating space-charge layer near the surface of the semiconductor. If the semiconductor is *p* type, the electrons come from the valence band, leaving behind excess holes to form the compensating charge. In the first case an "exhaustion layer" or a "depletion layer" arises at the surface—a region near the surface from which the current carriers have been exhausted. In the second case an "accumulation layer" arises at the surface, where added carriers (holes) have been supplied by the surface state. In both cases a double layer forms between the excess positive charge inside the semiconductor and the negative surface charge—a double layer that repels electrons from the surface, giving the surface a more negative potential. A more negative potential in a region leads to a higher electron energy in that region. On the band picture (Figure 7) this is represented by an upward bending of the bands near the surface. Figure 7(a, b) indicates the band picture before charge transfer, and Figure 7(a', b') shows the picture after charge transfer, when equilibrium conditions have been reached and the acceptor surface states A_1 or A_2 have acquired the appropriate negative charge.

Figure 8(a, b) shows the equivalent behavior with a positively charged surface with the surface state indicated by D for donor, an

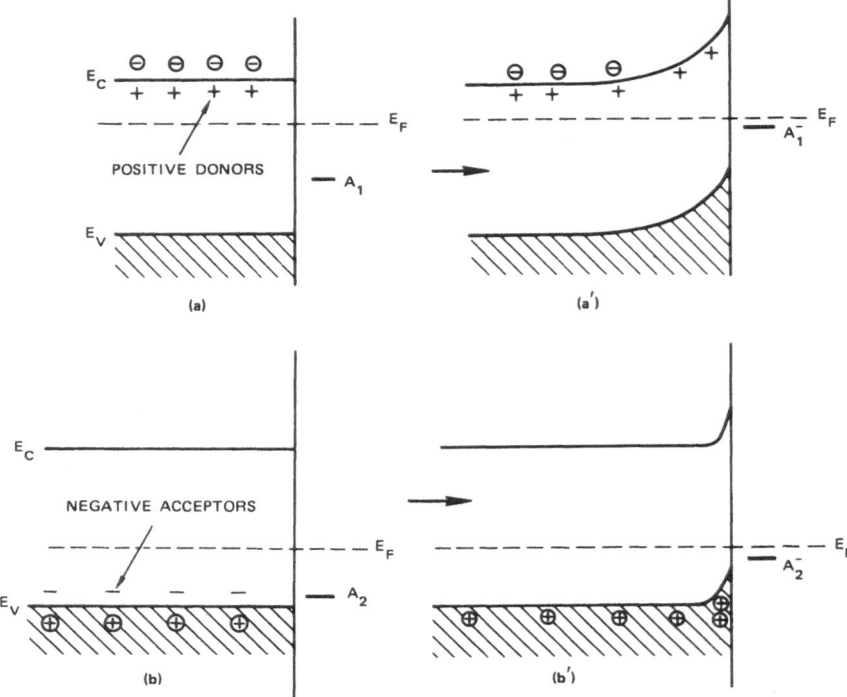

Fig. 7. Band bending with a negative charge on the surface states. The left-hand diagram in each case represents no charge transfer to surface states, and the right-hand diagram represents the equilibrium configuration.

adsorbate that tends to inject electrons. Here the positive charge at the surface is compensated by a negative charge in the semiconductor, and the bands bend down near the surface. If the semiconductor is *n* type (Figure 8a), the compensating negative charge is in the form of free electrons, and one has an accumulation layer. If the semiconductor is *p* type (Figure 8b), the holes are depleted near the surface, and the compensating negative charge is due to the acceptor ions, indicated by a minus sign in the band gap. This again is a depletion or exhaustion layer.

The third, rarer type of surface double layer, the "inversion layer," can be described with the help of Figure 7(a). If the sample is *n* type, with acceptor surface states, the bands bend up as indicated. The Fermi energy, the electrochemical potential for electrons, is, however, constant throughout the system at equilibrium. With more and more surface charge, the bands can bend to the point where the Fermi energy at the surface approaches the valence band. At some

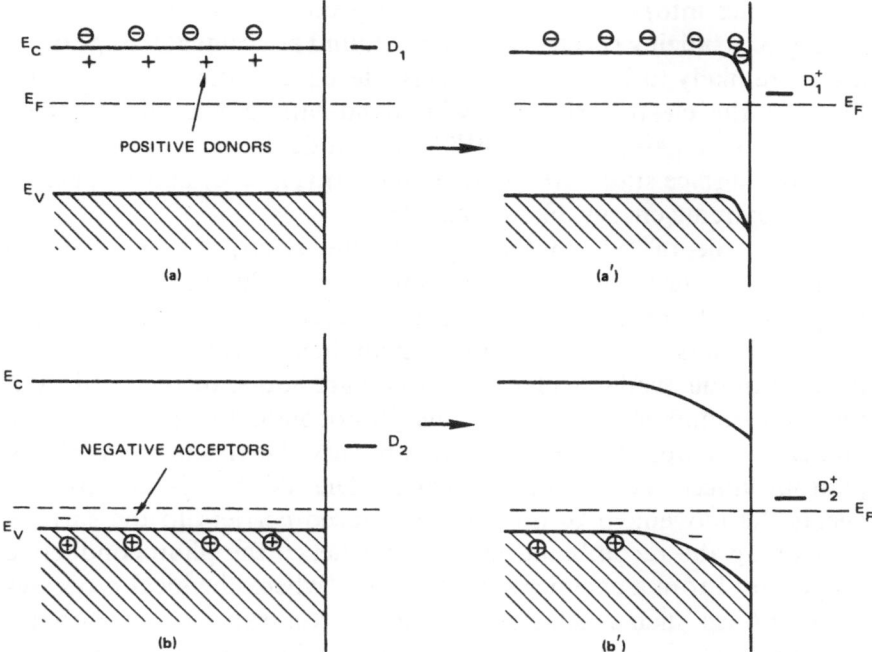

Fig. 8. Band bending with a net positive charge in the surface states.

point while charging the surface states the electrons will start coming dominantly from the valence band, leaving holes in the valence band. The surface becomes "inverted" and appears as a *p*-type conducting region. This might happen, for example, if a strong acceptor-type adsorbate, such as indicated by A_2 (Figure 7b), were deposited on the surface of the *n*-type material (Figure 7a).

A *p*-type semiconductor can also have an inversion layer where the compensating charge is due not only to negatively charged acceptors in the bulk, but also to electrons that have been injected into the conduction band near the surface. This could occur if a strong donor such as D_1 were deposited on a *p*-type semiconductor such as is shown in Figure 8(b). Here the semiconductor could behave as a *p*-type material with an *n*-type surface layer.

Of the various types of surface layer, the depletion layer is of most interest. It is qualitatively clear from Figures 7 and 8 that the depletion layer can be expected to be the most common for the wide-band-gap semiconductors that are of greatest interest in chemical studies. Very strong reducing agents D_1 or oxidizing agents A_2 are necessary to inject electrons into the conduction band or extract electrons from (inject holes into) the valence band, respectively. More common reducing or oxidizing species, such as are found as accidental impurities, are more likely to have median energy levels as indicated by A_1 and D_2. On the clean surface of wide-band-gap semiconductors also (above ~ 1 eV gap), it is found[10,26] that a depletion layer develops, since the surface states are of medium energy, like A_1 and D_2, rather than of extreme energy, like A_2 and D_1.

The role of the double layer in the electrical and chemical properties of the surface can be summarized as follows. The electrical properties of the semiconductor are influenced directly, because charge carriers (and for a thin semiconductor wafer a substantial fraction of the available charge carriers) are added or removed from the semiconductor. In addition, the photoconductive properties are influenced because the surface states are very effective in recombining photoproduced holes and electrons. The double layer strongly affects the movement of the photoproduced carriers to the surface, controlling the recombination rate and hence the photoconductive properties. From a chemical point of view, where chemical processes at the surface, such as adsorption and electron transfer to adsorbates (or from adsorbates) are of most interest there are two effects to anticipate, one equilibrium and one kinetic. As can be seen in Figures 7 and 8, comparing the diagrams on the left to those on the right, the surface states associated with the adsorbed species move relative to the Fermi energy[1,2] due to the double layer. At equilibrium, where the occupancy of the levels is described by the Fermi energy [Eq. (15)], this shift will dominate the surface coverage calculation.[48] The double layer also affects the kinetics of the carrier capture at the surface. It is seen from Figs. 7a' and 8b' that the band bending, particularly with a depletion layer, is actually an activation energy which must be overcome to move electrons (with n-type material) or holes (with p-type material) to the surface. Thus the double layer dominates both the kinetics and the equilibrium of charge exchange with surface species.

3.2. The Solution of Poisson's Equation for the Double Layer

In order to understand the character of the various types of double layer formation at the semiconductor surface, so that their influence upon adsorption and the semiconductor electrical properties can be better understood, it is of value to discuss the mathematical analysis.

The assumption will be made that the potential is a function only of x, the distance from the surface, and independent of y and z, distance parameters parallel to the surface. Poisson's equation,

$$d^2V/dx^2 = -\rho/K\varepsilon_0 \tag{29}$$

where ρ is the space charge, is solved from $x = 0$ (the surface) to $x = x$, with the boundary conditions that $V = 0$ at $x = x_0$ and $V = V_s$ at $x = 0$, and with $dV/dx = 0$ when $V = 0$ and $dV/dx = N_s/\varepsilon\varepsilon_0$ at $x = 0$, the latter arising from Gauss' theorem. Here x_0 is the inside boundary of the space-charge region, where the electric field $(-dV/dx)$ vanishes. With the bulk electron and hole densities denoted by n_b and p_b, a net space charge arises due to the potential V in the surface region (Figure 7) because the electron and hole densities are changed by $n_b(1 - e^{qV/kT})$ and $p_b(1 - e^{-qV/kT})$, respectively. Then

$$d^2V/dx^2 = (q/K\varepsilon_0)(n_b - p_b - n_b e^{qV/kT} + p_b e^{-qV/kT}) \tag{30}$$

This equation can be put in a very elegant mathematical form[2] utilizing dimensionless variables, suitable for computer solution. Here, however, since emphasis will be on the physical picture and solutions will be given for extreme conditions only, the above, more easily visualized variables in Eq. (30) will be used without change.

The above equation when multiplied by $2\,dV/dx$ can be integrated once with the boundary condition that $V = 0$ when $dV/dx = 0$:

$$(dV/dx)^2 = 2(q/K\varepsilon_0)[(n_b - p_b)V - kTn_b(e^{qV/kT} - 1) - kTp_b(e^{-qV/kT} - 1)] \tag{31}$$

Thus, for example, the surface electric field is given by the above expression with $V = V_s$. With a depletion layer the first term, $(n_b - p_b)V$, dominates—for a depletion layer on n-type material, $n_b \gg p_b$, of course, and the reverse holds on p-type material. When the second term dominates, that is, if $V_s > 0$ for n-type material or if $V_s \gg kT$ for p-type material, there are excess electrons at the surface—an accumulation layer on n-type material or an inversion layer on p-type material, respectively. Similarly, if the third term dominates, we have excess holes at the surface—an accumulation layer on p-type material or an inversion layer on n-type.

The solution of Eq. (31) with the first term dominating (a depletion layer) is found by neglecting the exponential terms. A double integration of the resulting expression with the boundary conditions above yields the Schottky relation:

$$V_s = -q(n_b - p_b)x_0{}^2/2K\varepsilon_0 \tag{32}$$

The other case of most interest is an accumulation layer (found when an electron-donating species is adsorbed on an *n*-type semiconductor). If we specify an *n*-type semiconductor with $qV/kT > 3$, for example, and use the boundary condition that $V = V_s$ at $x = 0$, Eq. (31) can be integrated to yield

$$x = (2kTK\varepsilon_0/qn_b)^{1/2}(e^{qV_s/2kT} - e^{qV/2kT}) \tag{33}$$

For n_b about $10^{18}/cm^3$ the coefficient is about 50 Å, so the potential falls off rapidly—the surface layer is very thin. This is still a much deeper surface layer than that on a metal, which will approach a few angstroms, or a Helmholtz double layer at a metal/electrolyte interface, which can also approach angstrom thickness for a concentrated electrolyte. However, the accumulation double layer at a typical semiconductor surface is much thinner than the typical depletion layer, which can be shown with Eq. (32) to be of the order of 1000–5000 Å. The relative thickness of the various types of surface layer is of interest because one can qualitatively estimate the charge involved in the double layer and thus anticipate the density of adsorbed ions or density of surface carriers for the various cases. By analogy with a simple parallel plate capacitor formulation

$$Q = CV, \quad \text{where} \quad C = AK\varepsilon_0/r_0 \tag{34}$$

where Q is the charge on each plate, V is the potential difference, and C is the capacity, given in the second equation as proportional to the area A and the inverse of the plate separation r_0, it is seen that the thinner the double layer, the greater the charge to be expected, for a similar potential difference. A reasonable double layer potential for any of the above cases is a few tenths of a volt, so with an exhaustion layer surface states will accept the order of 10^{11} or 10^{12} electrons/cm^2.

3.3. Electron Transfer to Surface States

In order to analyze a particular system and determine both the charge transfer to surface states and the associated surface double layer, one must combine the solution of Eq. (31) with the analysis of Eq. (15). Equation (31) can be solved to give the surface potential V_s as a function of the net charge in the surface states N_s. Equation (15) is the second relation between the charge transferred to the surface N_s and the surface barrier height V_s. When two relations are combined to eliminate V_s, which is done by numerical methods,[1,2] in principle a solution yielding the charge in surface states as a function of the surface state distribution and the bulk semiconductor properties is obtained. In the next section the quantitative interpretation of experimental results is discussed for systems with a fixed surface state density. However, a semiquantitative discussion of the behavior is of value. Because it is of particular interest, the case of volatile surface states (adsorbed species) will be used as an example in this semiquantitative discussion. Mark[49,50] has solved the equations for various fixed surface state densities and provides numerical solutions. However, normally the surface states due to adsorbed species are not fixed in density and/or an appropriate distribution is not known.

The influence of the double layer on the equilibrium and kinetics of ionosorption will be examined for the case when a depletion layer [Eq. (32)] develops. An n-type material with ionized donors of density N_D will be specified, with the energy of the surface states designated E_{eff}, as in Eq. (17). In the discussion Figure 7(a, a') will be used for illustration.

From Figure 7(a, a') it is seen that the position of the Fermi energy moves relative to the surface states as the double layer develops. The band bending simply reflects an increase in the electrical potential in the area, so clearly all electronic energy levels, including the conduction band edge E_{cs}, and the surface states, including the surface state energy E_{eff}, move together. Thus $E_{cs} - E_{eff}$ is fixed, and from Eq. (32) and by inspection from Figure 7(a')

$$|eV_s| = E_{cs} - E_F - \mu = q^2 N_D x_0{}^2/2K\varepsilon_0 \qquad (35)$$

where E_{cs} is the energy of the conduction band edge at the surface and μ is the energy difference between the conduction band and the Fermi energy deep in the bulk semiconductor. The total charge brought out to the surface is that originally contained in the depletion layer, so

$$N_s = N_D x_0 \qquad (36)$$

and Eq. (35) can be rewritten

$$|V_s| = qN_s{}^2/2K\varepsilon_0 N_D \qquad (37)$$

With the terminology of Section 2.5, the redox couple includes a neutral adsorbate X, which could, for example, be oxygen O_2, and a reduced form, X^-, which would then be $O_2{}^-$. Equations (15) and (17) yield

$$E_F = E_{eff} + kT\ln\{[S_2 \cdot X^-]/[S_1 \cdot X]\} \qquad (38)$$

in order that the surface state be in equilibrium. Combining this with

Eqs. (35) and (37) to eliminate V_s and (E_F) and solving for the amount adsorbed vs. the concentration of neutral adsorbed species, we find, assuming $N_s = [S_2 \cdot X^-]$,

$$kT \ln\{[S_2 \cdot X^-]/[S_1 \cdot X]\} = -q^2[S_2 \cdot X^-]^2/2N_0 K\varepsilon_0 + E_{cs} - \mu - E_{eff} \tag{39}$$

This equation is only soluble graphically, but has the following character.[4] At very low coverage of neutral adsorbed species the first term on the right is negligible, and the coverage by charged adsorbate ions is proportional to the coverage of neutral atoms. As the coverage of charged species increases (in Figure 7a' when the "surface state" E_{eff} approaches the Fermi energy) the first term on the right dominates, and the coverage by charged species abruptly saturates, the log term becoming insensitive to the amount of neutral species present, leading to

$$[S_2 \cdot X^-] = (2N_D K\varepsilon_0/q^2)^{1/2}\{E_{cs} - \mu - E_{eff}$$
$$- kT\ln([S_2 \cdot X^-]/[S_1 \cdot X])\}^{1/2} \tag{40}$$

In summary, the double layer affects the equilibrium adsorption by moving the surface states toward the Fermi energy and, as first pointed out by Weisz,[48] imposing an abrupt upper limit on the charge transfer. Inserting reasonable values into Eq. (40), it can be seen that this limit will be the order of 10^{-3}–10^{-2} monolayer of net charged species.

The calculation has been done for a depletion layer for simplicity, but the principle holds for the other type of surface layer, although the mathematics becomes more complex. For all surfaces, transfer of charge to the surface will result in a double layer resisting the further transfer. Qualitatively, because the double layer thickness is much smaller, the saturation value is higher if an inversion or accumulation layer is present. On a metal the saturation value will be still higher.

Thus it is seen that the double layer is expected to dominate the capture of carriers on surface states and the equilibrium quantity of charged adsorbed species.

Kinetic effects of electron transfer to surface states can also be controlled by the presence of the surface double layer. Much of the quantitative analysis is postponed to Section 5, where nonequilibrium effects on surface states are explored more carefully. In particular, the theory of electron capture and hole capture at nonvolatile surface states leading to recombination of excess carriers, which has been derived in detail and amply proved by experiment, will be reviewed.

Here we will discuss kinetics of adsorption. The effect of the surface barrier on irreversible charge capture, where irreversible adsorption is the most important example, was suggested at a much earlier date but still is understood only qualitatively, since other complicating factors, such as surface heterogeneity, are difficult to avoid.

As early as 1952 it was shown[51] that ionosorption, through surface double layer effects, gives rise to an activation energy increasing with coverage, and thus can "pinch itself off." With zero surface coverage (left side in Figure 7a) there is no surface barrier, since there is no double layer. But with the formation of the double layer the carriers must cross a potential barrier to reach the surface, which increases in height as the surface coverage θ increases. With an increasing surface potential barrier the rate of electron capture decreases, and it can easily occur at low temperature that the rate of carrier exchange becomes negligibly slow before equilibrium is reached. Melnick[52] pointed out that this "pinchoff" effect will lead to the Elovich equation, a rate equation often observed in adsorption phenomena:

$$d(\Delta\theta)/dt = ae^{-b\Delta\theta} \tag{41}$$

The Elovich equation arises by the following argument: If it is assumed that the rate of electron transfer to the surface species is proportional to the density of electrons at the conduction band at the surface, n_s, then it becomes exponentially dependent on the Fermi energy at the surface,

$$dN_s/dt = c_1 n_s = c_1 N_c \exp\{-(E_{cs} - E_F)/kT\} \tag{42}$$

where N_c is the effective density of states in the conduction band and c_1 is a rate constant. But from Eqs. (35) and (37), $E_{cs} - E_F$ equals a constant plus a term proportional to N_s^2,

$$E_{cs} - E_F = A + \alpha N_s^2 \tag{43}$$

Now if surface states are present on the surface prior to the adsorption measurements, then ΔN_s is no longer the total surface charge, but

$$\Delta N_s = N_s - N_{s0} \tag{44}$$

and if $\Delta N_s \ll N_{s0}$, Eqs. (43) and (44) yield

$$E_{cs} - E_F = B + b\,\Delta N_s \tag{45}$$

Inserting this into Eq. (42), it is clear that a form similar to the Elovich equation is obtained. According to Melnick, the adsorption measurements leading to the Elovich equation are often performed under conditions such that the above model may hold, i.e., there is an initial surface barrier. In the opinion of the present author, apparent experimental agreement with Elovich kinetics cannot be considered evidence for double layer pinchoff. Other effects, particularly surface heterogeneity, will also contribute. It is one of the features of the Elovich equation, with its logarithmic behavior and its two arbitrary constants, that it can be fitted reasonably well to many decay or rise curves that otherwise would be represented using multiple-time-constant analysis.

4. Measurement Methods and Observations

4.1. Measurement of the Surface Double Layer

There have been several measurement techniques, based on electrical measurements,[53-55] capacity measurements,[56] and recently optical measurements,[57] which have been used to examine the predictions of the theory of the double layer at the semiconductor surface. In general, it can be said that for semiconductors where the bulk properties are understood and where diffusion of imperfections in the space-charge region does not occur the results are fairly consistent with the double layer theory. The major unknown factors in surface phenomena on semiconductors are associated with the surface groups (diffusion, adsorption, reaction, and even identification of species at the surface).

Probably the simplest and most direct verification of the theory of the double layer arises in experiments at the semiconductor/electrolyte interface[56] when measurements of capacity vs. surface potential are made. Figure 9 shows a schematic of the simplest form of experimental setup for such a measurement. A voltage is applied between a semiconductor electrode and a platinum working electrode and because the capacitive impedance at the platinum electrode is low, a capacity bridge in this arm of the circuit measures the capacity across the high-resistance semiconductor depletion layer. A potential V_R between a reference electrode and the semiconductor electrode is measured, and changes in the potential can be ascribed to changes in the double layer at the semiconductor surface. The solution of Eq. (31) can be expressed in terms of the change of the charge in the double layer (the differential capacity) as a function of the surface potential V_s. In particular, it was shown by Dewald[58] that with a simple depletion layer the differential capacity C is related to the surface potential by

$$C^{-2} = (V_s - kT/q)(2/qN_D A^2 K\varepsilon_0) \tag{46}$$

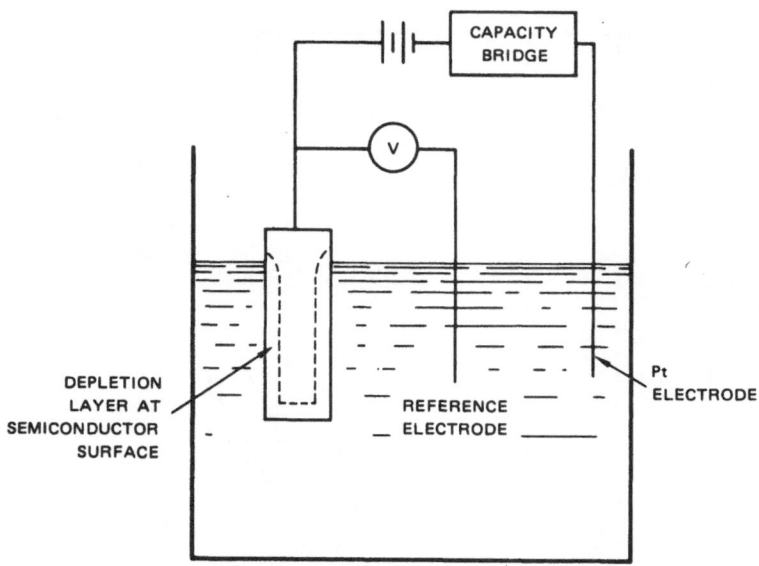

Fig. 9. Schematic of a simple electrochemical cell for the study of surface
layers on semiconductors.

where A is the surface area. Thus in regions where the semiconductor
space charge dominates and where changes in applied potential
appear across the semiconductor space charge a plot of C^{-2} vs. V_R
($V_R = V_s + \text{const}$) is predicted to be linear and inversely proportional
to the donor density. Dewald[58] found excellent agreement between
the theory and experiment with ZnO, not only for the depletion layer
region as discussed above, but extending into the accumulation layer
region. Figure 10 shows an example of such an experimental test. The
apparent consistency of Eq. (46) with experiment confirms the original
assumption that with ZnO the Helmholtz double layer potential on the
liquid side of the interface does not vary with applied voltage or current.

Brattain and Boddy[59] made similar measurements on germanium, a material of particular interest because of the ease of forming
an inversion layer at the surface and thus studying all three types of
double layer. It was found again that the capacity/potential curve
agrees with theory if care is taken to maintain a constant potential
difference across the Helmholtz double layer. To maintain a constant
Helmholtz potential, it is necessary to prepolarize the system by
holding the applied potential at a certain value, then abruptly change
the potential to the value of interest and make a rapid capacity
measurement before the Helmholtz potential can change.

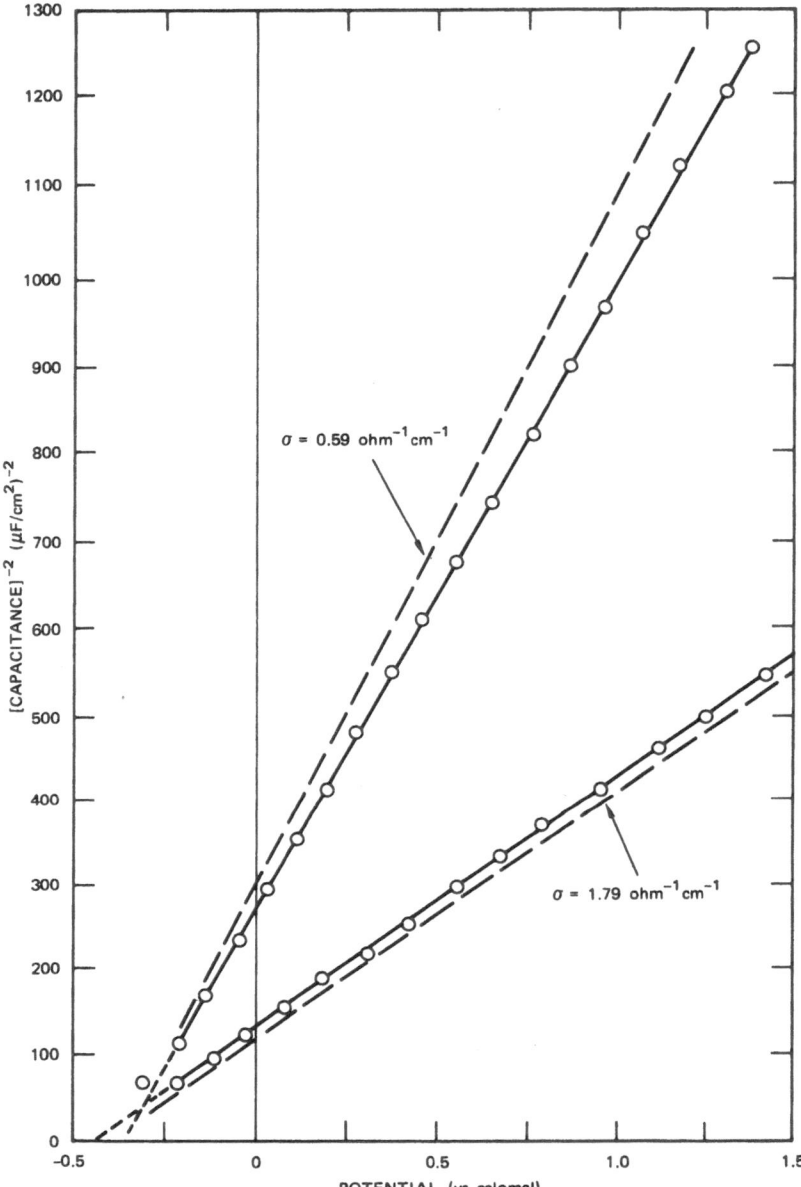

Fig. 10. Measurements of capacity vs. voltage on ZnO, suggesting agreement with double layer theory. The dashed lines show the theoretical slopes. From Dewald.[58]

If one accepts the mathematical theory of the surface double layer, then by analyzing a few electrical measurements one can in principle obtain a fairly complete picture of the surface. For example, a complete description of the surface double layer can be obtained from the measurement of the conductance of the sample, analyzed using a numerical solution of Eq. (31). This technique of surface layer analysis depends upon the identification of a minimum in conductance corresponding to a unique position of the Fermi energy approximately at midgap position at the surface. If the bands bend higher than the amount needed to produce this conductance minimum, surface conductance due to holes begins to increase appreciably, and if the bands bend lower, surface conductance due to electrons increases rapidly. If one can calibrate the conductance measurement by identifying the conductance at this minimum value, corresponding to what one may call an "intrinsic surface," one can calculate from theory the band bending at any other value of conductance. The minimum value can be reached by varying the surface double layer using gas adsorption or a normal electric field. A second method for analyzing the double layer, at least for obtaining the conductance at one value of V_s (and thus calibrating the conductance) is electroreflectance,[57,60,61] which identifies the point where the bands are flat out to the surface. This is particularly suitable for wider band-gap materials, where an "intrinsic surface" cannot be obtained. The principle here is that certain reflection maxima near the absorption edge of the crystal are sensitive to the electric field. The electric field [from Eq. (31) with $x = 0$] due to the surface double layer is modulated by a small field, and the reflected beam shows corresponding modulation. The phase of the modulation depends, however, on the direction of the band bending. If the band bending is upward, a negative potential to the field electrode increases the electric field, whereas if the band bending is downward, a negative potential decreases the electric field. The sign changes at "the flat band" condition, where the field goes to zero, permitting calibration of the conductance/surface potential relationship at this point.

4.2. Measurement of Surface State Energy Levels

The measurement of the energy and density of surface states at a solid/solid interface (such as the Ge/GeO_2 interface) is done by several relatively reliable methods. A few of the techniques used to study interface states will be mentioned, since these illustrate well many

239

of the electrical measurements used and useful for surface studies. Following this, measurements of surface states on the clean surface and measurements of adsorbate surface states will be described in somewhat more detail. In no case will the quantitative analysis be reviewed in detail; for this the reader is referred to the original articles or to more lengthy surveys.[1,2]

The reason that studies of interface states could be interpreted quantitatively even in early measurements is that often there are few surface states at such an interface. This is the case for Ge/GeO_2, the first system studied extensively. With a low density of states, measurements such as the field effect measurement are highly useful. Thus, for example, if an electric field of modest strength is applied normal to the surface, it induces a charge the order of 10^{11} electrons/cm^2. If the density of surface states is low, not all of this charge will be captured by surface states; some of the charge will remain in the conduction band of the semiconductor.[53] This remaining charge is measured by the change in conductance that it causes, so one can determine the charge captured by the surface states as a function of the total charge added. The measurement is termed the field effect measurement. Working back by theoretical analysis,[1,2,62] a plot of the "field effect conductance" vs. applied field can provide the energy and density of the surface states. Obviously, if the surface state density is extremely high, absorbing effectively all the induced electrons, there will be no measurable conductance change to interpret. This is usually the case if the charge is permitted to leak out to adsorbed species[43] or other surface states of high density.

Usually it is convenient when determining the two parameters, energy and density of surface states, to measure two transport properties simultaneously. Conductance measurements are often used, combined with measurements of other transport properties[1,2] such as field effect, Hall effect, surface recombination velocity, surface photovoltage, or work function. From conductance measurements the band bending at the surface is determined as discussed in the last section. With band bending known from such conductance measurements, analysis of field effect to yield information about the density of surface states is mathematically messy, requiring numerical solution of the double layer equations, but is basically straightforward. Similarly, one can use other surface properties, such as the surface recombination velocity (with the appropriate theories as discussed in Section 5.1 relating surface recombination velocity to the double layer potential) and analyze for the surface state density and energy.

Measurement of surface states at the clean surface is usually more difficult because the density of states is higher, but such measurements are of greater theoretical interest because (a) many chemical problems associated with sorbates are avoided, and (b) the surface state distribution is more reproducible than in the case of interface states. Cleavage in vacuum is one of the preferred ways to produce a clean surface. Early work[63,64] on the cleaved surface of germanium and silicon utilized an approach conceived by Statz and his co-workers[65] for interface surface states. Positive results by the Statz method are obtained only if there are donor states near the conduction band or acceptor states near the valence band, so that an inversion layer can be obtained. On germanium the cleaved surface is always p type,[63] so the surface states on n-type germanium can be examined. A three-layer, transistor type structure is constructed, with the three regions p type, n type, and p type, respectively. The conductance between the two p-type regions is very low under normal conditions because of the junctions, but after cleavage the p-type inversion layer across the surface of the n region provides a conducting path between the two p-type regions. This conductance allows one to characterize the surface properties. When a large bias is applied across the p–n junction(s) the bias appears across the "junction" between the n region and its surface p-type layer, and one can induce large changes in the surface state charge. By using very heavily doped germanium, to press the technique to the limit, Palmer and his co-workers were able to move the Fermi energy across the surface states from a position 0.34 eV below the midgap position (near the edge of the valence band) to a position 0.2 eV below the midgap position, and found that the concentration of electrons in the acceptor surface states remained unchanged at about $2 \times 10^{12}/\text{cm}^2$. Thus the Fermi energy is above the acceptor states throughout this range. It is thus concluded that the acceptor states dominating the clean surface of germanium are all at or below the valence band edge with a density at least the above value. Oxygen adsorption lowers the measured density, suggesting a local interaction between the oxygen and the germanium surface groups passivating the surface states. A curve showing density of surface states as a function of oxygen history is given in Figure 11.

Unfortunately, the approach is not applicable for silicon, since the surface states are too close to the midgap position. It has not been attempted with other materials. Henzler and Heiland[66,67] used combined conductance and field effect to investigate the surface states on germanium. Assuming a continuous distribution of surface

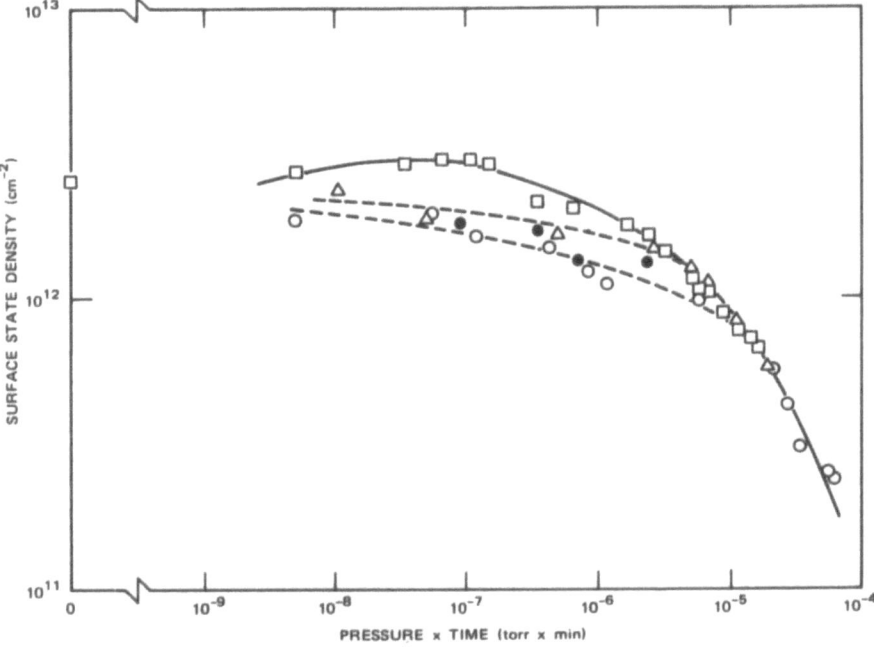

Fig. 11. The variation in surface state density as oxygen is adsorbed on a clean surface. From Palmer and co-workers.[64]

state energy, they find 1.4×10^{14} acceptor states/cm²/eV, decreasing to 2×10^{13}/cm²/eV with heat treatment of the germanium to 300°C. Their conductance results confirmed that admission of oxygen lowers the surface state density further. The results are thus qualitatively similar but may represent a higher density than the results above.

Another approach to the study of the surface states of medium high density as found on clean surfaces was developed by Gobeli and Allen.[68,69] They varied the bulk Fermi energy (by using a spectrum of samples of different impurity content) and measured the surface Fermi energy. The surface Fermi energy is measured by work function and photoelectric threshhold. If there are no surface states, the surface Fermi energy will be the same as the bulk. If surface states are present, capture of charge on these surface states will cause the surface Fermi energy to shift toward the surface state energy. If the surface state energy is extremely high, the surface Fermi energy will be controlled entirely by the surface states, and changes in bulk doping will have no effect on the surface Fermi energy. Surfaces prepared by cleavage in ultrahigh vacuum are the only ones that can be used, for a requirement of the method is that the surface state distribution be the same for all samples studied, so surface states due to adsorption, which will depend on doping, must be avoided. It is assumed that the clean, cleaved surface has the same spectrum of surface states independent of doping.

Gobeli and Allen found that the surface state density is substantial, and doping could only vary the surface Fermi energy with very high impurity densities, both on germanium[68] and silicon.[69] Results for silicon, where some effect was observed, are shown in Figure 12. The abscissa shows the variation of the bulk Fermi energy and the ordinate shows the two measurements which reflect the surface Fermi energy. Over much of the range the curves are horizontal lines, indicating no variation in surface Fermi energy as the bulk Fermi energy

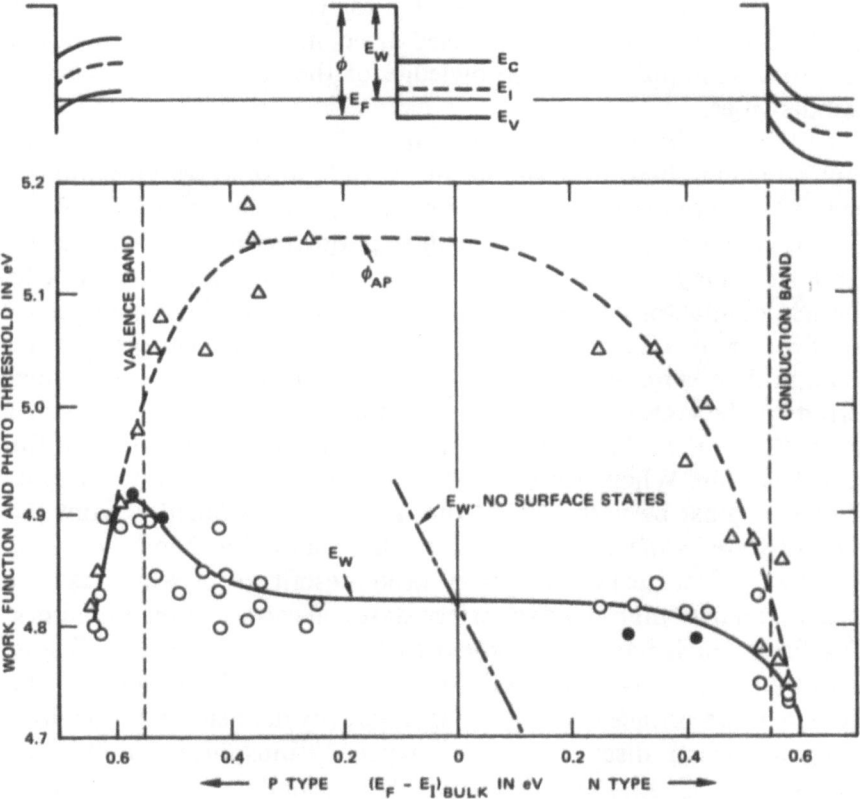

Fig. 12. The variation of work function and photoelectric threshold on cleaved silicon with changes in bulk Fermi energy. The curves can be interpreted in terms of surface state energy and density. From Allen and Gobeli.[69]

varies. Thus the surface state density is high enough to control the surface Fermi energy in this central region. Such data can be interpreted, as summarized by Davison and Levine,[12] in terms of a surface state distribution consisting of two maxima, a donor type at the lower energy an acceptor type at a somewhat higher energy, as predicted by the clean surface theories of Section 2.1.

The measurement techniques such as field effect so useful for low surface state densities only provide studies of transient effects[43] when dealing with high surface state densities, and other methods must be used. The problem of high surface state density is further compounded by the fact that, in contrast to interface states and perhaps to surface states at a clean surface, in the case of adsorbed or deposited species the chemical complications described in Section 1.4 become very important.

Measurement techniques suitable for foreign sorbates are thus much harder to find, and those used are compromises which lead at best to a semiquantitative knowledge of the surface state energies and densities.

One technique which has been used[70,71] but must be used with caution is illumination of the surface with light of less-than-bandgap energy. If there are sufficient surface states and the extinction coefficient is sufficiently high, effects such as photoconductance will be observed and may be associated with excitation from the surface state to the semiconductor conduction band. However, clearly in many cases (see Section 5.4 on electron injection from dyes) the electron is raised to an excited state of the surface complex, and from there it is transferred to the conduction band. Thus the results reflect the internal chemistry of the surface species, not the surface state location in the band diagram. When this method of obtaining surface state energies is used it must be verified that the transition examined is from the ground state of the surface group to the conduction band.

Two other techniques have been described[72] which permit measurement of high-density surface states on certain semiconductors. The first, which has been applied to the semiconductors ZnO and recently to TiO_2, is a method based on electrostatic charging of the surface. If the surface is charged negatively (by depositing oxygen ions using a corona discharge in an oxygen atmosphere, or chloride ions from a CCl_4 + Ar atmosphere), one finds with these n-type semiconductors that a high electrostatic voltage can be induced, of the order of 50 V. Then any surface states will tend, from a simple energetic point of view, to inject their electrons into the conduction

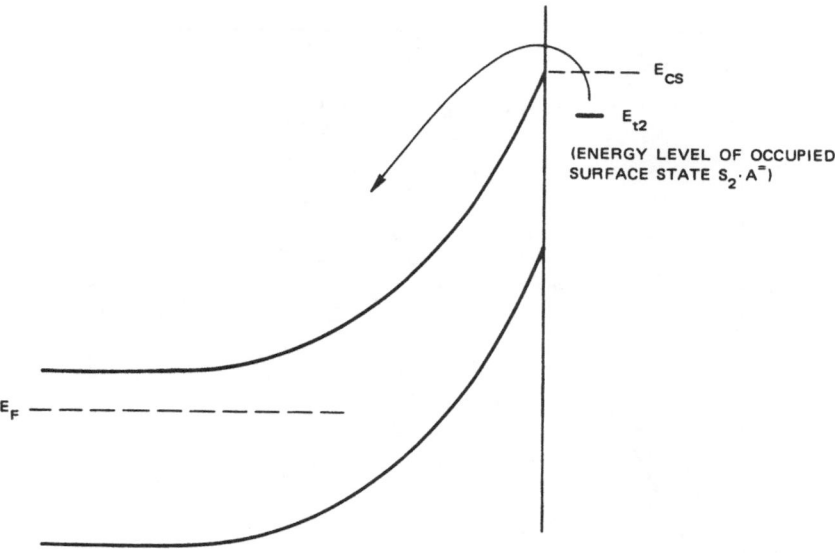

Fig. 13. Band bending on an *n*-type semiconductor with a large negative electrostatic surface charge. Donor surface states will tend to inject electrons as illustrated.

band of the solid, as illustrated in Figure 13. The reaction assumed is the forward direction of Eq. (9). The electrons will flow into the semiconductor from the surface states, neutralizing the electrostatic potential, at a rate governed by the energy necessary to thermally excite the electrons into the conduction band, which is the depth of the occupied surface state below the conduction band, $E_c - E_{t2}$. By measuring the electrostatic potential, the rate of electron injection is determined as a function of temperature, and an Arrhenius plot provides the surface state energy relative to the conduction band.

The technique, unfortunately, does not discriminate between activation energies derived from electron transfer processes [route of Eq. (9)] and activation energies associated with chemical processes [electron injection from O^- by the reverse direction of Eqs. (27) and (26), for example]. Such discrimination is desirable for a complete picture, but presently can only be suggested based on the known chemistry of the surface species and its likely kinetic behavior.

The other technique is an equilibrium technique, so is less dependent on kinetics. Here the current flow of electrons across the surface barrier (Figure 7a) is measured, and since the surface barrier is determined by surface states, the current flowing can be interpreted in terms of the surface state energy. To force current flow across the

245

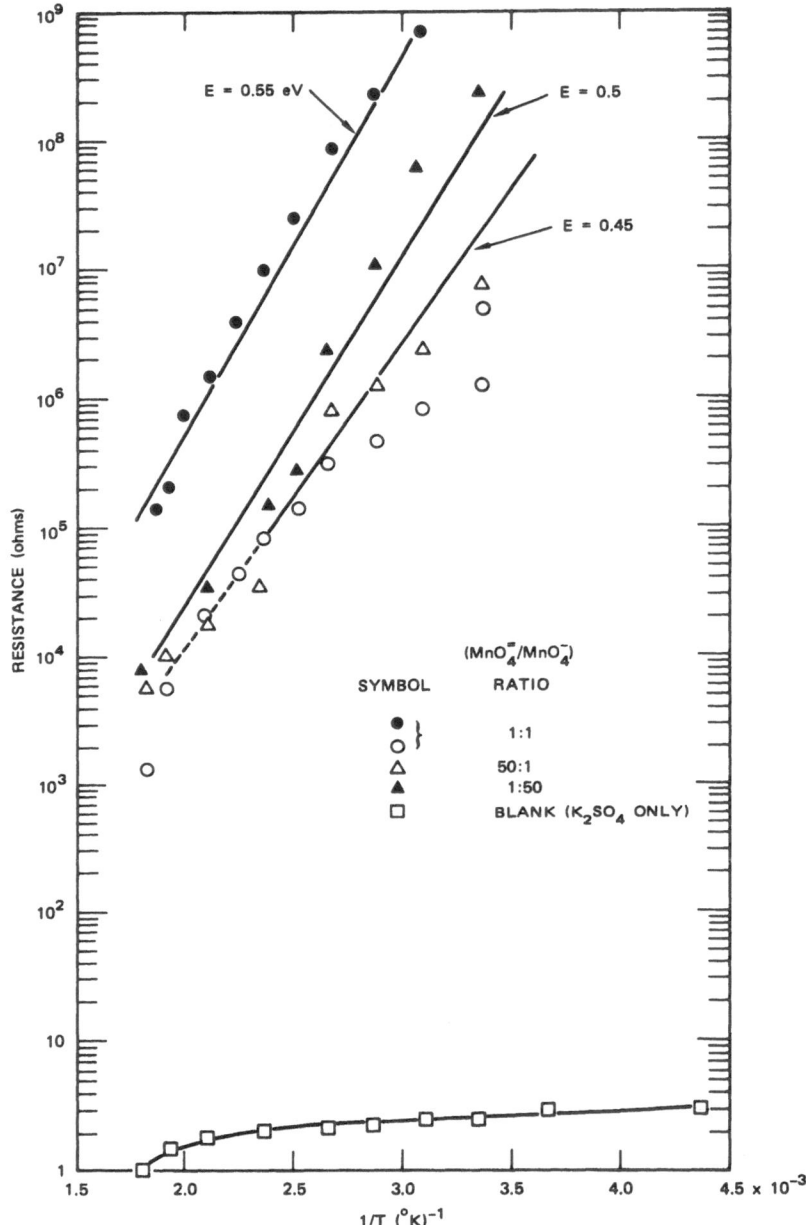

Fig. 14. Arrhenius plot of the conductance of a pressed pellet of ZnO powder with MnO_4^-/MnO_4^{2-} surface states. From Morrison.[72]

surface barrier, the sample is measured in the form of a compressed powder, where the surface barrier dominates each intergranular contact.

Results of such measurements are shown[72] for example in Figure 14, showing the conductance of ZnO pressed powder sample with the manganate/permanganate redox couple (potassium salts) forming the surface state. The details of the analysis will not be discussed here, but from the plot of conductance vs. inverse temperature, it can be concluded that the surface state [E_{eff} from Eq. (17)] is about 0.5 eV below the ZnO conduction band.

Another technique for measuring high-density surface states has been suggested based on field emission of electrons, which has been found[73] applicable at the (0001) face of ZnO. As the electric field was increased a high emission current was observed, which was interpreted as emission from a surface state. After the surface state was exhausted of electrons the current decreased to a lower value. Analysis according to a Fowler–Nordheim plot suggested a surface state at 0.056 V under the construction band edge. Such behavior has not been observed on other materials.

5. Kinetics of Electron and Hole Reactions

This section is concerned with nonequilibrium conditions, not only the case where there is a surplus of electrons and holes in the semiconductor and the surface acts to recombine them and restore the system toward equilibrium, but also the case where the surface chemistry itself is not at equilibrium and injection or extraction of carriers by surface states occurs. Either of these processes can lead to chemical changes (oxidation or reduction of surface species). Surplus electron–hole pairs often arise due to optical effects or transistor effects. The action of surface states to recombine such electron/hole pairs will thus affect such bulk semiconductor properties as photoconductivity or fluorescence on the one hand and gain of a transistor or reverse current of a diode on the other. However, in the present discussion the interest is centered more on the surface recombination process itself. Such recombination can be simple nonchemical surface recombination or can involve chemical changes leading to photocatalysis or photolysis. The most important examples of the second case, the case of nonequilibrium chemistry, are electrochemical or catalytic systems, where redox reactions on semiconductors will often involve surface capture or injection of carriers.

5.1. Electron–Hole Recombination

When radiation of band gap energy or greater falls on a semi-conductor, electron–hole pairs are produced. A valence electron in the valence band is excited to the next highest state, the conduction band (ignoring localized states due to flaws). The excited electrons are free to move in this almost empty band, and the unoccupied level in the valence band, the hole, can also move freely through the crystal. Thus both these photoproduced carriers can move to the surface and react with surface states. The reaction is usually exothermic (see Figure 15) both when an electron from an occupied surface state moves to occupy the empty level in the valence band (hole capture) and also when an electron in the conduction band is captured by an unoccupied surface state (electron capture). If the surface states are the reduced and oxidized forms of the same chemical group, as in Figure 15, there is no net chemical change due to this series of steps:

$$p + A^{2-} \rightarrow A^{-} \tag{47}$$

$$e + A^{-} \rightarrow A^{2-} \tag{48}$$

and the photon energy, which initially was converted to the energy of the hole–electron pair, is dissipated as heat energy.

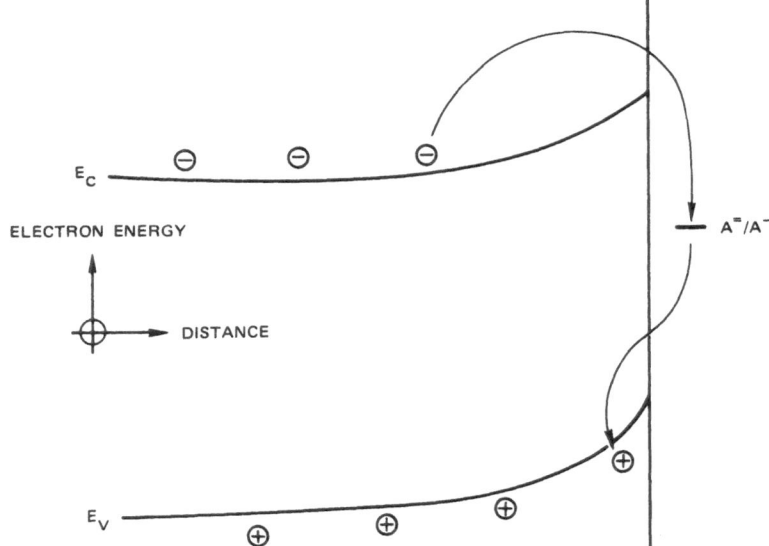

Fig. 15. Electron capture by a surface state A^- followed or preceded by hole capture by A^{2-} (an electron from the surface state annihilates a hole), illustrating surface states as recombination centers.

If we assume the surface state A^{2-}/A^{-} is ideal, that there are no chemical or Franck–Condon effects to split the occupied and unoccupied levels (Section 2.4), the kinetics can be analyzed[1,2,74] to provide a model system upon which to base later development. The assumption of no Franck–Condon effects is, of course, valid for surface states at a solid/solid interface, but, as discussed in earlier sections, must be justified for adsorbates.

Adopting the terminology which appears most general,[2] we define N_t as the total density of surface states of interest ($[A^{-}] + [A^{2-}]$), n_s^* and p_s^* as the densities of mobile electrons and holes at the surface (the asterisks are a reminder that these are nonequilibrium values), n_b^* and p_b^* as the values just under the surface double layer, and n_t^* as the density of occupied surface states (the density of A^{2-} species under kinetic conditions). First-order kinetics of Eqs. (47) and (48), both in forward and reverse directions, is assumed. The rate of hole capture is assumed to be proportional to p_s^* and to n_t^*; and the rate of hole injection is assumed to be proportional to the number of holes in the surface states $N_t - n_t^*$ (the number of unoccupied surface states) and to p_1, a hole emission constant made up of a Boltzmann factor reflecting the energy difference from the valence band to the surface state and the effective density of (occupied) states in the valence band N_v:

$$p_1 = N_v \exp\{-(E_t - E_{vs})/kT\} \tag{49}$$

where E_t is the energy of the surface state and E_{vs} the energy of the valence band at the surface. Then the net rate of hole capture is given by U_p:

$$U_p = K_p p_s^* n_t^* - K_2 p_1 (N_t - n_t^*) \tag{50}$$

where K_p is the rate constant for the forward direction of Eq. (49) and $K_2 p_1$ is the rate constant for the reverse direction. From the principle of detailed balance we can set $U_p = 0$ at equilibrium, leading to

$$U_p = K_p\{p_s^* n_t^* - p_1(N_t - n_t^*)\} \tag{51}$$

An analogous expression can be derived for the rate of electron capture U_n:

$$U_n = K_n\{n_s^*(N_t - n_t^*) - n_1 n_t^*\} \tag{52}$$

where

$$n_1 = N_c \exp\{-(E_{cs} - E_t)/kT\} \tag{53}$$

in complete analogy with the expression for p_1, with N_c the density of states in the conduction band.

At steady state the rates of electron and hole capture are equal, and if the density of surface states is low, the occupancy of the states will be the dominant variable. Then the condition

$$U_p = U_n = U \tag{54}$$

defines n_t*. Elimination of n_t* from the above equations yields for the rate of hole or electron capture

$$U = K_n K_p N_t (n_s^* p_s^* - n_b p_b)/[K_n(n_s^* + n_1) + K_p(p_s^* + p_1)] \tag{55}$$

Now the surface densities of holes and electrons equal the bulk densities p_b* and n_b*, respectively, multiplied by the appropriate Boltzmann factor $\exp(-eV_s/kT)$ and $\exp(eV_s/kT)$, respectively, if recombination in the space-charge region is neglected, and if the current of holes and electrons across the space-charge region is assumed low so as not to disturb the Boltzmann distribution:

$$p_s^* = p_b^* \exp(-eV_s kT), \qquad n_s^* = n_b^* \exp(eV_s/kT) \tag{56}$$

In the absence of bulk trapping effects we can assume that the increase in bulk hole density $\delta p_b = p_b^* - p_b$ is the same as the increase in the bulk electron density $\delta n_b = n_b^* - n_b$, and we will assume that the fractional increase of the carrier densities is small:

$$\delta n_b < n_b, \qquad \delta p_b < p_b \tag{57}$$

With Eq. (57) the rate of recombination becomes proportional to the bulk density of photoproduced hole–electron pairs δn_b, the proportionality constant is denoted as s, termed the "surface recombination velocity." The value of s can be calculated from Eqs. (55)–(57):

$$s = U/\delta n_b$$
$$= [(K_p K_n) N_t (n_b + p_b)/2n_i]$$
$$\times \{K_n n_b \exp(eV_s/kT) + K_p p_b \exp(-eV_s/kT)$$
$$+ K_n n_b \exp[(E_t^0 - E_F)/kT] + K_b p_b \exp[(E_F - E_t^0)/kT]\}^{-1}$$
$$\tag{58}$$

where E_t^0 is the energy of the surface state with no band bending.

A much more elegant form of this expression is available[2] for further mathematical processing. The above form is shown here to emphasize the variation of the surface recombination velocity with changes in the surface barrier V_s and with the relative positions of the Fermi energy and the surface state level under flat band conditions (no net charge in surface states). With a given Fermi energy and surface state energy there is [from Eq. (58)] a region or value of V_s where surface recombination is most efficient; s is maximum. If V_s is much higher than this value, then the first term in the denominator is large. If V_s is much lower, then the second term in the denominator is large. With V_s intermediate, both holes and electrons can reach the surface at a high rate, and s is largest. The same applies to E_t, the surface state energy. Figure 16 shows a typical curve[75] for the variation of s with surface potential $U_s (= eV_s/kT)$ for germanium.

The model is found to be in agreement with experiment for simple surface states such as states at the interface between a semiconductor and an oxide. It has also been shown[76] to hold for surface states at the germanium/liquid interface. It thus provides a firm basis on which to build as the more complex chemical phenomena are included.

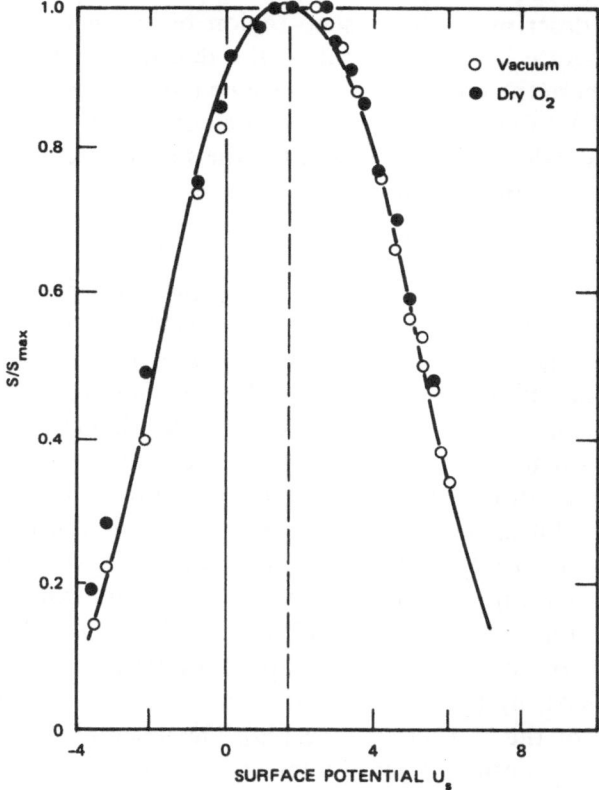

Fig. 16. Relative surface recombination velocity as a function of the surface potential for an *n*-type germanium sample. From Many and Gerlich.[75]

Another case of simple recombination, which perhaps may be applicable more often with adsorbates at the surface, has been developed by Morrison.[77] The preceding analysis was based on a low density of surface states, assuming the dominant variable under illumination is the change of occupancy of these surface states. In the other extreme, where the surface state density is extremely high, the photoprocesses cannot cause a large change in the fractional occupancy of the levels. As was discussed with respect to Eq. (46), there is an effective upper limit on charge transfer to the surface of about 10^{-3} monolayer, because greater transfer produces impossibly high V_s variation. There is, however, no upper limit to the density of surface states. Thus one can easily deposit a redox couple as a surface state, to densities of $10^{15}/cm^2$. Then the surface potential V_s can be changed significantly by electron transfer without appreciable effect (only a few percent change) on the surface state occupancy.

In this description there is no region of interest where the re-combination rate is proportional to the density of photoproduced carriers, so here the concept of the surface recombination velocity is not useful. The reader is referred to the original literature for the details of the calculation; the form for the rate of recombination of excess carriers is found to be:

$$(d/dt)(\delta n_b) = B(1 - e^{\beta \delta n_b}) \qquad (59)$$

where B and β are constants. Essentially the form arises from the following physical model: With Eqs. (56) and (57) it is easily shown that if V_s is negative, then $\delta p_s > \delta n_s$, so the surface becomes more positively charged and V_s tends toward zero. This trend is normally magnified if capture by surface states is included. If, on the other hand, V_s is positive, then $\delta n_s > \delta p_s$, and again V_s tends toward zero. Thus the effect of illumination is always to reduce the surface barrier. Now it is assumed to a linear approximation that the reduction in V_s is proportional to the excess surface charge. Then the rate of carriers crossing the barrier, varying as $\exp(-|eV_s|/kT)$, shows an exponential dependence on the surface charge and this rate is described by the last term of Eq. (59). The rate of carrier capture of carriers which need not cross a barrier is in certain cases independent of V_s, and this capture rate is given by the first term of Eq. (59). The resemblance of Eq. (59) to the Elovich equation (41) is of interest. Unfortunately, with the two constants, comparison with experiment is difficult here as in the Elovich case: Fitting the expression to experimental data does not provide a sufficiently stringent test.

5.2. Hole Capture, Oxidation of Molecules by Holes

In recent years there has been increasing interest in the chemical effects of electron and hole reactions with surface adsorbates. Much of this work has been done by electrochemical studies, where the products can be carried from the surface into the solution and electrical neutrality maintained by the current flow through the external circuit. Thus, for example,[39] the photocatalysis of formic acid at the ZnO surface could be studied in its separate steps of (a) hole capture by formate ions as an anodic process, (b) electron capture by oxygen as a cathodic process, and (c) hole and electron capture together (no external current) as a catalytic process. Here we discuss such electron and hole capture reactions as observed in solution[30,78] as well as at the gas/solid interface.

In this section hole capture is examined. As discussed earlier, holes from some semiconductors can be extremely powerful oxidizing agents, so hole capture is of substantial chemical as well as physical interest. Electrochemical studies of the kinetics of hole capture by foreign species have been made by Gomes and co-workers[79] on ZnO. The holes are photoproduced and the rate at which they reach the surface to provide anodic current is controlled by the light intensity. Hole capture on some simple two-equivalent species is easily detectable because after hole capture the resulting unstable free radical immediately injects an electron (as discussed in Section 2.4). If the concentration of such a two-equivalent species is held constant and a one-equivalent ion is slowly added to the solution, a greater and greater fraction of the hole current goes to the one-equivalent species. This is observed experimentally by the electron current going down, while the photoproduced hole current is held constant. We can thus calculate the relative hole capture cross sections (rate constants) of the surface states R and X according to:

$$p + R \rightarrow R^+ \tag{60a}$$

$$R^+ \rightarrow e + R^{2+} \tag{60b}$$

$$p + X \rightarrow X^+ \tag{61}$$

where R is the two-equivalent reactant and X the one-equivalent reactant. The total current $J_p + J_e$ due to the reaction is measured, and with the hole current J_p constant, changes in the electron current J_e are related to reactant concentrations. If a linear adsorption isotherm is assumed and the reactions (60a) and (61) are first order in the concentration of reactants, it is easily shown that

$$J_p/J_e = 1 + (K_X/K_R)([X]/[R]) \tag{62}$$

where the square brackets indicate concentrations in solution. Experimentally it was found that the form of Eq. (62) was followed. If there is specific adsorption on the solid, the K's are the products of the rate constants and sorption constants (assuming linear, or Henry's law, adsorption). If an assumption of negligible specific adsorption is made, which is most reasonable *a priori* assumption, one can obtain the relative cross sections for hole capture on reducing agents. For example, if I^- is designated unity to establish a reference value, Br^- shows a hole reactivity of 0.012 and Cl^- is zero, to the experimental accuracy. Methanol has a hole reactivity of about 0.09 and ethanol of about 0.17 on the same scale. The results are intuitively satisfying, following some sort of reasonable scale of ease of oxidation or redox potential, but a quantitative model is not yet available.

Other studies of hole capture have been made at the semiconductor/electrolyte interface with current doubling as the indicator. In particular, the work of Gerischer and Haberkorn on CdS should be noted (see Ref. 80).

One of the most important reactions of photoproduced holes in semiconductors, from a strictly commercial point of view, is the oxidation of organic molecules on the semiconductor surface. The importance arises because such a semiconductor/organic system appears so often in coatings, such as paints, and in filled polymers. Here the organic polymer is present as the binder and the structural material and the semiconductor is present as the white or colored pigment or as an ultraviolet absorber to protect the polymer against direct photolysis. However, the indirect photolysis occurring by hole–polymer interaction leads to chalking, cracking, and embrittlement of the polymers.

Relatively little work has been done in this field to date, from a fundamental point of view, to study such interactions. An empirical approach is used commercially: The pigment particles are coated with a silica/alumina mixture of the order of a few hundred angstroms thick, and this presumably physically blocks the holes from the surface, and is found reasonably effective.

Some effort has been made[77] to prevent these reactions by introducing simple recombination centers to compete for the holes. In particular, on ZnO the couple $Fe(CN)_6^{4-}/Fe(CN)_6^{3-}$ was used as a recombination center to prevent photolysis of a ZnO-based coating.

Some of the basic studies of hole oxidation of polymers is included in the above discussion of Gomes' work, where measurements of the hole capture cross section of methanol and ethanol in solution were discussed.

Recently Sancier and Morrison[81] compared this approach and that of electron spin resonance to provide information regarding oxidation of organic molecules by holes and its prevention by using recombination centers (to compete with the organic molecule for the holes). It was found, for example, that neopentane was stable against oxidation, whereas dioxane was readily oxidized.

Although the data even on ZnO are sparse, some suggested generalizations can be made from these and unpublished results. Organic molecules, if oxidizable, usually, but not always, show current doubling on ZnO, indicating the free radical formed by hole capture has an occupied energy level above the conduction band (Section 2.4). The molecular groups found to be oxidizable by holes at room temperature cannot be well classified at this time, since there are insufficient studies, but appear consistent with the ease of oxidation by catalytic processes using gas-phase oxygen.

A particularly interesting hole/surface state interaction from both a practical and fundamental point of view is that of the oxidation of OH^- to OH by photoirradiated TiO_2. This process has been shown to occur by Völz and his co-workers[82] and to be the dominant cause of degradation of outdoor paint. There is little doubt that oxidation by photoproduced holes is the basic process, and this primary action is followed by oxidation of the paint binder to CO_2 and water by the OH radicals.

These powerful oxidizing species—holes from TiO_2—will also oxidize mercury vapor to mercurous ions to mercuric ions. Kaluza and Boehm[83] have shown that TiO_2 irradiated in mercury vapor photocatalyzes the formation of mercuric ions in the presence of oxygen. Presumably the oxygen is required to absorb the electrons and maintain electrical neutrality on the surface.

Photolysis of crystals by hole capture at the surface has been studied[84] at the gas/solid interface for the case of vacuum photodecomposition of ZnO, by the reaction

$$p + ZnO \rightarrow \tfrac{1}{2}O_2\uparrow + Zn_i^+ \tag{63}$$

where the oxygen is lost to the gas phase and the zinc is dissolved into the lattice as an interstitial zinc donor. Because such donors are produced (or more accurately, because the electrons are now permanently in the conduction band), a conductance measurement enables the photolysis reaction to be monitored. Collins and Thomas[84] made the first quantitative measurements of this photolytic process and showed that photolysis, as opposed to photodesorption of oxygen, must be occurring. The conclusion was based on the observation that far more donors were produced than could be accounted for by photodesorption. As discussed with regard to Eq. (40), oxygen adsorption on ZnO should be limited by double layer effects. On a crystal where the limit was calculated to be 4×10^{10} ions/cm^2 they observed an increase in conductance corresponding to 10^{13} charges/cm^2.

At the semiconductor/liquid interface quantitative measurements of photolysis due to holes were made as early as 1955 by Brattain and Garrett.[85] They were able to show with a germanium electrode in an electrochemical cell that a current of holes reaching the surface caused (a) dissolution of the germanium and (b) electron injection into the germanium conduction band. The reaction could be written as

$$Ge + (4x)p + 6OH^- \rightarrow (GeO_3)^{2-} + 4(1-x)e + 3H_2O \quad (64)$$

where x is the fraction of total current carried by holes. The value of x ranges from 0.6 for very low hole currents to 0.8 for high hole currents. The reason suggested for the electron injection is related[30] to the two-equivalent surface state concept, namely, that as the germanium ion is oxidized from a valence of zero to a valence of plus four, energy levels for four electrons are involved, and for some valences (plus one, for example) the species may be unstable, represented by a high energy level, and the oxidation will occur by electron injection rather than the ion maintaining its unstable valence state until the arrival of a hole. Unfortunately, the reaction is somewhat complicated for quantitative analysis. Other materials that have been examined with band gap greater than germanium require holes[86] for anodic dissolution but do not inject electrons.

5.3. Electron Capture, Reduction of Molecules by Electrons

The reduction of adsorbate molecules by electrons has been known and studied for many years but surprisingly little quantitative information has been available. Presumably the reason has been the complexity of the interpretation. For example, adsorption of oxygen by electron transfer is a well-known phenomenon and has been studied using electron spin resonance, work function, conductance, and gas pressure measurements. However, the phenomenon, as discussed in Section 2.5, is both chemically and physically complex, even if one considers only adsorption and excludes the formation of oxide phases. Thus, although oxygen adsorption is by far the most important and the most studied case of electron capture at the surface by a foreign species, we will illustrate the principles with a far less important case, the capture of electrons by ferricyanide ions in solution.[87] In this case both the chemistry and the physics are relatively simple, so the example provides a basis for building up to the more complex electron capture (reduction) reactions.

In the study of electron capture at the ZnO electrode by ferricyanide ions in solution the reaction is

$$e + Fe(CN)_6^{3-} \rightarrow Fe(CN)_6^{4-} \tag{65}$$

which is a reaction ideal in its chemical simplicity, for the six stable cyanide ligands enclosing the iron ion make it unlikely that there will be any strong local chemical interaction between the iron and the solid. The stability of the iron cyanide complex and the fact that only two oxidation states for iron cyanide are possible almost preclude any reaction involving iron but the above.

If the above reaction rate can be described as first order in the density of electrons and first order in the density of available ferricyanide ions, then the current J produced by the above reaction will be given, with the help of Eq. (52), by

$$J = qU_n = qK_n n_s^* N_{Fe} \tag{66}$$

where the reverse direction of Eq. (52) is neglected on the assumption $n_t^* = 0$, namely that no $Fe(CN)_6^{4-}$ ions are present. (It is found experimentally that even with such ferrocyanide ions present there is no measurable anodic current.) Here N_{Fe} ($= N_t - n_t^*$) is the concentration (per unit area) of accessible ferricyanide ions.

With measurements of n_s^* made by capacitance measurements and N_{Fe} taken proportional to the concentration of ferricyanide ions in solution, Freund and Morrison[87] were able to show that electron capture by such foreign ions was indeed described by Eq. (66), thus showing that it can be expected that the laws found for simple recombination should hold for more complex chemical reactions if the complications due to chemical transformations can be accounted for.

An electron capture process involving more complicated chemical steps is that of iodine, as examined by Gerischer and Meyer[88] and by Cardon and Gomes.[89] From the observation that the capture rate is first order in electron and in iodine concentrations it is concluded that the first reaction step is

$$I_2 + e \rightarrow I_2^- \qquad (67)$$

The next step in reduction is still open to question. Cardon and Gomes claim on the basis of electrical noise measurements that

$$I_2^- \rightarrow 2I^- + p^+ \qquad (68)$$

(i.e., hole injection) is excluded, but cannot distinguish between the reaction route

$$2I_2^- \rightarrow I_3^- + I^- \qquad (69a)$$

followed by

$$I_3^- \rightarrow I_2 + I^- \qquad (69b)$$

and by reduction of the I_2, or the other reaction route, the direct electron capture I_2^-:

$$e + I_2^- \rightarrow 2I^- \qquad (70)$$

The difference is important from a surface state discussion, for by the first mechanism only one surface state is involved, followed by a chemical reaction (disproportionation), whereas by the second mechanism two surface states must be considered.

More recently Memming and Möllers[37] have studied electron capture by quinones on various semiconductors. They analyzed the current/voltage characteristics in accordance with Marcus' model[90] for polarization "broadening" of energy levels in solution (Section 2.4) and concluded that the energy levels of unoccupied surface states broaden to a normal dist.ibution with a standard deviation of the order of 0.4 eV.

Current doubling effects can be observed with electron capture on p-type materials. The chemistry is analogous to that of Eq. (60), except that electron capture by an oxidizing agent forms a free radical which extracts an electron from the valence band (hole injection):

$$e + X \rightarrow X^- \qquad (70a)$$

$$X^- \rightarrow X^{2-} + p \qquad (70b)$$

where X^- is an unstable radical. For example, current doubling was observed in studies of electron capture by Gerischer,[80] who studied capture of photoproduced electrons by bromine on GaAs, and by Memming,[91] who observed it by H_2O_2 on GaP.

5.4. Electron and Hole Injection from Foreign Species

Electron injection at the gas/solid interface has been found by Mollwo and Schreiber[92] for hydrogen atoms on ZnO, and provides one of the few well-studied cases at the solid/gas interface when electron injection may relate to a simple, easily understood chemically driven process, that of a strong reducing agent (high surface state energy) injecting electrons. In most experiments studying electron and hole extraction and injection caused by changes in the gaseous ambient, polar molecules such as water or ammonia have been an integral part of the gas cycle. Because of the complications described in Section 2.4, such measurements are not easily interpretable, since the contribution of the polar properties are difficult to separate from the chemical and electronic properties of the system. Electron injection at the gas/solid interface caused by a shift of electronic conditions from equilibrium was discussed in Section 4 (with reference to Figure 13) as a method of surface state measurement.

Studies of electron or hole injection from species in solution can provide a much better overview of injection processes, because there are many more active species available for examination. The many ions soluble in water can be compared with respect to their ability to inject electrons and holes, and can be classified according to their chemical properties. Although the work is associated with a polar medium (water), the polar properties are better defined and are the same in all measurements.

For example, the rate of reduction of various species on germanium by valence electron capture (hole injection) was studied by Pleskov[93] (and reported in English by Myamlin and Pleskov[94] and Turner[95]). Expressed as the ratio y of hole injection to total reduction, his results for various species are shown in Table 1, together with the redox potential $E°$.

From the discussion of Section 1.3 (see Figure 4) it is expected that the more negative the redox potential, the lower the energy level will be. If the energy level is below the valence band edge of germanium, one expects the dominant reduction mechanism to be hole injection, with y large. Unfortunately for the simple theory, the results in Table 1 are almost completely contrary to this expectation. Possible reasons for this apparent discrepancy have been suggested.[30] In particular, it seems that the simple theory must be modified for many of these ions since they are two-equivalent, and thus require two energy levels for their description, and that the experimental work was done

TABLE 1
Fraction y of Current Carried by Hole Injection*
during Reduction of Various Ions in Solution

Surface species	y	E°
H_2O_2	0	-1.77
$K_2Cr_2O_7$	0.03–0.08	-1.33
Quinone	0.38	-0.70
KI_3	0.42	-0.54
$K_3Fe(CN)_6$	0.66–0.80	-0.36
$KMnO_4$	0.78–0.88	-0.56

* Electrons captured from the valence band.

before the germanium/electrolyte interface was well understood, and the work should be redone, with proper control of the Helmholtz double layer and the density of conduction band electrons, and taking into account adsorption.

Electron injection into ZnO has been studied by Freund,[96] who restricted the study in general to one-equivalent materials. He found that in the case of redox couples with a redox potential greater than about zero the reducing agent injected electrons; in the case of redox couples with a redox potential less than about zero the reducing agent did not inject electrons. This study shows better agreement with the expectations of the simple model.

More recently in electron injection studies Van den Berghe and Gomes[36] studied a series of one-equivalent reducing agents as surface states injecting electrons from solution into the conduction bands of CdS, ZnO, and CdSe. Surprisingly, it was found that if a species injected electrons into one semiconductor, it was able to inject into all. They found that Cr^{2+}, V^{2+}, Ti^{3+}, and Fe(II)-EDTA inject, whereas Fe^{2+} (aquo) and ferroine do not. The relative standard redox potentials are consistent with expectations, the more positive redox potentials of the first group suggesting a higher surface state energy. From calculations of the potential of the conduction band edge and the potential of the average "surface state" energy from the standard redox potential in solution they conclude that the injecting species have an average "surface state" energy well below the conduction band in each case, so the energy of the occupied surface state is substantially below the conduction band edge. They conclude that polarization broadening of the levels (Section 2.4) spreads the levels to overlap the conduction band. However, this does not easily explain the sharp cutoff between the injecting and noninjecting surface states. With such broadening all the species studied should inject to a measurable amount, at least in the semiconductor with the lowest conduction band energy, ZnO. The behavior of ions in solution as surface states is thus much better understood than the behavior of adsorbed species as surface states, but the models are not entirely clarified as yet.

Another system of high technical importance involving carrier injection is the semiconductor/organic dye system. Such organic dye reactions have been important through the years in photography, where silver salts are used, but have achieved recent renewed impetus because of their importance in electrophotography, where the solid is photoconducting ZnO.

From a fundamental point of view the system is particularly interesting, since it represents an easily accessible system illustrating optical absorption by "surface states" which then, in their excited state, are capable of injecting electrons (or holes) into the semiconductor.

The chemical details of the reactions are not yet satisfactorily understood, so a quantitative picture is not available. Thus we will only discuss briefly the qualitative picture available at the present time.

Photoinduced electron injection from dyes shows almost the same wavelength dependence as the dye absorption spectrum.[97–99] Thus the possibility of a direct electronic transition from the ground state of the dye molecule to the conduction band can be dismissed. However, there are two possibilities which can describe the electron injection. One is direct electron injection from the excited state of the dye to the conduction band of the solid; the other is energy exchange, where the energy of the excited dye molecule is transferred to a second surface state, and the second surface state injects its electron. The actual mechanism for any case has not yet been clarified, although there has been considerable discussion and experimentation.

Several reviews are available discussing dye sensitization.[100–102] In future work, photoproduced carrier injection from surface states, both from dyes and other surface states, will doubtless provide copious quantitative information about carrier/surface state interactions.

References

1. D. R. Frankl, *Electrical Properties of Semiconductor Surfaces*, Pergamon Press, New York (1967).

2. A. Many, E. Harnik, and Y. Margoninski, in *Semiconductor Surface Physics* (R. H. Kingston, ed.), Univ. of Pennsylvania Press (1957), p. 85.

3. F. F. Volkenstein, *The Electronic Theory of Catalysis on Semiconductors*, Macmillan, New York (1953).

4. S. R. Morrison, *J. Catalysis* **20**, 110 (1971).

5. V. J. Lee, *J. Chem. Phys.* **55**, 2905 (1971).

6. J. K. Dixon and J. E. Longfield, in *Catalysis*, Vol. VII (P. H. Emmett, ed.), Reinhold, New York (1960).

7. S. Z. Roginskii, in *Proc. 2nd Int. Congr. Catalysis* (Paris, 1960).

8. M. Green, J. A. Kafalas, and P. H. Robinson, *Semiconductor Surface Physics* (R. H. Kingston, ed.), Univ. of Pennsylvania Press, Philadelphia (1957), p. 349.

9. S. R. Morrison, *J. Phys. Chem. Solids* **14**, 214 (1960).

10. D. R. Frankl, *J. Electrochem. Soc.* **109**, 238 (1962).

11. P. J. Boddy and W. H. Brattain, *J. Electrochem. Soc.* **109**, 812 (1962).

12. S. G. Davison and J. P. Levine, *Solid State Physics*, Vol. 25 (F. Seitz and D. Turbull, eds.), Academic Press, New York (1970).

13. T. B. Grimley, *J. Phys. Chem. Solids* **14**, 227 (1960).

14. S. B. Davison and J. Koutecky *Proc. Phys. Soc.* **89**, 237 (1966).

15. D. Kalkstein and P. Soven, *Surface Sci.* **26**, 85 (1971).

16. I. Alstrup, *Surface Sci.* **20**, 335 (1970).

17. W. Shockley, *Phys. Rev.* **56**, 317 (1939).

18. J. Koutecky, *Adv. in Chem. Phys.* **9**, 85 (1965).

19. J. Koutecky, *J. Phys. Chem. Solids* **14**, 241 (1960).

20. J. D. Levine, *Phys. Rev.* **171**, 701 (1968).

21. J. Koutecky, *Adv. in Chem. Phys.* **9**, 85 (1965).

22. M. Henzler, *Surface Sci.* **25**, 650 (1971).

23. M. Green and M. J. Lee, in *Solid State Surface Science*, Vol. 1 (M. Green, ed.), Dekker, New York (1969).

24. J. D. Levine and P. Mark, *Phys. Rev.* **144**, 751 (1966).

25. P. Mark, *J. Phys. Chem. Solids* **29**, 689 (1968).

26. P. Mark, *Catalysis Rev.* **1**, 165 (1967).

27. S. R. Morrison, *Surface Sci.* **15**, 363 (1969).

28. R. K. Swank, *Phys. Rev.* **153**, 844 (1967).

29. F. Beck and H. Gerischer, *Z. Electrochem.* **63**, 943 (1959).

30. S. R. Morrison, in *Progress in Surface Science* (S. G. Davison, ed.), Pergamon Press, New York (1971).

31. G. C. Bond, *Catalysis by Metals*, Academic Press, New York (1962).

32. L. Pauling, *The Nature of the Chemical Bond*, 3rd ed., Cornell Univ. Press, Ithaca, New York (1960).

33. W. H. Weinberg and R. P. Merrill, *Surface Sci.* **33**, 493 (1972).

34. R. A. Marcus, *J. Chem. Phys.* **24**, 966 (1956).

35. H. Gerischer, *Adv. Electrochem. Electrochem. Eng.* **1**, 139 (1961).

36. R. A. L. Van den Berghe and W. P. Gomes, *Ber. Bunsenges.* **76**, 481 (1972).

37. R. Memming and F. Möllers, *Ber. Bunsenges.* **76**, 609 (1972).

38. S. R. Morrison (to be published).

39. S. R. Morrison and T. Freund, *J. Chem. Phys.* **47**, 1543 (1967).

40. S. R. Morrison, Surface Sci. **10**, 459 (1968).

41. W. H. Brattain and C. G. B. Garrett, *Bell Syst. Tech. J.* **34**, 1 (1955).

42. W. H. Brattain and J. Bardeen, *Bell Syst. Tech. J.* **32**, 1 (1953).
43. S. R. Morrison, in *Semiconductor Surface Physics* (R. H. Kingston, ed.), Univ. of Pennsylvania Press (1957).
44. R. H. Kingston and A. L. McWhorter, *Phys. Rev.* **103**, 534 (1956).
45. S. N. Kozlov, V. F. Kinselev, and Yu. F. Novototskii-Vlasov, *Surface Sci.* **28**, 395 (1971).
46. H. Chon and J. Pajares, *J. Catalysis* **14**, 257 (1969).
47. S. R. Morrison, *Surface Sci.* **27**, 568 (1971).
48. P. B. Weisz, *J. Chem. Phys.* **21**, 1531 (1953).
49. P. Mark, *Surface Sci.* **25**, 192 (1971).
50. T. A. Goodwin and P. Mark, *Prog. Surf. Sci.* **1**, 1 (1971).
51. S. R. Morrison, *Adv. Catalysis* **7**, 259 (1955); Tech. Rep. #4, ONR Contract N6 onr 24914 (1952).
52. D. A. Melnick, *J. Chem. Phys.* **26**, 1136 (1957).
53. W. L. Brown, W. H. Brattain, C. G. B. Garrett, and H. C. Montgomery, in *Semiconductor Surface Physics* (R. H. Kingston, ed.), Univ. of Pennsylvania Press (1957), p. 111.
54. W. A. Albers, Jr. and J. E. Thomas, Jr., *J. Phys. Chem. Solids* **14**, 181 (1960).
55. J. W. Zemel, *Phys. Rev.* **112**, 762 (1952).
56. Yu. V. Pleskov, *Croatica Chemica Acta* **44**, 179 (1972).
57. H. Gobrecht, R. Thull, F. Hein, and M. Schaldach, *Ber. Bunsenges.* **73**, 68 (1969).
58. J. F. Dewald, *J. Phys. Chem. Solids* **14**, 155 (1960).
59. W. H. Brattain and P. J. Boddy, *J. Electrochem. Soc.* **109**, 574 (1962).
60. B. Hoffman, *Z. Physik* **219**, 354 (1969).
61. B. O. Seraphin, *Surface Sci.* **8**, 399 (1967).
62. J. R. Schrieffer, in *Semiconductor Surface Physics* (R. H. Kingston, ed.), Univ. of Pennsylvania Press (1957).
63. D. R. Palmer, S. R. Morrison, and C. E. Dauenbaugh, *J. Phys. Chem. Solids* **14**, 27 (1960).
64. D. R. Palmer, S. R. Morrison, and C. E. Dauenbaugh, *Phys. Rev.* **129**, 608 (1963).
65. H. Statz, G. A. deMars, L. Davis, Jr., and A. Adams, Jr., *Phys. Rev.* **101**, 1272 (1956).
66. M. Henzler and G. Heiland, *Solid State Comm.* **4**, 499 (1966).
67. M. Henzler, *Surface Sci.* **9**, 31 (1968).
68. G. W. Gobeli and F. G. Allen, *Surface Sci.* **2**, 402 (1964).
69. F. G. Allen and G. W. Gobeli, *Phys. Rev.* **127**, 150 (1962).
70. J. Lagowski, C. L. Balestra, and H. C. Gatos, *Surface Sci.* **29**, 203 (1972).
71. J. Lagowski, E. S. Sproles, Jr., and H. C. Gatos, *Surface Sci.* **30**, 653 (1972).
72. S. R. Morrison, *Surface Sci.* **13**, 85 (1969); *Surface Sci.* **27**, 586 (1971).
73. J. Marlen, R. Leysen, and H. van Hove, *Phys. Stat. Sol.* **5**, 121 (1971).
74. D. T. Stevenson and R. J. Keyes, *Physica* **20**, 1041 (1954).
75. A. Many and D. Gerlich, *Phys. Rev.* **107**, 404 (1957).
76. R. Memming and G. Neumann, *Surface Sci.* **10**, 1 (1968).
77. S. R. Morrison, *J. Vac. Sci. Tech.* **7**, 84 (1970).
78. H. Gerischer, *Surface Sci.* **13**, 265 (1969).
79. W. P. Gomes, T. Freund, and S. R. Morrison, *J. Electrochem. Soc.* **115**, 818 (1968).
80. H. Gerischer, *Surface Sci.* **18**, 97 (1969).
81. K. M. Sancier and S. R. Morrison, *Surface Sci.* (in press).

82. H. G. Völz, G. Kämpf, and H. G. Fitzky, *Farbe u. Lack* **78**, 1037 (1972).
83. U. Kaluza and H. P. Boehm, *J. Catalysis* **22**, 347 (1971).
84. R. J. Collins and D. G. Thomas, *Phys. Rev.* **112**, 388 (1958).
85. W. H. Brattain and C. G. B. Garrett, *Bell Syst. Tech. J.* **34**, 129 (1955).
86. H. Gerischer, *Surface Sci.* **13**, 265 (1969).
87. T. Freund and S. R. Morrison, *Surface Sci.* **9**, 119 (1968).
88. H. Gerischer and E. Meyer, *Z. Physik Chem.* **74**, 302 (1971).
89. E. Cardon and W. P. Gomes, *Z. Phys. Chem.* **78**, 205 (1972).
90. R. A. Marcus, *J. Chem. Phys.* **24**, 966 (1956).
91. R. Memming, *J. Electrochem. Soc.* **116**, 785 (1969).
92. E. Mollwo and H. Schreiber, *Sol. St. Comm.* **8**, 1011 (1970).
93. Y. V. Pleskov, in *Proc. Int. Conf. on Semiconductor Physics, Prague*, Academic Press, New York (1961).
94. V. A. Myamlin and Y. V. Pleskov, *Electrochemistry of Semiconductors*, Plenum Press, New York (1967).
95. D. R. Turner, in *The Electrochemistry of Semiconductors* (P. J. Holms, ed.), Academic Press, London (1962).
96. T. Freund, *J. Phys. Chem.* **73**, 468 (1969).
97. K. Hauffe, V. Martinez, H. Pusch, J. Range, R. Schmidt, and R. Stechemesser, *Appl. Optics* (*Suppl.* 3) **1969**, 34.
98. S. Larach and J. Turkevich, *Appl. Optics* (*Suppl.* 3) **1969**, 45.
99. W. Bauer and G. Heiland, *J. Phys. Chem. Solids* **32**, 2605 (1971).
100. H. Meier, *Spectral Sensitization*, Focal Press, New York (1968).
101. H. Meier, *J. Phys. Chem.* **69**, 719 (1965).
102. E. Inoue, in *Current Problems in Electrophotography* (W. F. Berg and K. Hauffe, eds.), De Gruyter, New York (1972).

4

The Role of the Solid in Electrochemical Phenomena

*Wolf Vielstich, Karl H. Hamann, Joachim Heitbaum,
Wolfgang Schmickler, Helmut Schmidt, Eberhard
Schwarzer*
Institut für Physikalische Chemie der Universität Bonn
and
John M. Hale
Orbisphere Corporation, Geneva

1. Introduction

In this chapter we treat only those aspects of electrochemistry
in which the solid plays a part. If it reads much like an introduction to
all of electrochemistry, then this shows that electrochemical processes
—with very few exceptions—cannot take place without the participation of a solid.

We speak of an *electrochemical reaction** whenever in the course
of reaction an electronic charge transfer occurs between ionically and
electronically conducting phases.

Ionic conductors can be (1) liquids, namely electrolyte solutions
or electrolyte melts, or (2) solids possessing movable ions in their
crystal lattice which can conduct the current (solid electrolytes).
Chapter 6, Volume 4 deals with solid electrolytes, so we can restrict
the following discussion to liquid electrolytes.

Electronic conductors include metals, which exhibit an especially
good conductivity, and also several nonmetallic substances with

---

* For introduction and further reading see Refs. 1 and 2.

metallike conductivity, such as carbon in the form of graphite or active carbon, which are important in many technical applications. Finally, semiconductors exhibit both electron and hole transport; though their conductivity is much poorer, they are gaining more and more electrochemical importance, especially in photoelectrochemistry (see Section 6).

We can make the above characterization of electrochemical reactions more precise by stating: In an electrochemical reaction we are always concerned with heterogeneous processes which proceed at a liquid–solid phase boundary. For completeness some exceptions should be noted: solid electrolytes, which have been mentioned above, and liquid mercury electrodes, which are of technical importance. In these cases we still have phase boundaries, namely solid/solid and liquid/liquid respectively. Only in a very special case when the electrolyte contains solvated, and hence relatively stable, electrons can an electrochemical reaction proceed homogeneously. At present solvated electrons are of purely scientific interest and therefore only mentioned in passing here.

If one of the reacting species is gaseous, the formation of a three-phase boundary, solid–liquid–gaseous, is necessary for the reaction to proceed.

First we present a *static picture* of the phase boundary, i.e., we wish to establish what kinds of interactions occur between ionic (liquid) and electronic conductors (solid) even in the absence of electrochemical reactions, when no currents are passing through the boundary. These interactions result in an orientation or adsorption of ions or dipoles, respectively, and produce a zone of inhomogeneous charge and potential distribution, whose form depends on the nature and the concentration of the adjacent media.

Subsequently we consider the case of an electrochemical reaction with charge carriers passing the liquid/solid phase boundary (*charge transfer reaction*). In this case there are far-reaching analogies with purely chemical reactions: We shall see that there corresponds to the well-known chemical equilibrium an electrochemical equilibrium, which is established by two charge transfer reactions of equal rate and opposite direction. Further, we will see that also in electrochemical reactions the rates depend on the concentration of the reacting species and that they can be varied within wide ranges through changes in temperature and catalysts. For electrochemical reactions we have an additional experimental parameter, the *electrode potential*, with which we can influence the reaction rate.

After these fundamental considerations of phase boundaries without and with current flow, we look at several characteristic examples of electrochemical reactions. We shall find it convenient to distinguish between two cases: (1) The electrode material does not take part directly in the reaction but merely acts as a catalyst (*electrocatalyst*); (2) the electrode material is directly involved, i.e., it changes its valence in the course of the reaction (*metal-ion electrode*).

These studies will show that the charge transfer reaction is the characteristic step, but not the only step, in an electrochemical reaction. As in all heterogeneous reactions, mass transfer plays an important role; further, homogeneous chemical reactions such as dissociation or recombination can precede or follow the charge transfer. Finally, heterogeneous chemical reactions, adsorption and crystallization, can take place at the electrode surface and influence the overall reaction. Within such a reaction sequence, which is often long and branched, we find at least one, and possibly several, charge transfer reactions. Thus the role of the solid in electrochemistry is not confined to the promotion and catalysis of the charge transfer, but the electrode material has a determining influence on all the heterogeneous steps in the reaction sequence.

The current–potential characteristics of semiconductors differ considerably from those of metal electrodes. We have therefore devoted a special section to the electrochemistry of semiconductors.

In conclusion the dominant role of solids in the course of electrochemical reactions will be demonstrated in several technically important processes.

2. The Electrode/Solution Interface

2.1. The Inner and Outer Potentials

Before we discuss the charge distribution at the solid electrode/solution interface we have to clarify the concepts of a "potential" and a "potential difference across an interface." Let us consider a charged body placed in a vacuum. The *outer potential* ψ of the body is defined as the work required to bring a unit test charge from infinity to a point *just outside* the body (the test charge is supposed to have no influence on the charge distribution in the body). In contrast the *inner potential* ϕ is defined as the work required to take a test charge from infinity to a point *inside* the body. How do the two potentials differ?

Inside a solid there are a number of allowed and forbidden energy bands for the electrons. At the surface, there exist in addition localized electronic surface states, which do not necessarily lie within the allowed energy bands. The center of charge of these surface states generally does not coincide with the center of charge of the surface layer of atomic nuclei. As a result the surface of a solid is covered by a *dipole layer*, which generates a change in potential, the *surface potential* χ, between the inside and the outside:

$$\chi = \phi - \psi \tag{1}$$

At the surface of a fluid body, a dipole layer can also be generated by a polarization of the surface molecules.

The outer potential ψ can be measured by an electrostatic voltmeter, at least in principle. It is clear, however, that there is no direct way of measuring the inner potential. Indeed, the following argument shows that not even the difference in the inner potential of two adjacent phases can be measured. Consider two phases, an electrode M and a solution S, in contact. When we want to measure the potential difference ${}_M\Delta_S = \phi_M - \phi_S$ between M and S, we have to connect both to the terminals T_1 and T_2 of a voltmeter. In addition to the interface M–S we have thus also the interfaces T_1–M and S–T_2, and the instrument measures the total potential difference

$$\Delta\phi = (\phi_{T_1} - \phi_M) + (\phi_M - \phi_S) + (\phi_S - \phi_{T_2})$$
$$= {}_{T_1}\Delta_M + {}_M\Delta_S + {}_S\Delta_{T_2} \tag{2}$$

By choosing the same metal for the terminals T_1 and T_2 as for M, we can eliminate the term ${}_{T_1}\Delta_M$,* but there is no way of eliminating ${}_S\Delta_{T_2}$. In fact the terminal T_2 serves as a second electrode, or *counter electrode*. We can sum up our analysis by saying that the *half-cell potential* ${}_M\Delta_S$ cannot be measured, but only the potential difference across a complete cell. For practical convenience, the half-cell potential of a certain reference electrode, the *standard hydrogen electrode* (s.h.e.) is set equal to zero. The *electrode potential* is then defined as the potential difference between the electrode under consideration and a s.h.e. Practical ways of measuring the electrode potential are discussed in Section 3.

When we defined the inner and outer potentials, the test charge was supposed to interact only with the electrostatic field of the body. But a real charge carrier, an electron or an ion, interacts also chemically with its surroundings. The corresponding potential energy (per

---

* In fact, $\Delta\phi$ is independent of the material of the terminals.

mole) is the *chemical potential* μ. The total, or *electrochemical potential* $\bar{\mu}$ of a charge carrier is defined as the sum of the chemical potential μ and the electrostatic potential energy:

$$\bar{\mu} = \mu + zF\phi \tag{3}$$

where z is the charge number and F is Faraday's constant. In the case of an electron in a solid, the electrochemical potential is identical with the Fermi energy E_F. With the aid of the electrochemical potential we can easily formulate the condition of equilibrium at an interface between an electrode M and a solution S. If c is a charge carrier which can pass through the boundary between M and S, we have

$$_M\bar{\mu}^c = {}_M\mu^c + zF_M\phi_0 = {}_S\mu^c + zF_S\phi_0 \tag{4}$$

at equilibrium. The corresponding electrode potential (with respect to the s.h.e.) is the *rest potential* ϕ_0. When all the substances that take part in the charge transfer reaction have unit activities the corresponding rest potential is the *standard potential* ϕ_{00} of the half-cell; we are thus led to the well-known thermodynamic scale of standard potentials. If we impose a potential ϕ, different from the rest potential ϕ_0, on the electrode from without, the electrochemical potential in the electrode changes by an amount

$$_M\bar{\mu}^c(\phi) - {}_M\bar{\mu}^c(\phi_0) = zF(\phi - \phi_0) = zF\eta \tag{5}$$

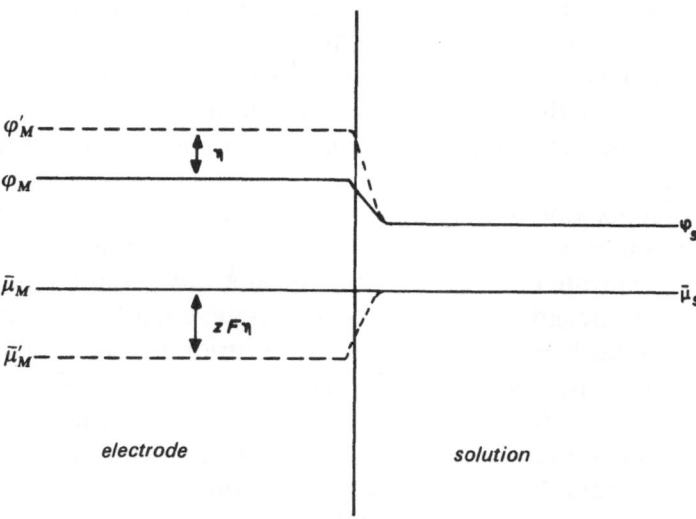

Fig. 1. Electrochemical potential $\bar{\mu}$ and inner potential ϕ at the electrode/solution interface at equilibrium and after application of an overpotential η (dashed lines).

and equilibrium is disturbed. The expression $\eta = \phi - \phi_0$, which is responsible for the deviation from equilibrium, is called the *overpotential*. The shift in electrochemical potential when an overpotential is applied is shown graphically in Figure 1.

2.2. Potential Distribution and Capacity of the Interface

In the remainder of this section we shall suppose that there is no current flowing through the interface electrode/solution, or that the currents are so small that their influence on the structure of the interface can be neglected. We have seen in Section 2.1 that generally, even at equilibrium, there is a potential difference (inner or outer) across the interface. Both surfaces carry excess charges which, since the system as a whole is electrically neutral, are of equal magnitude and opposite sign; the interface is therefore also known as the *double layer*.*

In *metal electrodes* the inner potential is constant because of the high conductivity. Consequently the charge density is zero inside the electrode, and all of the excess charge is concentrated on the surface. In *semiconductors* the excess charge extends further into the bulk of the electrode. Ions that are specifically adsorbed form a second layer of charge on the electrode surface. In a schematic representation (see Figure 2) the center of the adsorbed ions defines the *inner Helmholtz plane* (i.H.p.). While the ions in the i.H.p. are not or only partially solvated, the next layer of ions, which is situated on the *outer Helmholtz plane* (o.H.p.) is assumed to be fully solvated. The higher the conductivity of the solution, the more of the excess charge we expect to be concentrated on the o.H.p. (a quantitative treatment will be given below). Beyond the o.H.p., in the *diffuse layer*, the charge density falls off to zero.

In practice one usually adds a surplus of an indifferent electrolyte to the solution to minimize Ohmic losses. All of the excess charge is then concentrated in a layer about 10–20 Å thick, and the potential drops off over the same distance. When the potential difference between the electrode surface and the bulk of the solution is 100–200 mV, as it often is under experimental conditions, the electric field at the interface is about 10^8 V/m. At such high electric field strengths nearly all the polar solvent molecules are oriented along the field, and dielectric saturation occurs. For water, the effective dielectric constant in the double layer varies between six and 70.

---

* For more extensive treatments of the electrical double layer see Refs. 3–6.

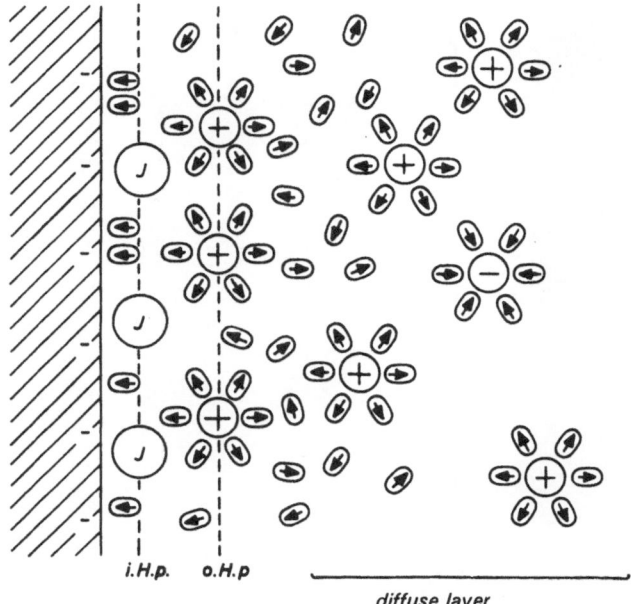

Fig. 2. A microscopic picture of the interface (schematic); J denotes a specifically adsorbed anion or cation.

To calculate the potential distribution at the interface, we introduce a coordinate system as shown in Figure 3. For simplicity we assume that there is no specific adsorption on the electrode; the general case is treated in the cited literature. Let ϕ_e be the inner potential at the electrode surface. The space between the electrode and the o.H.p. is free of charge in the absence of contact adsorption; accordingly the potential changes linearly:

$$\phi(x) = \phi_e - [(\phi_e - \phi_1)/a]x \qquad \text{for} \quad 0 \le x \le a \qquad (6)$$

where $\phi_1 = \phi(a)$ is the potential at the o.H.p. In the diffuse layer we must calculate ϕ from Poisson's equation:

$$\Delta\phi(x) = (\partial^2/\partial x^2)\phi(x) = -(4\pi/\varepsilon)\rho(x) \qquad (7)$$

$\rho(x)$ is the charge density and ε is the dielectric constant of the solution in the diffuse layer. Using a continuum model for the charge distribution, we have for the case of a z–z electrolyte

$$\rho(x) = ez[n_+(x) - n_-(x)] \qquad (8)$$

where $n_+(x)$ and $n_-(x)$ are the densities of the cations and anions, respectively. According to Boltzmann statistics, these are given by

$$n_\pm(x) = n_0 \exp(\mp ze\phi/kT) \qquad (9)$$

271

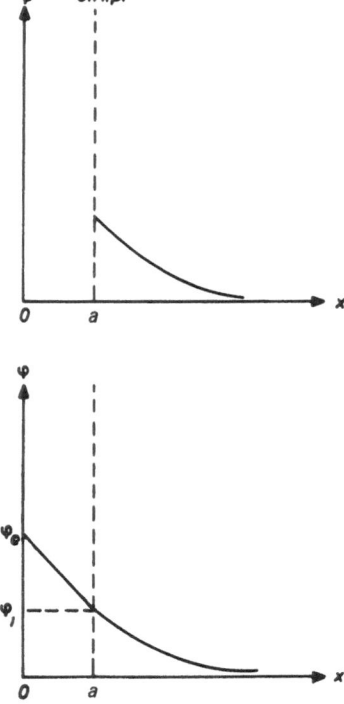

Fig. 3. Potential ϕ and charge ρ distribution at the metal electrode/solution interface in the absence of contact adsorption.

n_0 is the ion density in the bulk of the solution, where the inner potential has been set equal to zero for convenience.

Equations (6)–(9) have the following solution[7]:

$$\phi(x) = \frac{4kT}{ze} \tanh^{-1}\left[\tanh\frac{ze\phi_1}{4kT} \exp - \frac{x-a}{L_D} \right] \quad \text{for} \quad x \geq a \quad (10)$$

where $L_D = (kT\varepsilon/8\pi n_0 e^2 z^2)^{1/2}$ is the *Debye length*. For small values of ϕ_1, Eq. (10) can be approximated by

$$\phi(x) \simeq \phi_1 \exp[-(x-a)/L_D] \quad \text{for} \quad x \geq a \quad (11)$$

which shows that the potential decays roughly exponentially within the diffuse layer, the Debye length L_D acting as a characteristic decay parameter. We note that L_D decreases with n_0 and hence with the conductivity of the solution as predicted above, but increases with temperature: The Brownian movement tends to disperse the charge accumulation at the boundary.

The potential ϕ_1 at the o.H.p. can be calculated from Eqs. (6) and (10) by requiring that $\phi(x)$ be continuous and that the charge on the electrode surface be of equal magnitude and opposite sign to the total charge in the double layer.[8]

Since the inner potential cannot be measured, there is no direct way of verifying the above theory experimentally. It can be indirectly tested, however, by measuring the *differential capacity* $\tilde{C}$, the effective capacity for alternating current. We can regard the interface, for the case of a metal electrode, as two capacitors in series:

$$1/\tilde{C} = (1/\tilde{C}_1) + (1/\tilde{C}_2) \tag{12}$$

where $\tilde{C}_1$ is the differential capacity of the o.H.p. with respect to the electrode and $\tilde{C}_2$ is that of the diffuse layer. For semiconductor electrodes, which we shall treat below, we get a third contribution from the diffuse charge distribution in the electrode. For $\tilde{C}_1$ we have from Eq. (6)

$$\tilde{C}_1 = \varepsilon' A/4\pi a \tag{13}$$

where A is the electrode surface and ε' is the effective dielectric constant for the part of the solvent that lies between the electrode and the o.H.p. From Eq. (10) we get

$$\tilde{C}_2 = (\varepsilon/4\pi L_D) \cosh(ez\phi_1/2kT) \tag{14}$$

So $\tilde{C}_2$, and hence $\tilde{C}$, has a minimum for $\phi_1 = 0$; at this point, which is known as the *point of zero charge* (pzc), the inner potential is constant throughout the solution and both surfaces are free of any excess charge. Consequently the outer potentials of electrode and solution are equal at the pzc. The predicted minimum of the differential capacity is indeed observed experimentally (see Figure 4) for not too high electrolyte concentrations and can be used for an accurate determination of the pzc. Most experimental studies of the pzc have, however, been performed at liquid mercury electrodes, where the pzc coincides with the maximum of the surface tension (capillary rise).

For detailed comparison of Eqs. (6)–(11) with experiment a constant value for $\tilde{C}_1$ is usually chosen to fit the differential capacitance at the pzc.[3] For dilute solutions the agreement is good near the pzc, but at higher potential differences deviations occur. These can be partially explained by dielectric saturation effects on ε', which tend to decrease $\tilde{C}_1$. It is also known that the combination of the nonlinear Boltzman distribution with the linear Poisson equation leads to

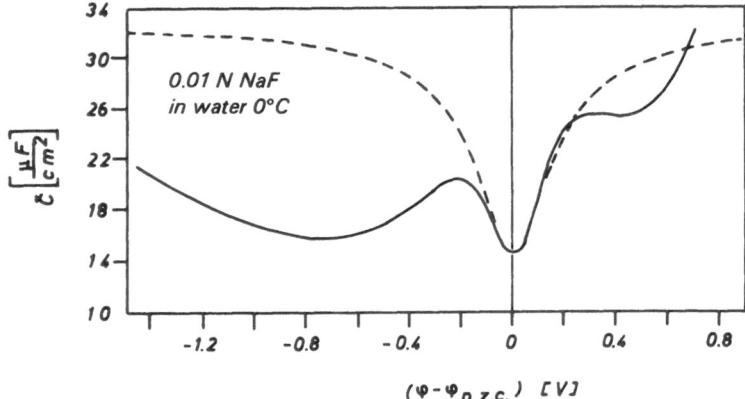

Fig. 4. The differential capacity for an unadsorbed electrolyte. Dashed line: theoretical prediction; solid line: Grahame's[9] data.

inconsistencies, which become especially important at high ionic concentrations. A more consistent theory is based on the method of cluster expansions,[10] but it also gets into difficulties at high concentrations.

So far we have only been concerned with the potential distribution in the solution. What about the electrode? For metal electrodes the potential is constant, but for a semiconductor we expect to have a diffuse charge layer just as in the solution. Indeed, the potential distribution in a semiconductor electrode is very similar to Eq. (10); the exact details depend somewhat on the band structure.[11,12] The main difference is that usually the conductivity of the semiconductor electrode is several orders of magnitude smaller than that of the solution, and the Debye length accordingly greater, so that the diffuse layer extends further into the electrode (see Figure 5). For details see Section 6.

2.3. The Potential of Zero Charge for Metal Electrodes

If no specific adsorption occurs at a metal electrode, an interesting relation can be derived between the electrode potential at the pzc and the electronic work function of the metal. Consider a metal electrode M in contact with an electrochemically inert solution S; the electrode potential is measured by means of a voltmeter with terminals T and T', both of the same material, with respect to a reference

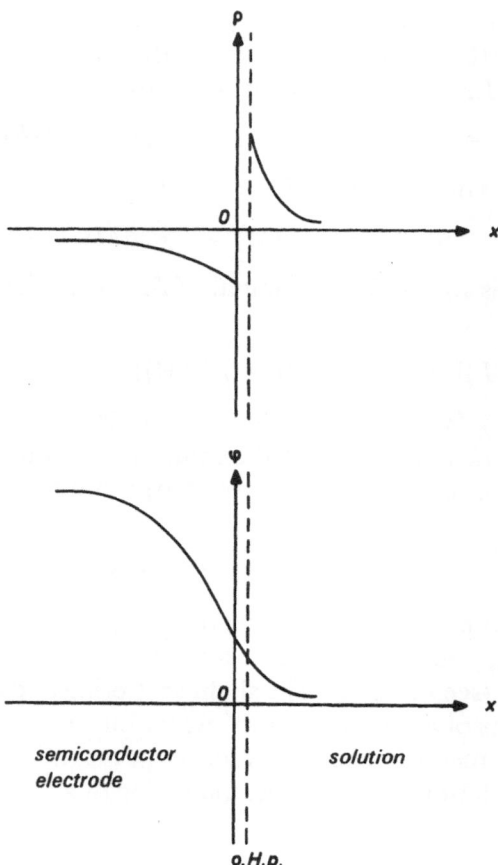

Fig. 5. Potential ϕ and charge ρ distribution at the semiconductor electrode/solution interface in the absence of contact adsorption.

electrode R. The potential difference E measured by the voltmeter is

$$E = \phi(T) - \phi(T') = [\phi(T) - \phi(M)] + [\phi(M) - \phi(S)]$$
$$+ [\phi(S) - \phi(R)] + [\phi(R) - \phi(T')] \qquad (15)$$

At the pzc we have $\psi(M) = \psi(S)$, and hence $\phi(M) - \phi(S) = \chi(M) - \chi(S)$. The conditions of equilibrium are

$$\mu(T) - F\phi(T) = \mu(M) - F\phi(M) \qquad \text{for the interface } T\text{--}M$$

$$\mu(S) - F\phi(S) = \mu(R) - F\phi(R) \qquad \text{for the interface } S\text{--}R$$

$$\mu(T') - F\phi(T') = \mu(R) - F\phi(R) \qquad \text{for the interface } T'\text{--}R$$

where we have used the fact that $\mu(T') = \mu(T)$, since the chemical potential μ of the electrons depends only on the substance under consideration. Inserting these relations into Eq. (15), we obtain

$$\phi_{pzc} = -(1/F)\mu(M) + \chi(M) - \chi(S) + (1/F)\mu(S) \tag{16}$$

The electronic work function of the metal is

$$W(M) = -\mu(M) + F\chi^{vac}(M) \tag{17}$$

where $\chi^{vac}(M)$ is the surface potential of M when placed in vacuum. Thus we have

$$\phi_{pzc} = (1/F)W(M) + [\chi(M) - \chi^{vac}(M)] - \chi(S) + (1/F)\mu(S) \tag{18}$$

The difference $\chi(M) - \chi^{vac}(M)$ should, to a first approximation, not depend on the metal but only on the solution. We thus obtain a linear relation between the potential of zero charge and the electronic work function

$$\phi_{pzc} = (1/F)W(M) + c(S) \tag{19}$$

where $c(S)$ is a function of the solvent only. A plot of $W(M)$ in electron volts versus ϕ_{pzc} should thus give a straight line with slope one, as is indeed the case (see Figure 6). The slight scattering of data is caused by different degrees of orientation of the water molecules at the interface due to specific metal–solvent interaction; Trasatti[13] has suggested an improved relationship which accounts for this effect.

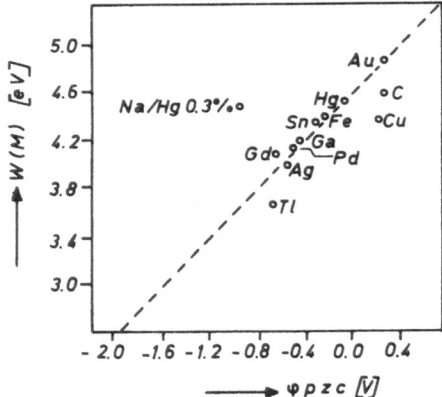

Fig. 6. The potential of zero charge vs. electronic work function for several metals. Measurements were performed in unadsorbed electrolytes.

3. Current–Potential Relation of an Electrode under Load

In this section the disturbance of the electrochemical equilibrium at the interphase electrode–electrolyte is described. For the experimental situation this implies the application of a current i or an overpotential η (see Section 2.1). The overall electrochemical reaction is determined not only by the charge transfer as a basic step but by the mass transfer of the active species and/or by chemical steps preceding or following the electrode reaction.

3.1. The Dynamic Equilibrium at the Interface; the Charge Transfer Current–Overpotential Relationship

In chemical kinetics the overall process of a homogeneous or heterogeneous reaction can be described by the equation

$$nA + mB \underset{k_b}{\overset{k_f}{\rightleftharpoons}} pC + qD \tag{20}$$

the reaction rates in the forward and backward directions being

$$v_f = k_f[A]^n[B]^m \qquad \text{and} \qquad v_b = k_b[C]^p[D]^q \tag{21}$$

The rate constants k_f and k_b depend on the température and the pressure only. If $v_f = v_b$, we have chemical equilibrium, thermodynamically described by $\Delta G_{pT} = 0$. An overall reaction in either direction takes place if we disturb the equilibrium by experimental change in concentrations, temperature, or pressure. Equations (20) and (21) demonstrate that usually the chemical equilibrium is a dynamic one, forward and backward processes occurring simultaneously. A disturbance of the ratio v_f/v_b from the equilibrium value results in a net flux of the chemical reaction in one direction.

The most characteristic difference between a chemical and an electrochemical reaction is the presence of a new important parameter: the electrode potential ϕ. Furthermore, an electrochemical reaction sequence involves the change from a current carried by electrons to a current carried by ions at the electrode surface. This charge transfer can occur by two different modes:

(a) Passing of metal ions through the Helmholtz double layer (see Section 2); the case of deposition and dissolution of metals will be discussed in Section 5.

(b) Passing of electrons through the Helmholtz layer; the inert electrode acts as an *electrocatalyst* in donating or accepting electrons.

This case of a so-called *redox electrode* will be taken as an example here.

The charge transfer at a redox electrode with n electrons can be described as follows:

$$O + ne^- \underset{k_{ox}}{\overset{k_{red}}{\rightleftharpoons}} R \qquad (22)$$

k_{red} and k_{ox} being the rate constants for reduction and oxidation, respectively. They depend, as stated above, not only on temperature and pressure, but also on the electrode potential ϕ. In addition, the k's are not characteristic for a special type of electrode reaction, as, for instance,

$$Fe^{3+} + e^- \rightleftharpoons Fe^{2+}$$

or the reduction of Cl_2 to Cl^- ions

$$Cl_2 + 2e^- \rightleftharpoons 2Cl^-$$

but they usually depend very much on the catalytic activity of the electrode material used (see Section 4).

As in chemical kinetics, in the case of a so-called *reversible electrode reaction* we have a dynamic equilibrium with the same charge transfer rate in either direction; related to 1 cm^2 of surface the rate of this electrochemical reaction is the *exchange current density j_0*. The j_0 value is a measure of the activity of a given electrocatalyst at given working conditions.

In order to prove the reality of the exchange current assumption, the rate of exchange between metal atoms and metal ions in the case of dynamic equilibrium

$$Ag \rightleftharpoons Ag^+ + e^-$$

has been followed by using a radioactive silver ion solution (^{110}Ag). Starting with a pure ^{109}Ag metal electrode, after dipping the electrode in radioactive electrolyte solution for a certain length of time, ^{110}Ag atoms are observed in the upper metal layers.

Due to the fact that the rate constants k_{red} and k_{ox} depend on the potential ϕ, one has, by changing from the equilibrium potential ϕ_0, a resulting net current i through the interface, cathodic or anodic depending on the applied *overpotential $\eta = \phi - \phi_0$*,

$$i = SF(k_{ox}c^\circ_{red} - k_{red}c^\circ_{ox}) \qquad (23)$$

with S the electrode surface, F Faraday's constant, and c°_{red} and c°_{ox} the bulk concentrations of the reduced and oxidized species, respectively.

The assumption that the bulk concentrations $c°$ are established at the electrode surface also, implies that mass transfer processes are fast relative to the electrode reaction (see below). Equation (23) demonstrates that the net current i can be understood throughout the studied potential range as a sum of an anodic and a cathodic partial current $i = i_+ - i_-$, or for 1 cm^2 of surface $j = j_+ - j_-$. In the case of equilibrium potential $\phi = \phi_0$, we have

$$|j_+| = |j_-| = j_0, \qquad \text{but} \quad j = 0 \tag{24}$$

From chemical kinetics we know that the rate constants in Eq. (23) can be expressed as the product of a collision factor k' and an exponential term depending on the free enthalpy of activation $k = k' \exp(-\Delta G^{\neq}/RT)$.

For the charge transfer step the enthalpy of activation depends on the electrode potential. To a first approximation we have

$$\Delta G^{\neq} = \Delta G^{\neq}(\phi_0) - \alpha F\eta$$

with $\eta = \phi - \phi_0$ (overpotential) and $\alpha = -\partial(\Delta G^{\neq})/\partial(F\eta)$. The *transfer coefficient* α lies between zero and one.* In accordance with Figure 7 we obtain for the partial current densities j_+ and j_-

$$\begin{aligned}
j_+ &= Fc^{\circ}_{red}k'_{ox} \exp[(-\Delta G_0^{\neq} + \alpha F\eta)/RT] \\
j_- &= Fc^{\circ}_{ox}k'_{red} \exp\{[-\Delta G_0^{\neq} - (1-\alpha)F\eta]/RT\}
\end{aligned} \tag{25}$$

From the form of Eq. (25) we conclude that only the part $\alpha|\eta|$ or $(1-\alpha)|\eta|$ of the overpotential is effective in accelerating the anodic or cathodic partial current. Figure 8 shows that this can be understood via the qualitative study of the appropriate Morse diagrams. The energy difference in Figure 8(b) is $F\eta$; at the maximum of the energy barrier, however, only the amount $\alpha F\eta$ or $(1-\alpha)F\eta$, respectively, is effective.[14]

Separating in Eq. (25) the terms depending on overpotential and introducing the exchange current density j_0 (for $\eta = 0$ one has $j_+ = j_0$), we obtain as the basic current–overpotential relation of a redox process

$$\begin{aligned}
j &= j_0(\exp[(\alpha F/RT)\eta] - \exp\{-[(1-\alpha)F/RT]\eta\}) \\
j_0 &= Fc^{\circ}_{red}k'_{ox} \exp(-\Delta G_0^{\neq}/RT) = Fc^{\circ}_{ox}k'_{red} \exp(-\Delta G_0^{\neq}/RT)
\end{aligned} \tag{26}$$

with the two characteristic parameters j_0 and α. The slope of the

---

* α is defined as the *anodic* transfer coefficient; in the English literature one often has α as the cathodic and $1 - \alpha = \beta$ as the anodic coefficient.

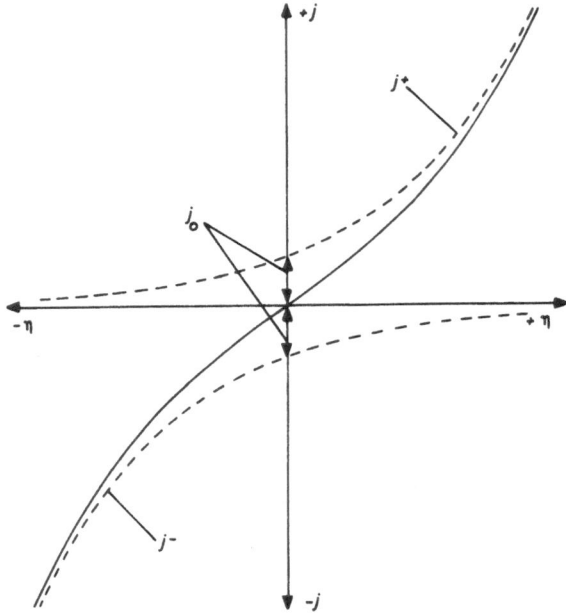

Fig. 7. Current density–potential relationship for the charge transfer step; the two partial current densities j_+ and j_- add up to the total current density j, j_0 is the exchange current density.

current–overpotential curve is determined by the dynamic exchange at the equilibrium; for $\eta \ll RT/F$ it follows from the expansion of the exponential functions in (26)

$$j \simeq (F/RT)j_0\eta \tag{27}$$

For high values of overpotentials we obtain in logarithmic terms

$$\ln j \simeq \ln j_+ = \ln j_0 + \alpha(F/RT)\eta = a + b\eta$$
$$\ln j \simeq \ln j_- = \ln j_0 - (1 - \alpha)(F/RT)\eta = a - b'\eta \tag{28}$$

The slope of these curves is determined by the α value. For most redox reactions the α value ranges from 0.4 to 0.6; i.e., the current–potential plot for the charge transfer shows a nearly symmetric behavior, as in Figure 7.

3.2. Summation of Current–Potential Plots of Single Electrodes; Current–Voltage Relations of Electrolytic Cells

In the foregoing subsection we considered the behavior of single electrodes (or better, of half-cells, see Section 2.1) under load. Practical

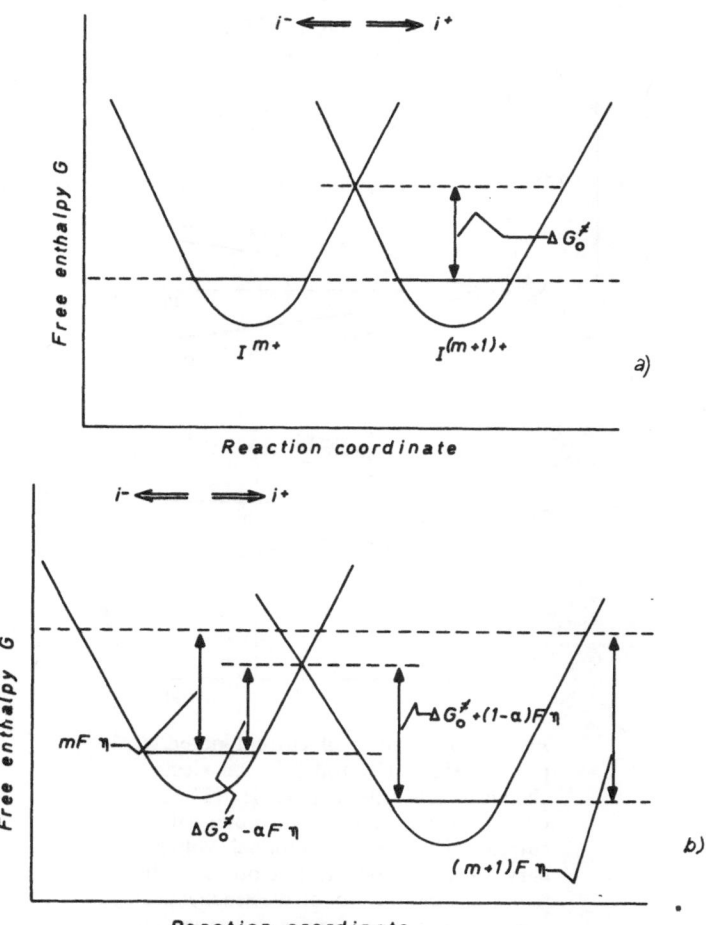

Fig. 8. Energy as function of the reaction coordinate for an electrode reaction of the type $I^{(m+1)+} + e^- \rightleftharpoons I^{m+}$. (a) At the equilibrium potential ϕ_0 with $|j_+| = |j_-| = j_0$ and equal concentrations of I^{m+} and $I^{(m+1)+}$. (b) At an anodic overpotential $\phi - \phi_0 = \eta$; α is the transfer coefficient.

experiments always imply the combination of at least two working electrodes in a complete cell. The potential difference between the two terminals of a cell we call the terminal voltage E_{term}. At open circuit, when no current is flowing, E_{term} is equal to E_0, the open-circuit voltage. The maximum value for E_0, which can be expected from thermodynamics at constant p and T according to

$$E_0 = -\Delta G/nF \tag{29}$$

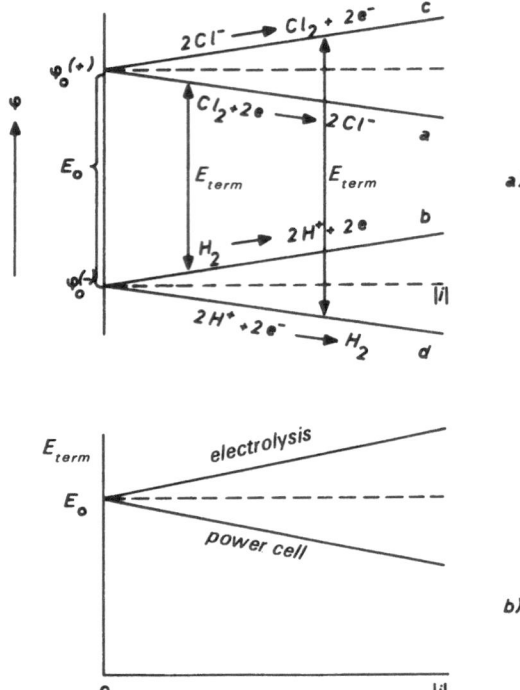

Fig. 9. Summation of current–potential plots of single electrodes for an electrochemical cell (the couple H_2/H^+ and Cl_2/Cl^- as an example). (a) potential vs. current; E_{term} is the terminal voltage; curves a and b denote potentials of the anode and cathode for an electrochemical power source, and c and d the potentials for anode and cathode for electrolysis. (b) terminal voltage E_{term} as function of the current i; E_0 is the open-circuit voltage (emf).

is observed only in exceptional cases (i.e., for the H_2/H^+, Cl_2/Cl^- couple in acid media). Independent of the difficulties, the summation of current–potential curves obtained in practical experiments can be made as in Figure 9. Here anodic and cathodic currents are plotted as absolute values. The equilibrium potentials ϕ_0 of the two electrodes are marked on the ordinate axis. The potential difference between the positive and the negative terminals of the cell is equal to E_0 at $i = 0$; at $i \neq 0$ one has $\phi_+ - \phi_- = E_{term}$.

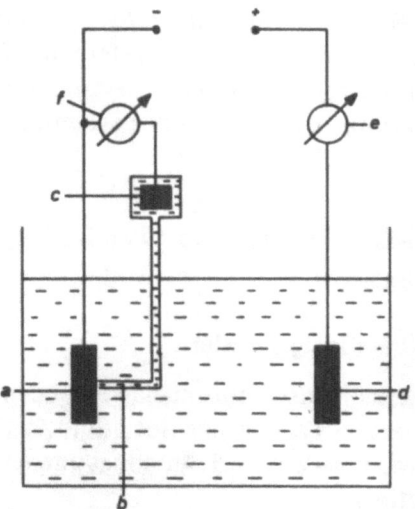

Fig. 10. Experimental setup for measuring current–potential characteristic of a single electrode; a, working electrode; b, Luggin–Haber capillary; c, reference electrode; d, counter-electrode; e, ammeter; f, high-resistance voltmeter.

The cell terminals have different names, depending on the direction of the current. A given electrode is called the *cathode* if a negative charge is transferred from the metal into the solution. Therefore in the case of an electrochemical power source the positive terminal is the cathode. The potential–current plot (Figure 9a) has a negative slope. The potential of the *anode* is increasing with current (curve b). In the case of electrolysis the negative terminal is the cathode, E_{term} is increasing with the current applied by a dc power source from the outside. For simplification the current–potential in Figure 9a are linear; the resulting current–voltage plots are given in Figure 9b.

As was discussed in Section 2, the potential difference between an electronic conductor and an electrolyte, i.e., the single electrode potential, cannot be measured as such. The potential of a single electrode therefore has to be determined against another half-cell as reference whose potential is kept constant throughout the investigation. In the experimental cell, we introduce a third electrode, the so-called *reference electrode* (Figure 10). In order to avoid the voltage drop at the Ohmic resistance of the electrolyte between anode and cathode, one touches the potential near the surface of the working

electrode by means of a *Luggin–Haber capillary*. The potential
between the working electrode and the reference electrode is recorded
as function of the current (between working electrode and counter
electrode). The potential of the reference system is not influenced, due
to the high-resistance voltmeter used.

3.3. Influence of Mass Transfer and Chemical Steps on the Current–Potential Relation

3.3.1. Concentration Overpotential

In most practical cases the surface concentrations c_i^s of the
electrochemically active species are not equal to the bulk concentration c_i^o as assumed in Section 3.1. In consequence, Eq. (26) has to be
transformed as follows:

$$j = j_0 \left\{ \frac{c_{red}^s}{c_{red}^o} \exp\left(\frac{\alpha F}{RT}\eta\right) - \frac{c_{ox}^s}{c_{ox}^o} \exp\left[-\frac{(1-\alpha)F}{RT}\eta\right] \right\} \qquad (30)$$

In order to obtain the complete current–overpotential relation, it is
necessary to calculate the surface concentrations c_i^s in the given cases.
The deviation of the c_i^s from the c_i^o values can be caused by limited
diffusion (in stagnant electrolytes) or by *convection*. Limited mass
transfer can easily be proved by varying the intensity of stirring. The
important interaction of mass and charge transfer will be discussed
later.

If the intensity of stirring has no influence on the reaction rate,
the deviation of c_i^s from c_i^o and the resulting limiting current are
caused by rate-determining *chemical steps*. In all three cases mentioned
the additional overpotential is called the *concentration overpotential*.

Typical preceding chemical steps are adsorption and homogeneous or heterogeneous dissociation. For the anodic hydrogen
oxidation one obtains at high overpotentials (100–300 mV) a limiting
current due to slow adsorption and/or dissociation:

$$H_{2,sol} \xrightarrow{\text{slow}} 2H_{ad}$$

$$2H_{ad} \xrightarrow{\text{fast}} 2H^+ + 2e^-$$

An example for a cathodic limiting current is the slow dissociation
of weak acid molecules during the hydrogen evolution out of the

respective electrolyte solutions:

$$HA \xrightarrow{\text{slow}} H^+ + A^-$$

$$H^+ + e^- \xrightarrow{\text{fast}} H_{ad}, \qquad 2H_{ad} \xrightarrow{\text{fast}} H_2$$

3.3.2. The Influence of Mass Transfer in Stagnant Electrolytes

In an electrolytic cell without convection the application of a current pulse results in a change of the concentration (decrease or increase) at the surface and in a depletion (or enrichment) of the nearby electrolyte (Figure 11a, curve 2). In the case of cathodic deposition, high overpotentials result in zero surface concentration (curve 3) and in a following decrease of the concentration profile (curve 4). The limiting current density j_{lim} as function of time is given by Fick's first law in the case of zero concentration at the surface ($c = 0$ at $x = 0$; x is the direction normal to the electrode)

$$j_{lim} = -nFD(\partial c/\partial x)_{x=0} = nFDc_0/\delta \qquad (31)$$

with

$$\delta = (\pi D t)^{1/2} \qquad (32)$$

The length δ of the so-called diffusion layer according to Eq. (31) is defined by the dashed-line construction for curve 3 in Figure 11a.

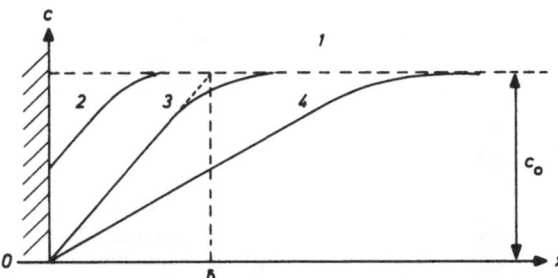

Fig. 11a. Concentration profile during the deposition of an electrochemically active species at the electrode surface in stagnant electrolyte; c_0, bulk concentration; x, distance from the electrode; δ, diffusion layer: 1, without current; 2, at a time t after applying a current; 3, as in 2 but under limiting current conditions; 4, as in 3 but later in time.

3.3.3. Mass Transfer under Convective Diffusion

Using convection as mode of mass transfer the current–potential and current–voltage curves after a transient period become independent of time. With a time-independent concentration profile we also have a constant thickness of the diffusion layer δ.

Such steady-state current–potential curves are presented in Figure 11b. Curve 1 can only be obtained at infinite rate of mass transfer and chemical reaction steps. The anodic parts of curves 2 and 3 refer to two different intensities of stirring the electrolyte and two respective stationary diffusion layers δ_2 and δ_3, $\delta_2 > \delta_3$. Mass transfer-controlled limiting currents can be increased to very high values if no other reaction step becomes rate determining. In the example of Figure 11b the adsorption and dissociation of molecular hydrogen finally limit the anodic current density (curve 4). At cathodic hydrogen evolution (left part of Figure 11b) so far no limiting current range has been observed. The stirring by gas bubbles, increasing with current, takes care of the mass transfer requirements.

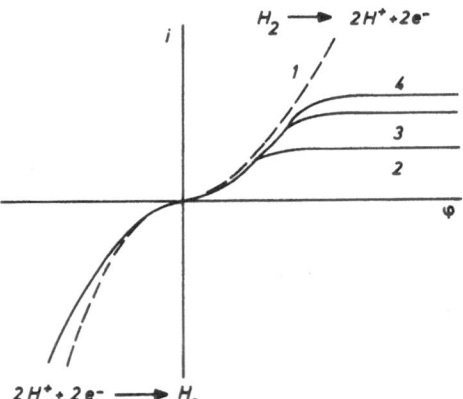

Fig. 11b. Current–potential curve showing a limiting current range at anodic potentials for different rates of convective diffusion (hydrogen oxidation at platinum in acid solution as an example): 1, charge transfer control at infinite rate of convection and preceding chemical reaction(s), in this example adsorption and dissociation; 2, medium rate of convection; 3, increased convection; 4, chemical reaction-limited current at high rate of convection.

4. The Electrode as a Catalyst

4.1. Electrocatalysis

In ordinary chemistry a catalyst is a substance which increases the rate of a reaction without being consumed. A catalyst increases the rates of both the forward and the backward reactions in the same proportion; it has no influence on the equilibrium constant, which, in condensed phases, is solely determined by the reaction free enthalpy.

In this section we shall consider electrochemical redox reactions such as

$$O_2 + 4H^+ + 4e^- \rightleftharpoons 2H_2O$$

—reduction of oxygen in acid solutions—where an electron transfer takes place between an ion or molecule in solution and a solid electrode. Since the electrode is left unchanged by the overall process, it acts as a catalyst, an *electrocatalyst*. The rates of electrochemical redox reactions vary widely with the electrode, while the rest potential ϕ_0, which corresponds to the equilibrium constant K in chemical kinetics, does not depend on the electrode material.

When we compare the catalytic properties of two different electrodes for the same redox reaction, we shall call that electrode a better catalyst which, at a given overpotential, gives higher current densities. When a charge transfer is rate determining, the current density at an overpotential η is given by the Butler–Volmer equation (26) (see Section 3):

$$j = j_0(\exp[(\alpha F/RT)\eta] - \exp\{-[(1 - \alpha)F/RT]\eta\})$$

If the transfer coefficient α is nearly the same for both electrodes, the electrode with the higher exchange current density j_0 is the better electrocatalyst at all potentials. This is usually the case when the reaction mechanism and the rate-determining step are the same. In some instances, however, the transfer coefficients differ widely; the current–potential characteristics of the two electrodes can then cross. Figure 12 shows Tafel plots ($\log j$ versus η) for the oxidation of ethylene at platinum and at an 80% Pt–20% Rh alloy (see also Section 4.3.2b). At potentials below 0.45 V (vs. s.h.e.) the alloy is a better catalyst, while at higher potentials pure platinum is better. This indicates that the reaction mechanism and/or the rate-determining step are different at the two electrodes.

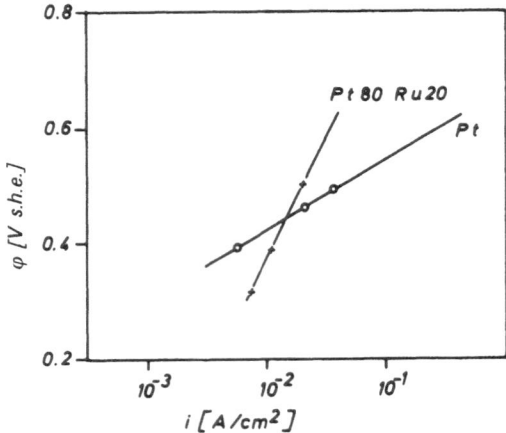

Fig. 12. Experimental current–potential relationship for the oxidation of ethylene on platinum and 80% Pt–20% Ru alloy (after Bockris and Reddy[1]; by permission of Plenum Press).

In the remainder of this section we shall have a more detailed look at redox mechanisms and consider the catalytic influence of the electrode material at individual reaction steps.

4.2. Redox Reactions without Adsorption

Electrochemical redox reactions are particularly simple when no adsorption of reactants or products occurs. While such reactions are rarely of practical importance, they have been the playground of theoreticians. This is, however, not the place to plunge into the depth of quantum electrochemistry; we shall restrict ourselves to a survey of activation enthalpies and exchange current densities.

According to current theories[15,16] the heat of activation $\Delta H^{\neq}$ of a simple redox reaction depends only on the energy of reorganization of the solvation sphere of the reactant; the electrode material should have no influence. This prediction agrees fairly well with the results of Bockris et al.[17] for the reaction

$$[Fe(H_2O)_6]^{2+} \rightleftarrows [Fe(H_2O)_6]^{3+} + e^-$$

in acid solution (see Figure 13)—the deviation of palladium could be caused by competitive hydrogen evolution.

The exchange current density does, however, vary with the electrode material. According to Bockris and Reddy (Ref. 1, pp. 1141–

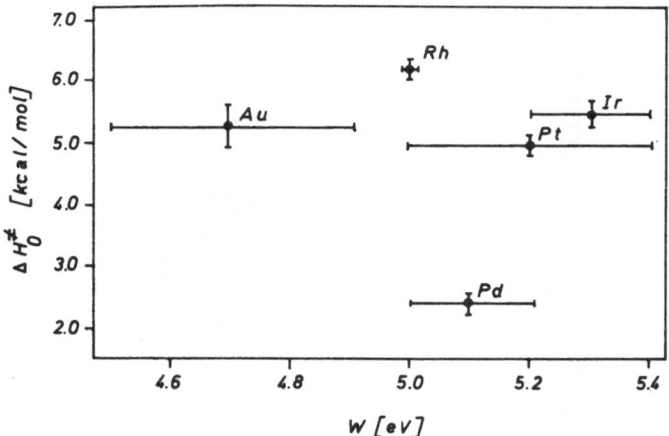

Fig. 13. The heat of activation $\Delta H_0{}^{\neq}$ as a function of the substrate work function W for the ferric–ferrous redox system on the noble metals (Bockris *et al.*[17]; by permission of American Institute of Physics).

1173), a plot of $\log j_0$ versus the electronic work function W of the metal gives a straight line, for the following reason: As we have seen in Section 2, the concentration c of the reactant in the o.H.p. differs from that in the bulk of the solution, unless the electrode is at the pzc; we have

$$c \sim c_0 \exp[-z(\phi - \phi_{pzc})/kT]$$

according to Boltzmann distribution [cf. Eq. (9)]. On the other hand, $j_0 \sim c$ [cf. Eq. (26)] and $\phi_{pzc} \sim W$ [cf. Eq. (19)], from which it follows that

$$\log j_0 \sim W$$

The experimental evidence (see Figure 14) is, however, not quite convincing.*

4.3. Redox Reactions with Adsorption

The majority of electrochemical redox reactions include an adsorption step. While in the simple redox reactions which we discussed in the previous section the electrode and the reactants did not come into direct contact and the electrode material had but little influence on the reaction rate, the interaction between an adsorbed

*Note Added in Proof. The matter has recently been reconsidered by Trasatti, *J. Electroanal. Chem.* **44**, 367 (1973); he has confirmed the linear relation between $\log j_0$ and W, but criticized the work of Bockris as giving the wrong slope.

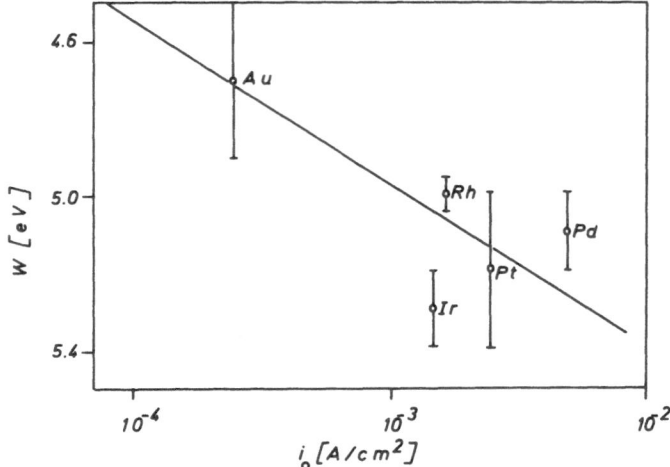

Fig. 14. The exchange current density i_0 for the Fe^{3+}/Fe^{2+} redox couple (0.01 M concentration) plotted against the work function W of the noble metal substrate (Bockris *et al.*[17]; by permission of American Institute of Physics).

reactant and an electrode is much stronger. We are therefore not surprised to find that, whenever adsorption is involved, the reaction rate and even the reaction mechanism vary considerably with the electrode material; indeed, some reactions will occur readily at a particular electrode and not at all at others.

Before we discuss the influence of adsorption on the reaction in several typical examples (Sections 4.3.2 and 4.3.3), we shall treat some theoretical aspects of adsorption bonds. Finally, we shall briefly describe three experimental methods for the study of adsorption layers.

4.3.1. Estimation of the Strengths of Adsorption Bonds

Most theoretical studies of adsorption phenomena have so far been done on adsorption from the gas phase, where the situation is much simpler. Electrochemists have usually adapted these results to adsorption from the liquid phase by introducing corrections for the influence of the solvent.

Let us consider the adsorption of hydrogen on metals:

$$H_2 \rightleftharpoons 2H_{ad}$$

which is so far the best-studied process. We can write the enthalpy of adsorption approximately as

$$\Delta H_{ad} = 2D_{MH} - D_{HH} \tag{33}$$

where D_{HH} is the dissociation energy of the hydrogen molecule and D_{MH} the energy of the metal–hydrogen bond. D_{MH} can be estimated from *Pauling's equation*, which for our case is

$$D_{MH} = \tfrac{1}{2}(D_{MM} + D_{HH}) + 23.06(x_M - x_H)^2 \tag{34}$$

x denotes the electronegativity; the numerical factor has been chosen to give the binding energy in kcal/mole. We obtain for the enthalpy of adsorption

$$\Delta H_{ad} = D_{MM} + 2 \cdot 23.06(x_M - x_H)^2 \tag{33a}$$

The metal binding energy D_{MM} is approximately one-sixth of the latent heat of sublimation; the electronegativities have been tabulated. For adsorption from the gas phase Eq. (33a) gives good results for most metals, e.g., Fe, Ni, W (see Table 1). while considerable discrepancies are observed in some cases. For electrochemical adsorption the experimental data are still uncertain, so that no comparison can be made at the present time.

On transition metal electrodes the adsorbed species are bonded predominantly via the outer d orbitals. The *d-band character*, i.e., the

TABLE 1
Calculated and Experimental Values of ΔH_{ad} for Hydrogen Adsorption

Metal	ΔH_{ad}, kcal mole	
	Exp.	Calc.
Co	24	31.0
Rh	26	32.3
Ru	26	38.1
Ir	26	38.1
Pd	27	22.5
Cu	28	25.6
Pt	28	22.6
Ni	29–32	28.6
Mo	40	42.9
Cr	45	24
Ta	45	49.6
Fe	32–36	31.6
W	45–52	45.6

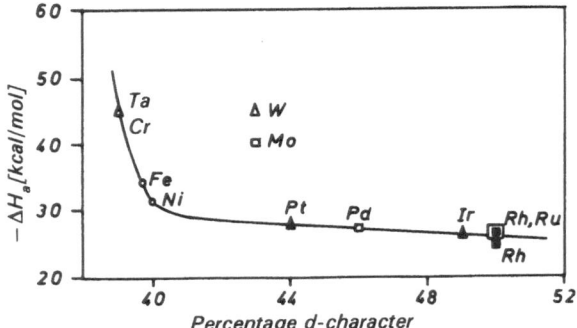

Fig. 15. Variation of the initial heat of adsorption of hydrogen as a function of percentage d character of various metals. Open symbols, evaporated metal films; solid symbols, silica-supported metals (after Bond[74]; by permission of Academic Press).

percentage of occupied outer d-electron orbitals, serves as a convenient characterization of the adsorption properties. Metals with a low d-band character, such as elements of the groups VIIIb and Ib, have many unpaired electrons and can thus form strong adsorption bonds with electron donors, while elements with a high d-band character have but few unpaired electrons and form only weak bonds. Thus the heat of adsorption decreases with increasing d-band character, as shown in Figure 15 for the adsorption of hydrogen.

More rigorous treatments of adsorption from the liquid state are not yet available, since our present knowledge of surface states of solids is still meager.

4.3.2. Reactions with Adsorption of Reactants and/or Products

We consider an adsorption process $X \rightarrow X_{ad}$ embedded in a reaction sequence. When the free enthalpy of adsorption is low, as is usually the case for electrode materials with a high d-band character, the adsorption process is slow and can become the rate-determining step in the sequence. The coverage θ_X of the electrode with the adsorbate is then low since the adsorbed species is consumed in one of the relatively fast following steps.

By using an electrode material with better adsorptive properties (e.g., with a lower d-band character) we can increase the reaction rate until another step becomes rate determining. A further increase in the energy of adsorption will not yield a greater current density; it can even result in a decrease of the reaction rate by slowing down the desorption of the products. This pattern will emerge again in the following examples.

a. Hydrogen Evolution

In the early days of electrochemistry, research centered about the evolution of hydrogen, which is probably still the best-studied electrochemical process, even though it is far from being completely understood. Two different reaction mechanisms are known in acid solutions:

(I) The Volmer–Tafel mechanism:

$$H^+ + e^- \rightleftharpoons H_{ad} \qquad \text{(Volmer reaction)}$$

$$2H_{ad} \rightleftharpoons H_2 \qquad \text{(Tafel reaction)}$$

(II) The Volmer–Heyrovsky mechanism:

$$H^+ + e^- \rightleftharpoons H_{ad} \qquad \text{(Volmer reaction)}$$

$$H_{ad} + H^+ + e^- \rightleftharpoons H_2 \qquad \text{(Heyrovsky reaction)}$$

The Tafel and Heyrovsky reactions can occur in parallel. Gerischer[18] has estimated the energy of activation for all three reactions as function of the enthalpy of adsorption ΔH_{ad}. He came to the following conclusions: For low ΔH_{ad} the Volmer reaction is rate determining; the coverage θ_H of the electrode with hydrogen is then low and the rate of

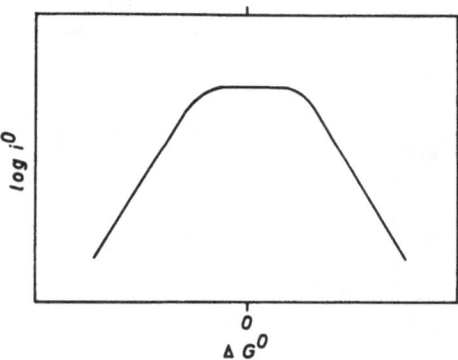

Fig. 16. Variation of the exchange current density with the standard free energy of adsorption of hydrogen. The horizontal segment corresponds to intermediate coverage and the logarithmic Temkin isotherm. The linear segments correspond to the Langmuir isotherm (Parsons[19]; by permission of the American Chemical Society).

293

the Tafel reaction, which depends quadratically on θ_H, can be neglected—the Heyrovsky reaction predominates. For intermediate ΔH_{ad} the Heyrovsky and the Tafel reaction can proceed at about equal rate, while for high ΔH_{ad} again the Heyrovsky mechanism is favored, and is also rate determining.

Parsons[19] has estimated the dependence of the exchange current density j_0 on the free enthalpy of adsorption ΔG_{ad}. Independent of which reaction step is rate determining, he arrived at the following conclusion: For low ΔG_{ad} the coverage θ_H of the electrode and hence j_0 are low. With increasing ΔG_{ad} both the coverage θ_H and j_0 increase. At $\Delta G_{ad} = 0$ the exchange current density reaches its maximum. A further increase in ΔG_{ad} results in a slowing down of the desorption process, and j_0 decreases again. A plot of $\log j_0$ versus ΔG_{ad} should thus give a "volcano"-shaped curve (see Figure 16). Parsons has substantiated this qualitative reasoning by a detailed mathematical analysis. Experimental determinations of ΔG_{ad} are very uncertain but such measurements as do exist seem to corroborate his theory (see Figure 17).

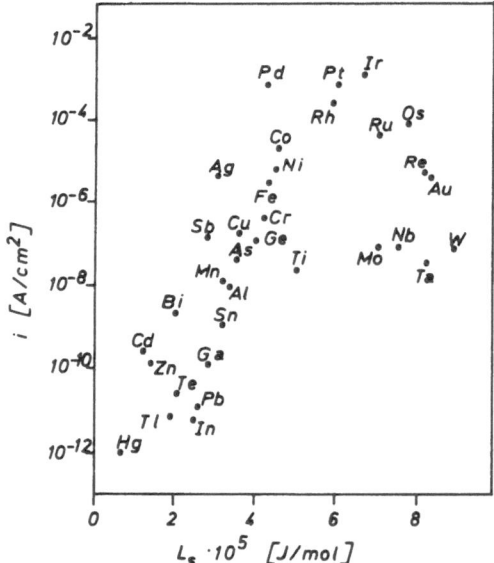

Fig. 17. Exchange current density i_0 for hydrogen evolution vs. latent heat of sublimation L_s; according to Eq. (33a) L_s is approximately proportional to ΔG_{ad} (after Kuhn et al.[75]; by permission of *Journal of Electroanalytical Chemistry*).

b. The Oxidation of Ethylene

In the anodic oxidation of hydrocarbons not only the reactant itself but also an oxygen donor is adsorbed at the electrode. For the oxidation of ethylene in aqueous solutions, which proceeds according to the overall reaction equation

$$C_2H_4 + 4H_2O \rightarrow 2CO_2 + 12H^+ + 12e^-$$

the oxygen is delivered by water. The reaction mechanism has not yet been completely elucidated, but the first steps are known to be

$$C_2H_4 \rightleftharpoons (C_2H_4)_{ad}$$

$$H_2O \rightleftharpoons OH_{ad} + H^+ + e^-$$

$$(C_2H_4)_{ad} + OH_{ad} \rightarrow \text{organic radicals}$$

Only noble metals are suitable as electrodes, since other metals dissolve at the required potentials.

As usual, we expect a volcano-type curve when we plot the current density versus d-band vacancies. At first the increase in the adsorption

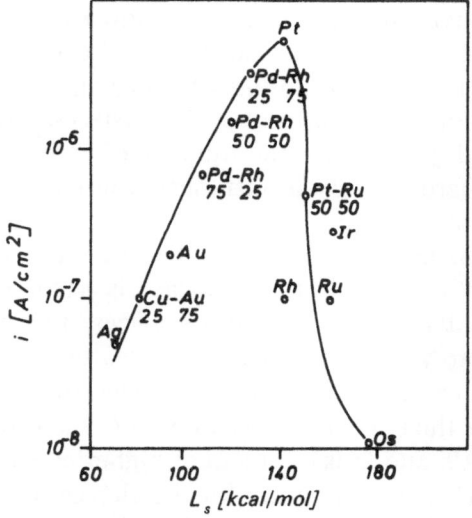

Fig. 18. The volcano relationship for the oxidation of ethylene upon various substrates, relating the reaction rate in amperes per square centimeters to the substrate heat of sublimation (after Kuhn *et al.*[76]; by permission of the American Chemical Society).

rate speeds up the overall reaction, while at still lower *d*-band character desorption is slowed down and becomes rate determining. There is, however, a reason why in this particular instance the descent of the volcano curve should be steeper than usual, and the peak more pointed: The adsorption of ethylene and that of OH radicals are competitive processes. The heats of adsorption of the hydroxyl radicals on noble metals are much higher than those of ethylene. We therefore expect a continuous increase of OH adsorption with decreasing *d*-band character, while the adsorption of ethylene should pass through a maximum, probably at platinum, since at metals with a still lower *d*-band character the prevailing OH-adsorption will reduce the adsorption of ethylene.

So we expect a sharp maximum of the reaction rate at platinum, which is borne out in Figure 18, where the current density is plotted, not against *d*-band vacancies, but against the latent heat of sublimation, which, according to Pauling's equation (cf. Section 4.3.1), is also a measure of the strength of adsorption bonds.

4.3.3. The Influence of Surface-Active Substances on the Catalytic Activities of Electrodes

So far we have only considered the adsorption of reactants or products. But adsorbed species need not take part directly in the reaction in order to have a profound influence on the catalytic activity of the electrode. In the trivial case such "surface-active substances," as they are called, just reduce the number of active sites and act as inhibitors; but there are several instances where they cause a significant increase of the reaction rate.

The anodic oxidation of formic acid at platinum in acid solutions is such a case. When the electrode surface is partially covered with lead the current density at potentials between 300 and 700 mV (vs. s.h.e.) is considerably greater than at a pure platinum electrode[20] (see Figure 19). The catalytic activity of the electrode varies with the coverage θ: First the current increases with θ, then reaches its maximum at $\theta_{max} \approx 0.9$, and falls off again at higher coverages.

Since the energy of activation for the charge transfer step is not found to be lowered, the following explanation seems plausible: The formic acid is adsorbed at lead atoms and oxidized at platinum; the reaction rate therefore increases with the number of active boundary sites lead/platinum, which reaches its optimum value at θ_{max}. Similar effects are observed with adsorbed sulfur and selenium.[21]

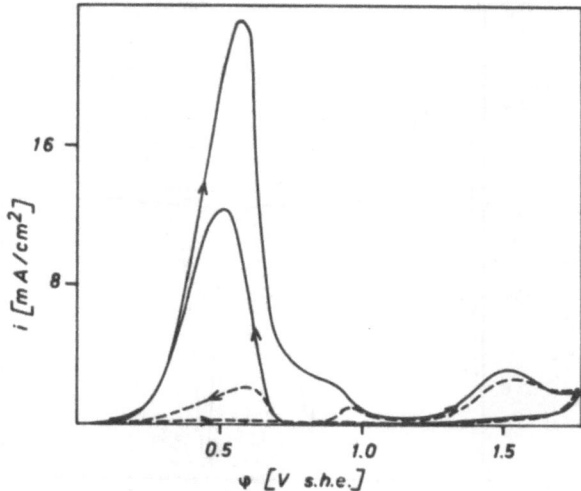

Fig. 19. Current–potential curves on platinum in 0.5 M $H_2SO_4 + 1 M$ HCOOH: broken line, without lead; solid line, with lead. Measurements were made by the triangular potential scan method, $v = 100$ mV/sec; see Section 4.3.4.

4.3.4. Experimental Methods for the Study of Adsorption

There are a number of experimental techniques for the study of adsorption processes: capacity measurement,[22] radiometric determination,[23] charging curves, triangular potential scan,[24] reflection spectroscopy, and recently mass spectroscopic methods. Of these we shall only treat the potential scan method, internal reflection spectroscopy, and the X-ray photoelectron spectroscopy (XPS) method; for the other techniques the interested reader is referred to the cited literature.

a. Triangular Potential Scan Method

In the triangular potential scan (or cyclic linear voltage sweep) method a periodic triangular potential is imposed on the electrode. In aqueous solutions the hydrogen evolution potential (0 mV s.h.e.) and the beginning of oxygen evolution (~ 1600 mV s.h.e.) are usually chosen as turning points, the scan rate being of the order of 0.01–0.1 cycle/sec.

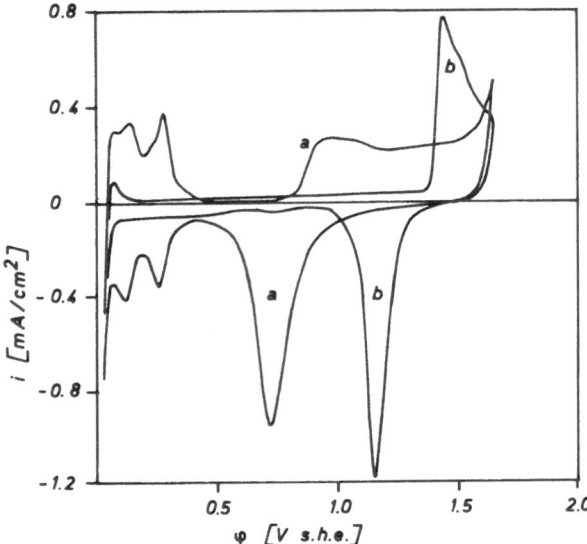

Fig. 20. Current–potential curves in 0.5 M H$_2$SO$_4$: a, on platinum; b, on gold. Measurements were made by the triangular potential scan method, $v = 100$ mV/sec; see Section 4.3.4.

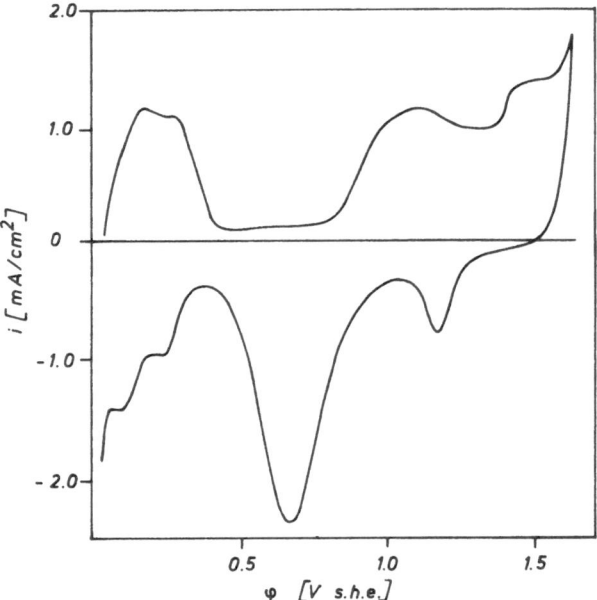

Fig. 21. Current–potential curves in 0.5 M H$_2$SO$_4$ on 60% Pt–40% Au alloy. Measurements were made by the triangular potential scan method, $v = 100$ mV/sec; see Section 4.3.4.

Figure 20(a) shows the resulting current–potential characteristic of a smooth platinum electrode in 1 N sulfuric acid. In the anodic direction, adsorbed hydrogen is oxidized between 0 and 350 mV; the two current peaks correspond to two different adsorption states of hydrogen on platinum.[27] Between 350 and 800 mV no electrochemical reaction occurs; the residual current is due to double layer charging. At about 800 mV, chemisorption of oxygen sets in, and at 1600 mV, oxygen evolution begins. In the backward (cathodic) scan the reverse processes are observed; we note that the reduction of the adsorbed oxygen layer requires an overpotential of several hundred millivolts. Figure 20(b) shows the corresponding diagram for a gold electrode; here the coverage with hydrogen is much smaller; the chemisorption of oxygen sets in more rapidly, but at a considerably higher potential. These two diagrams are additively superimposed for a gold/platinum alloy electrode (see Figure 21); an experienced experimenter can recognize the constituents of an alloy and also estimate their percentages.

b. Internal Reflection Spectroscopy

During the last decade several optical methods[28,70-72] have been developed for *in situ* study of adsorption layers. In internal reflection spectroscopy a glass slide is coated with a thin, conducting film—gold, platinum, and doped tin oxide have been found suitable. A light beam is incident from the electrode side, refracted at the glass/film boundary and reflected at the film/solution interface. During reflection the light ray penetrates several hundred Ångstroms into the solution, so the amplitude of the reflected ray depends critically on the optical properties of the solution. Usually an angle of incidence is chosen such that the beam is not propagated into the solution phase. The light beam is then reflected several times within the film (see Figure 22) and the intensity and spectral distribution of the emerging light are observed. When an adsorption layer develops on the film the optical properties of the film/solution interface and hence the amplitude and spectral distribution of the reflected rays change in a characteristic way as a result of light absorption in the metal film and the adsorbed layer. Examples of the use of the method can be found in Ref. 72. At present, optical methods are being developed rapidly and show great promise not only for the study of adsorption layers, but also for very fast charge transfer reactions.

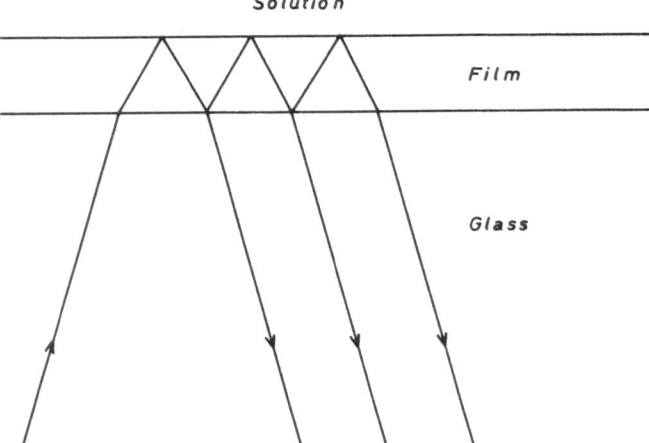

Fig. 22. Principle of internal reflection spectroscopy.

c. X-Ray Photoelectron Spectroscopy

In X-ray photoelectron spectroscopy (XPS) monoenergetic photons are incident on a sample surface; the generated electrons, which may be Auger, photo, or secondary electrons, are analyzed according to their kinetic energy and spin orientation. The photoelectron spectrum will reflect the electronic energy structure of the surface.

When the XPS method is used for analytical purposes, such as the study of adsorption, it is referred to as ESCA (electron spectroscopy for chemical analysis). When a species is adsorbed at a surface, the effective Coulomb potential of the inner electrons is changed. Correspondingly the energy spectrum of the inner electrons exhibits a shift, which varies with the strength of the bonds. In this way ESCA techniques distinguish, for instance, between adsorbed oxygen and chemically bound oxygen on an electrode surface. Examples of the use of ESCA in electrochemistry can be found in Ref. 73.

5. The Mechanism of Deposition and Dissolution of Metals

In the preceding section we have discussed electrochemical redox processes catalyzed on an inert metal electrode. In many electrochemical processes, however, the metal itself takes part in the electrode reaction concerned, e.g., metal ions in the solution are discharged on the electrode and incorporated in the metal lattice or, in the opposite direction, the metal undergoes dissolution, forming solvated metal ions in the liquid phase. It is clear that besides the kinetic phenomena of the charge transfer across the metal–solution interface, the crystallographic properties of the substrate play an important role in the rate of the overall process.

In the following we will concern ourselves first with the kinetics of metal deposition, second, with the crystallographic aspects of this subject, and finally, with metal dissolution. It should be mentioned that the corrosion of metals is a closely related subject, because the corrosion rate is determined by the rate of the anodic dissolution of the metal, by surface layer formation and inhibition, and by the properties of the electrolyte involved. But because of the significance of the corrosion phenomena, metal corrosion is discussed elsewhere in this treatise (Vol. 4, Chapter 9).

5.1. Kinetics of Deposition

5.1.1. Basic Steps of Metal Deposition

The overall reaction of metal deposition from an aqueous solution is represented by

$$[M^{z+}(H_2O)_n]_{sol} + ze^- \rightarrow M_{lattice} + nH_2O \qquad (35)$$

where $M^{z+}(H_2O)_n$ represents a hydrated metal ion in the bulk of solution and M is a metal atom in the crystal lattice. Such a reaction consists of a number of elementary steps which may proceed successively or in parallel with each other:

1. Transport of the hydrated ion from the bulk of solution to a position at the outer Helmholtz plane (cf. Section 2), from which it can be transferred to the electrode surface.

2. Charge transfer between the electrode and the ion (reduction of the ions) and its partial dehydration.

3. Lateral motion to a position of incorporation into the lattice or encounter with other particles (adions or metal atoms) to form a crystal nucleus.

4. Incorporation of the particle into a fixed lattice position and further dehydration.

Figure 23(a) illustrates the elementary steps within metal deposition mentioned above. Figure 23(b) shows that this need not be the only pathway of reaction; it may be that step 3 is not included in the reaction mechanism, but the metal ion is directly transferred from

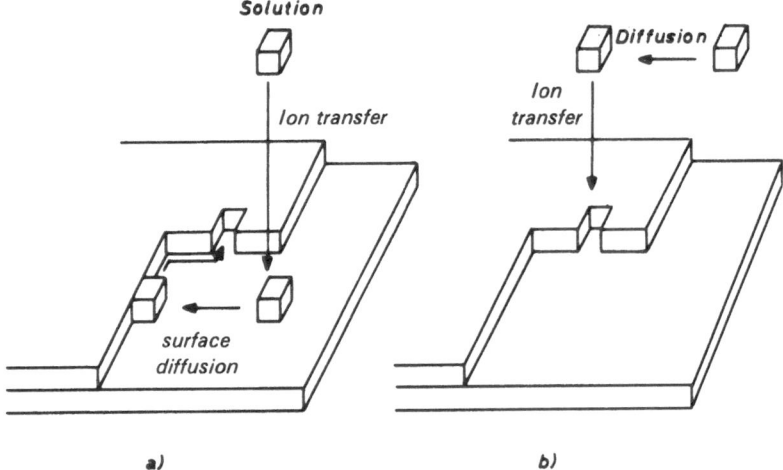

Fig. 23. Schematic illustration of the elementary steps in metal deposition:
(a) discharge followed by surface diffusion; (b) charge transfer limited to
points of incorporation.

the outer Helmholtz plane to the final place of incorporation into
the metal lattice.

The basic steps 2–4 make the difference between the metal–ion
electrode of interest in this section and the redox electrode at inert
electrocatalysts discussed in Section 4.

In detail, the surface diffusion has no effect on the kinetics of a
redox process and incorporation does not occur. On the other hand,
each of these steps may be rate determining in the case of the metal–ion
electrode. If surface diffusion or incorporation is rate determining,
this results in an additional overpotential, which is called the crystallization overpotential in order to distinguish it from the charge
transfer overpotential or the concentration (diffusion) overpotential
(cf. Section 3). It is clear that the crystallization overpotential strongly
depends on the crystallographic properties of the electrode surface.

As to the charge transfer, in the case of redox processes at inert
electrocatalysts there are electrons crossing the phase boundary [e.g.,
$(Fe^{2+})_{sol} \rightarrow (Fe^{3+})_{sol} + e^-$; reactant and reaction product both
remain in the solution]. But in the case of the metal–ion electrode the
metal ions themselves are transferred across the double layer [cf. Eq.
(35)]. As a consequence of this, charge transfer reactions with more
than one electron involved are possible at the metal-ion electrode,
while the simultaneous transfer of several electrons is extremely unlikely in the case of redox processes.

TABLE 2
Calculated Heats of Activation at the Zero-Charge
Potential for Direct Ion Transfer from the Undistorted
Hydrated Ions to Sites of the Metal Surface[30]

| Ion | Heat of activation from solution to site, kcal/mole | | |
	Plane surface	Edge	Kink
Ni^{2+}	130	190	190
Cu^{2+}	130	180	180
Ag^+ *	10	21	35

* Contrary to the predicted behavior, the deposition of silver preferentially
occurs by direct transfer to growth sites.

The transport phenomena involved in step 1 are essentially the
same as in the electrode processes discussed in the foregoing sections.

5.1.2. The Discharge Process

Two fundamental questions concerning the discharge steps of the
deposition process are: (1) Where does the discharge of the hydrated
ion take place, and (2) what is the molecular mechanism of discharge?

To answer the first question, Bockris and Conway[29] considered
potential energy diagrams for the deposition of cupric ions via
cuprous ions discharging at a plane surface, at a kink site, or at an edge
of the metal surface. Table 2 gives the calculated heats of activation*
for the different pathways.

From the data we conclude that by far the most frequently
occurring charge transfers are those onto crystal *planes*, because
transfer to kink sites and defects (e.g., direct deposition on growth
sites) is associated with much larger activation energies.

The second question, on the mechanism of discharge, cannot be
answered easily because the number of possible pathways very
strongly increases with the valence of the metal ions. In the case of
divalent ions four possible mechanisms of discharge can be foreseen:

$$M^{2+} + 2e^- \rightleftharpoons M \tag{36a}$$

---

* The heats of activation were calculated by taking the energy differences between the
initial state and the zero-point energy level of the activated state. They can be taken
to be, within the rather large limits of error of the calculation of ± 5 kcal/mole, equal
to the $\Delta G^{\neq}$ values of the different reaction paths.

$$M^{2+} + e^- \rightleftharpoons M^+; \qquad M^+ + e^- \rightleftharpoons M \qquad (36b)$$

$$M^{2+} + M \rightleftharpoons 2M^+; \qquad M^+ + e^- \rightleftharpoons M \qquad (36c)$$

$$M^{2+} + e^- \rightleftharpoons M^+; \qquad 2M^+ \rightleftharpoons M + M^{2+} \quad (36d)$$

Regarding these mechanisms of discharge, one can ask what the most probable pathway is. Concerning Eq. (36a), a direct two-electron exchange in most cases is unlikely because a high energy of activation is necessary and this consequently gives a low probability of two electrons tunneling simultaneously. On the other hand, one can argue that, except in the case of Cu^+, univalent ions are not known for the majority of divalent metals. They can exist, however, as strongly adsorbed, unstable intermediates. Therefore it should be concluded that multivalent ions in most cases are discharged in single-electron-transfer steps. Indeed, much experimental evidence has been accumulated in support of this conclusion for the discharge of Cu^{2+},[31] Zn^{2+},[32] and Fe^{2+}.[33] In the three cases mentioned, the mechanism of discharge follows the pathway (36b), and the rate-determining step is the divalent-to-monovalent transition, which is in agreement with energy considerations (i.e., transfer to a heavily hydrated divalent ion is more difficult than to a less hydrated monovalent one).

There are, however, divalent ions which do not follow the pattern described above. In the case of the discharge of Cd^{2+} the experimental results indicate a single-step two-electron transfer.[34] Figure 24

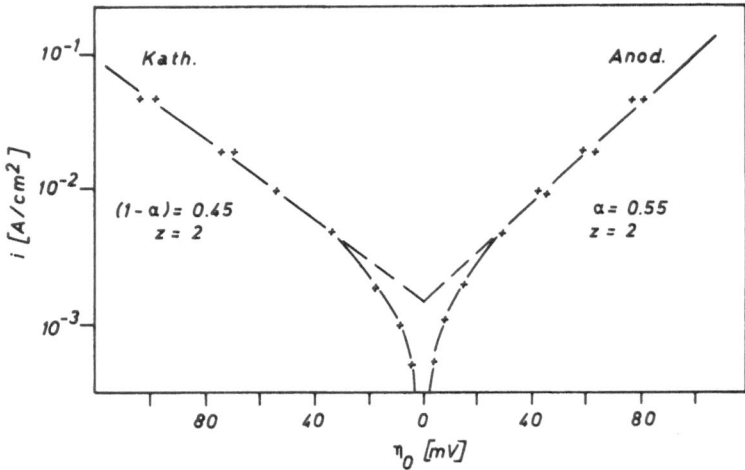

Fig. 24. Current density/overpotential curve for solid cadmium in 0.01 M Cd^{2+} + 0.4 M K_2SO_4 at 20°C (after Lorenz[34]).

shows the current density–overpotential curve for the system Cd/Cd^{2+}. The cathodic Tafel slope of RT/F is only explicable if a two-electron step is involved.

The examples given above show that in order to discriminate between the possible mechanisms for a given multivalent ion discharge, the orders of reaction, the transfer coefficients, and the stoichiometric numbers must be measured.

5.1.3. Intermediates

As already mentioned, univalent ions should be involved as adsorbed intermediates in the discharge process of multivalent metal ions. If this is true, the adsorbed species must exhibit pseudocapacitance effects in transient or ac phenomena. This is due to the fact, that, corresponding to the Nernst equation, the system tends to follow any change in potential by adjusting the concentration of the intermediate C_{M+} to the corresponding equilibrium value. This can be done only by exchanging charge with the electrode, i.e., by passing current. Mathematical computation shows that this current is proportional to the variation of the overpotential with time[30]

$$i_{PC} = C_{PC}(d\eta/dt)_{\eta \to 0} \tag{37}$$

The factor C_{PC} represents the pseudocapacitance acting in parallel with the capacitance of the double layer. C_{PC} is determined essentially by the equilibrium concentration of the intermediate. By measuring C_{PC}, the concentration of the monovalent ion at the surface can be computed.

In fact, the existence of monovalent intermediates at the metal surface can be demonstrated by such measurements. Furthermore, it

TABLE 3
Surface Adion Concentration at the Reversible Potential on Silver Electrodes as a Function of Surface Preparation

Surface preparation	$C_{ad,v} \cdot 10^{-11}$ mole/cm^2	Ref.
Quenched in H_2 atmosphere	90	35
Scraped in solution	15	36
Undefined	7	37
Quenched in He	3	38
Prepared *in situ* by anodic pulse	160	38

became evident that the equilibrium surface adion concentration is strongly dependent on the conditions of the surface, as Table 3 shows.*

5.1.4. Surface Diffusion and Incorporation

The adsorbed intermediates move by surface diffusion on the electrode surface and by reaching a growth point (kink, step, or dislocation) can be incorporated in the lattice. If the transport of the ions from the bulk of the solution is fast enough, the incorporation rate is represented by the equation[44]

$$i = i_{0,ad} \left\{ \exp\left(\frac{\alpha z F}{RT}\eta\right) - \exp\left[-(1-\alpha)\frac{zF}{RT}\eta\right] \right\} \frac{\lambda_0}{x_0} \tanh\frac{x_0}{\lambda_0} \quad (38)$$

This equation is the Butler–Volmer equation with superimposed terms containing λ_0 and x_0, which represent the surface conditions. In detail, λ_0, the "surface diffusion penetration," is a function of the surface diffusion coefficient D_S, the equilibrium adion concentration $C_{ad,0}$, and the exchange current density of the adions $i_{0,ad}$

$$\lambda_0 = (zF D_S C_{ad,0}/i_{0,ad})^{1/2} \exp[-(\alpha z F/2RT)\eta] \quad (39)$$

x_0 is the half distance between two steps at which incorporation can occur.

Two limiting cases of Eq. (38) can be distinguished:

1. If $\lambda_0 \gg x_0$, the tanh function can be replaced by its argument and Eq. (38) gives the simple current–potential relation. This case will be established if D_S and/or $C_{ad,0}$ produce a high value for λ_0, or if the surface has so many steps that x_0 becomes small. Consequently the adions reach points of incorporation in a very short time and the charge transfer step is rate determining.

2. If $\lambda_0 \ll x_0$, the tanh function can be replaced by 1 and Eq. (38) becomes

$$i_0 = i_{0,ad} \left\{ \exp\left(\frac{\alpha z F}{RT}\eta\right) - \exp\left[-(1-\alpha)\frac{zF}{RT}\eta\right] \right\} \frac{\lambda_0}{x_0} \quad (40)$$

Upon inserting Eq. (39) into Eq. (40), it can readily be seen that the slope of the cathodic current–potential curve, $\partial\eta/\partial(\log|i|)$, becomes

---

* The values given in Table 3 for the equilibrium surface adion concentration $C_{ad,0}$ are related to the pseudocapacitance C_{PC} by the formula
$$C_{PC} = (z^2 F^2/RT)C_{ad,0}$$

$-2RT/(1 - \alpha)zF$. This means that by the influence of the slow surface diffusion the slope of the current–potential relation is flattened by a factor of two in comparison to case 1. Furthermore, the absolute value of the current flow at a given overpotential in Eq. (40) is decreased by the factor $\lambda_0/x_0 < 1$.

The formulas given above indicate that the rate of metal deposition is largely influenced by the adion concentration, via λ_0, and by the number of growth points, represented by x_0. Both the adion concentration and the number of growth points are sensitive to the preparation of the surface. Table 3 indicates that the best activated electrode, in the sense that the incorporation rate is largest, is produced by anodic dissolution of some atomic layers from the old surface. Figure 25 shows the change of deposition rate with time, measured by the exchange current density. Evidently the rate of deposition is greatest in the case of the fresh electrode and the activation state of the electrode drops with time in several minutes.

One can ask why the high activity of the electrode, once established, does not remain constant. Several reasons can be given: (1) Smoothing of the surface during multilayer deposition (the number of growth points declines); (2) adsorption of impurities (the surface concentration of adions diminishes); and (3) tendency to form crystal faces of lower indices (the incorporation rate slows down).

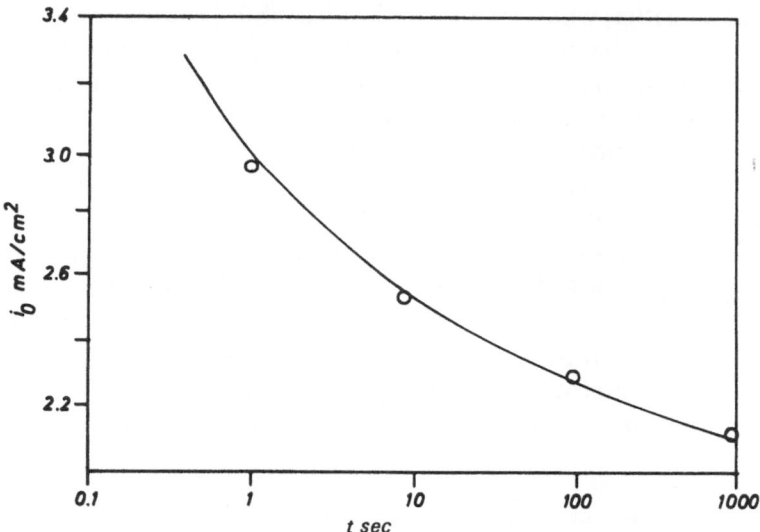

Fig. 25. Exchange current density i_0 of a copper electrode as a function of time after activation by anodic dissolution (after Bockris and Despić[30]; by permission of Academic Press).

5.2. Crystallographic Aspects of Metal Deposition

5.2.1. Brief Morphology of Deposits

Before the crystallographic aspects of electrodeposition are discussed, a short survey of possible forms of deposits shall be given. The morphologic properties depend largely on the conditions of electrocrystallization. Close analogies exist between the latter and the vapor-phase crystallization and ordinary crystallization from solutions and melts. However, considerable differences, arising from the presence of adsorbed layers (of ions and water molecules), the existence of the double layer field, and from potential effects on the electrodeposition, should also be expected.

Between the growth forms of deposits two effects can be distinguished: (1) *Amplification of surface roughness* occurs in electrode processes when deposition is carried out at low concentrations of ions in the solution. In this case, the transport of the ions from the bulk of solution is rate determining and electrocrystallization is favored at places of higher field strength, that is, on the top of elevations of the surface. This effect becomes extreme in the case of *dendritic growth*[39] when at some points of the surface needlelike or pinetreelike deposits grow deeply into the solution. A special case is the growth of *whiskers*,[40] i.e., threadlike deposits, which will be formed at fairly high current densities, when organic molecules inhibit the deposition at the plane parts of the electrode and only some points of the surface are free for electrocrystallization. (2) *Leveling*[41] is the phenomenon opposite the ones described above, when surface irregularities are ironed out at prolonged deposition. This effect is obtained when plating is carried out from concentrated metal salt baths. The leveling effect is intensified when strongly adsorbing substances of small concentrations inhibit the deposition at surface elevations. *Electropolishing*[42] is not a deposition effect, but shall be described here to complete this survey. It occurs when a metal surface is subjected to anodic dissolution in special baths (containing small concentrations of agents with a strong complexing affinity for the metal ions, in an electrolyte forming insoluble products with the same ions).

5.2.2. The Effect of the Potential on the Growth Forms

As can be seen in Figure 26, the growth forms in electrodeposition are dependent on the current density or on the overpotential of the

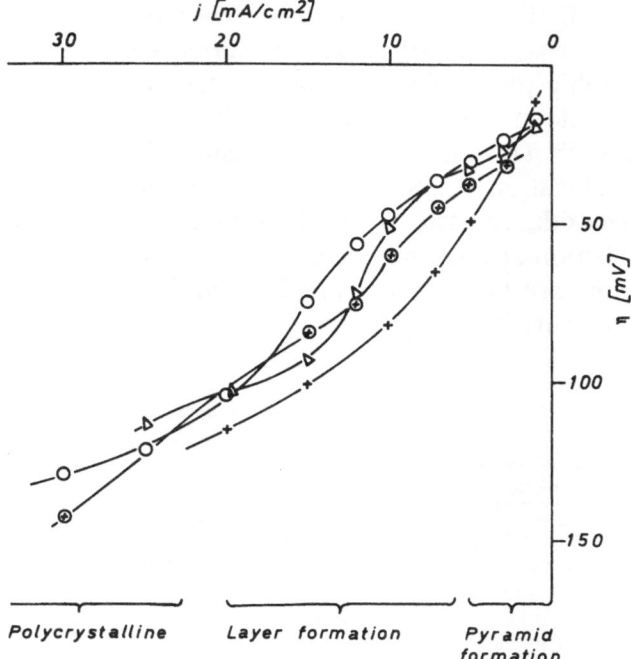

Fig. 26. Correlation between current density and observed growth forms in copper electrocrystallization (after Seiter and Fischer[43]; by permission of the Deutsche Bunsen Gesellschaft).

electrode, respectively. This can be understood if we imagine (1) that the adion concentration and the local current distribution are potential dependent[38]; (2) that in the case of local supersaturation of adions spontaneous nucleation occurs (supersaturation is an effect of adion concentration and therefore is potential dependent[44]); and (3) that adsorption of inhibitors largely influences the deposition forms (adsorption of organic molecules is potential dependent[45]).

5.2.3. The Mechanism of Electrocrystallization at Single Crystals

The final point of our discussion of electrochemical metal deposition is the mechanism of incorporation in the lattice. Usually this will be illustrated with the model of a single crystal.

As indicated in Figure 23, the incorporation should take place at monatomic steps on the surface which are reached by surface diffusion. The question is why monatomic steps are to be found on the surface.

First of all, the single crystal used as electrode can be cut in such a way that the electrode surface has a misorientation with respect to a crystallographic plane. The electrode surface therefore has a high crystallographic index, but electrocrystallization tends to form low-index planes.[46] This phenomenon is illustrated in Figure 27. Before electrocrystallization starts, the electrode surface, which may look optically smooth, consists of invisible microsteps (Figure 27a). By electrodeposition at these microsteps layers parallel to the crystallographic plane are built up (Figure 27b), and finally form a visible macrostep (Figure 27c).

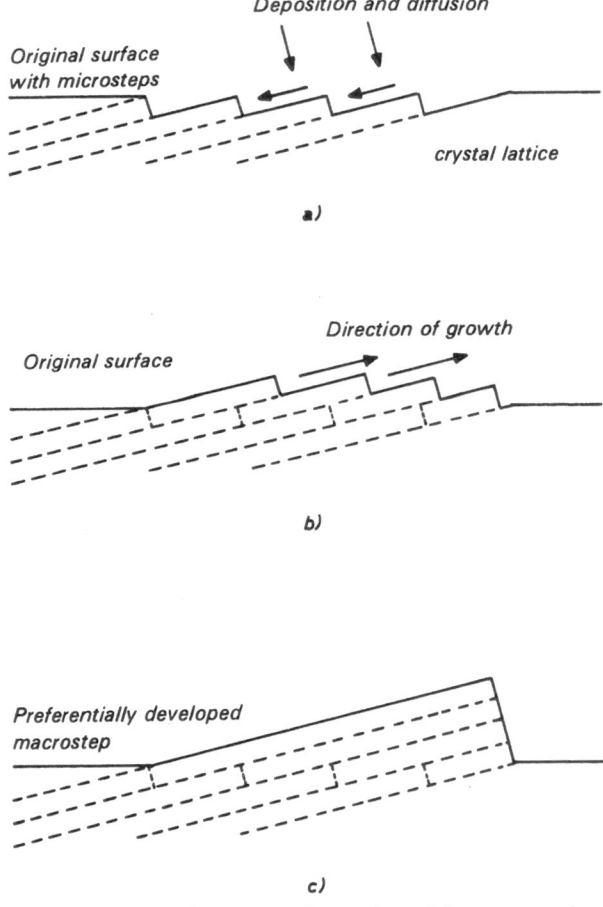

Fig. 27. Successive stages in preferential step or edge growth to produce surfaces of lower crystallographic index (after Bunn and Emmett[46]).

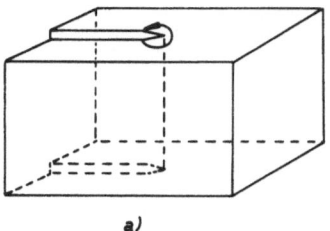 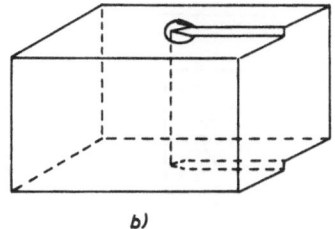

a)   *b)*

Fig. 28. Schematic illustration of screw dislocations: (a), right-handed screw; (b) left-handed screw.

The second reason for the appearance of monatomic steps on crystal planes is the fact that no crystal is ideal, but all show dislocations. Displaced planes (line dislocations) give the same effect as described above. The more frequent screw dislocations are schematically shown in Figure 28. The monatomic screw step propagates when electrocrystallization occurs by rotating itself with one end fixed at the point A where the screw dislocation emerges. A gradual sloping of the crystal surface occurs, so that at no point is there a discontinuity between the lattice layers. Thus a spiral with pyramidal shape develops[47] (Figure 29). The more hypothetical picture of crystal growth by an isolated screw dislocation can be generalized easily for the case of more than one screw dislocations. Figure 30 shows the simplest case of a pair of dislocations emerging sufficiently far apart and being of different sign (one rotating clockwise and one rotating counterclockwise). The result of both signs is a relatively smooth and

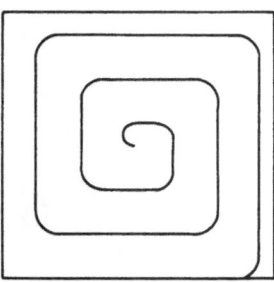

Fig. 29. Spiral growth with pyramidal shape at a single screw dislocation.

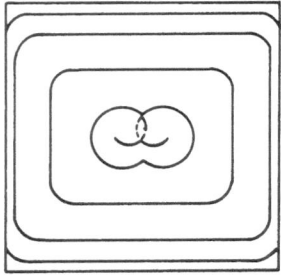

Fig. 30. Pyramidal growth due to a pair of screw dislocations.

even deposit. Screw dislocations are of special importance in crystal growth because the dislocation step never vanishes by continued deposition, in contrast to edge dislocations. Therefore two-dimensional nucleation producing new incorporation sites is not necessary.

Finally, if a crystal plane lacks steps and kinks, or if the incorporation is sufficiently inhibited, local supersaturation of adion concentrations can occur, so that the probability increases that new growing centers will emerge in the form of two-dimensional nuclei[48] (Figure 31). But at dislocation-free planes the electrocystallization is strongly

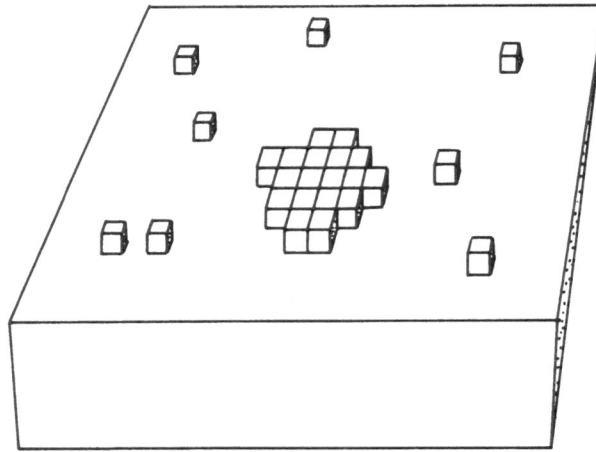

Fig. 31. Model of the formation of a two-dimensional
nucleus.

inhibited, as the large crystallization overpotential of such electrodes shows. Furthermore, at constant current, potential oscillations occur. This can easily be understood, if we consider that local supersaturations require a large potential as mentioned above. Therefore nucleation occurs with a large overpotential but the further growth at the steps of the nucleus continues with low overpotential until a new plane has completed.

5.3. Kinetics of Metal Dissolution

5.3.1. The Converse of Metal Deposition: Metal Dissolution

Applying an anodic potential to an electrochemically active metal electrode causes the metal to undergo anodic dissolution (oxidation of the metal atoms). Since the overall reaction of metal dissolution is the inverse process of metal deposition [cf. Eq. (35)], it can be assumed that the elementary steps of dissolution follow the same reaction path. However, this microscopic reversibility is limited to conditions of rate and potential not far from the equilibrium ones. There always exists the possibility of the reaction assuming an alternative parallel path and this may be favored by the potential being made anodic.

Often, the incorporation step is irreversible, so that the centers of most active growth by no means need also be centers of most active dissolution. This can be seen by applying alternating current to equilibrated metal surfaces. In most cases the surface structure will be changed (roughened or smoothed) which proves that dissolution and incorporation occur at different places on the electrode. This effect will be intensified if potential-dependent adsorption of inhibitors occurs. These surface-active substances may inhibit the deposition process at the required negative potential while desorbing and not affecting the anodic dissolution, and vice versa. Knowledge of the extent of irreversibility and how it is influenced by various factors is still very limited and of a qualitative nature.

5.3.2. Metal Dissolution with the Formation of Insoluble Products

A special case of metal dissolution occurs when the metal ions are insoluble in the electrolyte involved. Often, these insoluble products form a continuous film, which leads to passivation, i.e., the dissolution slows down. It can be shown[49] that the anodic current in the case of passivation film formation declines exponentially with

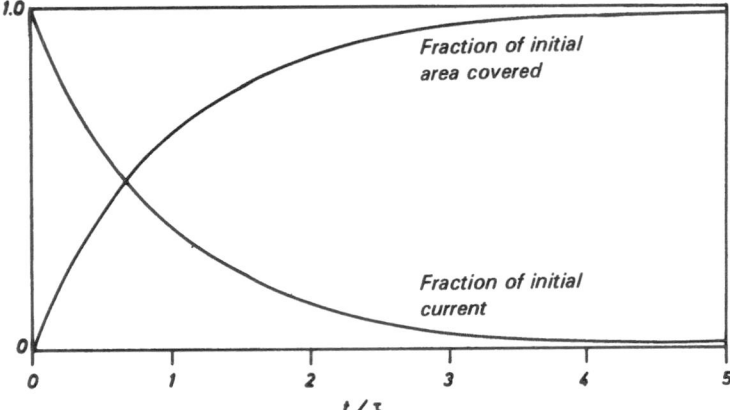

Fig. 32. Variation of current and area covered during film formation in metal dissolution (after Vermilyea[49]; by permission of John Wiley and Sons–Interscience).

time at constant overpotential, the dissolution process being the rate-determining step.

$$i = i_1 \exp(-t/\tau) \qquad (41)$$

Here i_1 is the current density on the uncovered surface and τ is a relaxation time depending on the valence of the metal ions formed and on the film thickness. The fraction of the initial area covered by the film increases with the same function, as Figure 32 indicates.

Continuous film formation is known; for example, in the anodic dissolution of tantalum a film of tantalum oxide forms and covers the electrode completely at a film thickness of a very few monolayers.[50] As soon as the electrode is covered, the overpotential required for continued current flow rises very rapidly and may reach hundreds of volts. The reason for this behavior is that the flow of current is controlled by the rate at which ions can be transported through the film.

On the other hand, many examples of the formation of non-continuous surface films are known. In this case surface reactions largely determine the electrode characteristic.

To give an example, the well-known reaction of the formation of lead sulfate on lead may be considered. This plays an important role in the lead accumulator. The overall reaction

$$Pb + SO_4^{2-} \rightleftharpoons PbSO_4 + 2e^- \qquad (42)$$

can proceed in the two steps

$$Pb \rightleftharpoons Pb^{2+} + 2e^- \tag{43}$$

$$Pb^{2+} + SO_4^{2-} \rightleftharpoons PbSO_4 \tag{44}$$

Both of the reactions (43) and (44) are apparently fast,[51] so that fairly high current densities can be passed through the electrode without much overpotential. In effect, the surface film of $PbSO_4$ has a very low and nearly constant resistance, of the order of ohms. Consequently the film must be discontinuous so that the current flows in between the $PbSO_4$ crystals. Therefore the film properties such as electronic resistivity or ionic mobility play almost no part in determining the electrode behavior.

6. The Electrochemistry of Semiconductors

6.1. Fundamentals and Terminology

The electrochemical properties of electrode/solution interfaces, such as the electrical capacity and the magnitude of the exchange current, ultimately depend upon the distribution of potential throughout the surface region and the distribution of available electronic energy levels on each side of the interface. These are just the properties which change most strikingly upon replacement of a metal by a semiconductor, hence it is instructive to study the electrochemistry of the semiconductor/electrolyte interface and to make a comparison with that of the metal electrode. The survey presented here is necessarily brief and the reader is referred to several reviews which develop the subject in more detail.[52−56]

We first consider the role played by the density of charge carriers in the solid. The density of conduction electrons in a metal $(n_m \sim 10^{22} \text{ cm}^{-3})$ is very high and the carrier density, electrostatic field, and potential change by an order of magnitude within the characteristic length

$$L_m = (\varepsilon_m E_F / 6\pi n_m e_0{}^2)^{1/2}$$

where ε_m is the permittivity of the metal and E_F is its Fermi energy. L_m is of the order of 0.5 Å for the noble metals, justifying the assumption of an ideal surface charge at a metal/electrolyte interface. Semiconductors have a concentration $n_s \sim 10^{13}–10^{18} \text{ cm}^{-3}$ of mobile charge carriers, and exhibit a characteristic length

$$L_s = (\varepsilon_s kT / 8\pi n_s e_0{}^2)^{1/2}$$

for spatial variations of the electrical variables in depletion and weak accumulation layers. L_s is of the order of 10^{-4} cm in typical semiconductors, such as intrinsic germanium, whence it follows that in this case the potential variations responsible for an established potential difference between an electrolyte and the semiconductor are principally located within the solid phase. In strong accumulation layers the carrier concentrations can become sufficiently high that surface charges may be formed as at metal interfaces.

It should be mentioned that certain complications frequently prevail at real semiconductor surfaces, due to one or more of the following effects:

(a) The atoms in the surface of molecular crystals may have un-saturated valences.

(b) There may be uncompensated ions in the surface of ionic crystals.

(c) There may be discrete surface energy states arising from the discontinuity in the potential energy–distance relationship at the surface.

(d) As at a metal interface, adsorption of species may occur from the solution onto the solid.

The establishment of equilibrium between a solution of a redox system and a solid is achieved by the transfer of charges across the interface, with the consequent creation of an interfacial potential difference $\phi_{s\text{-}e}$ given by

$$e_0 \phi_{s\text{-}e} = E_F - E_{red} + \alpha_{ox} - \alpha_{red} \qquad (45)$$

Here E_F is the Fermi energy of the solid, referred to the energy of the electron in vacuum, E_{red} is the vacuum energy level of the reduced species (the ionization potential), and α_i represents the real free energy of solvation of species i. Reducing agents, which have small ionization potentials, transfer electrons to the solid and cause downward bending of the energy bands, creating an accumulation layer in the conduction band and a depletion layer in the valence band of the semiconductor. In a similar fashion an oxidizing agent accepts electrons from the solid and causes the creation of an accumulation layer of holes in the valence band and a depletion layer of electrons in the conduction band. The resulting electrostatic energy diagrams are illustrated schematically in Figure 33; E_c represents the energy of the lower edge of the conduction band, E_v that of the upper edge of the valence band, and E_F that of the Fermi level. Continuous current flow through

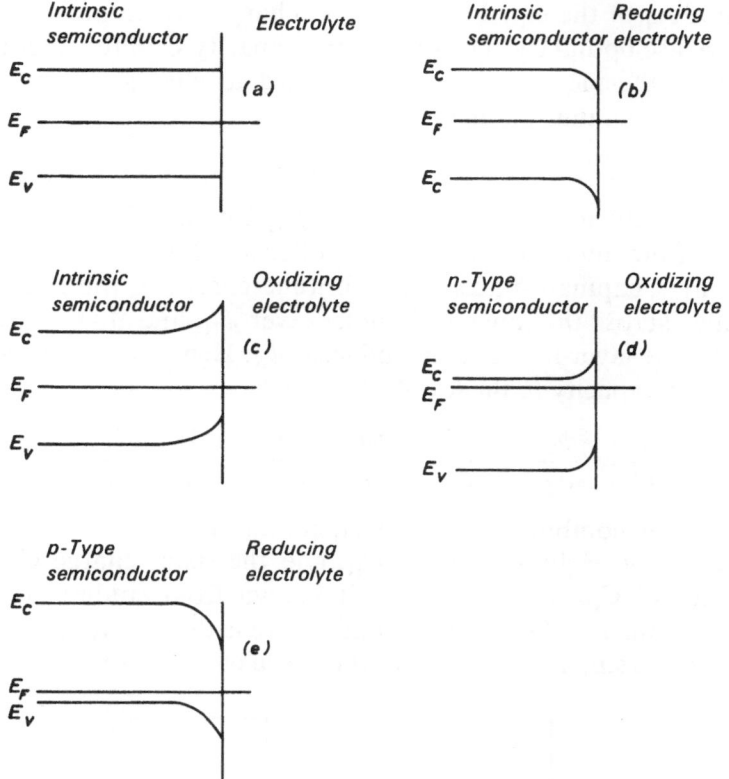

Fig. 33. Types of contact between a semiconductor and electrolyte solutions: (a) neutral contact between intrinsic semiconductor and inert solution; (b) ohmic contact for electrons between intrinsic semiconductor and a reducing agent; (c) ohmic contact for holes between intrinsic semiconductor and an oxidizing agent; (d) blocking contact for electrons; (e) blocking contact for holes.

ohmic contacts is possible when an appropriate field is applied to cause migration of carriers through the solid. Neutral and blocking contacts impede the flow of continuous current, and therefore resemble ideally polarizable metal electrodes.

6.2. Polarization Effects at Blocking Contacts; the Capacity of the Semiconductor/Electrolyte Interface

The application of a potential difference ϕ_{s-e} between a solid and an electrolyte causes a relative shift of the energy levels in the bulk of the two phases by the amount $e_0\phi_{s-e}$, and results in a change in the magnitude q of the equal but opposite charges which are separated by the phase boundary. The differential capacity $\tilde{C}$ of the interface is defined as the incremental charge needed to change the potential difference by an infinitesimal amount

$$\tilde{C} = \partial q / \partial \phi_{\text{s-e}}$$

At a semiconductor, the potential drop $\phi_{\text{s-e}}$ is divided between several regions of the interface, as was seen in Section 2, namely that across the Gouy–Chapman diffuse layer of ionic charge in the electrolyte ϕ_{GC}, that across the inner Helmholtz layer ϕ_{H}, and that across the space charge layer in the semiconductor ϕ_S. Hence it is necessary to express the capacity in the form[57]

$$\frac{1}{\tilde{C}} = \frac{\partial \phi_{\text{GC}}}{\partial q} + \frac{\partial \phi_{\text{H}}}{\partial q} + \frac{\partial \phi_S}{\partial q} = \frac{1}{C_{\text{GC}}} + \frac{1}{C_{\text{H}}} + \frac{1}{C_S}$$

for the series combination of the capacities of the Gouy–Chapman layer C_{GC}, the Helmholtz layer C_{H}, and the space charge C_S. The properties of C_{GC} and C_{H} are well known from studies at metal electrodes; the new feature to be noted here is that the total capacity may never be smaller than that of the space charge layer.

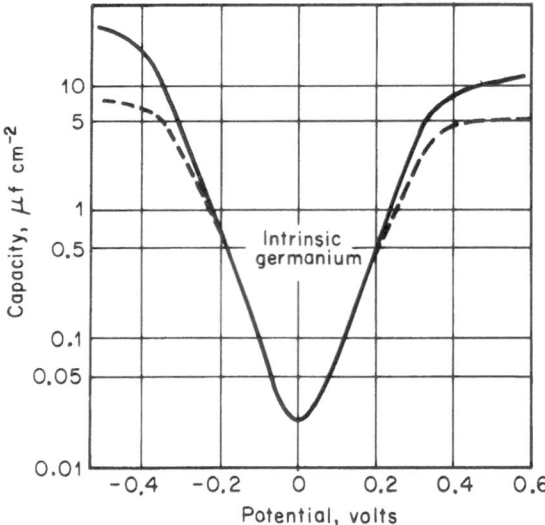

Fig. 34.. Differential capacity of an intrinsic Ge electrode with regard for degeneracy: solid line, space charge capacity; dashed line, space charge capacity in series with Helmholtz capacity[52] (by permission of Interscience Publishers).

The theoretical dependence[52] of the differential capacity of an intrinsic germanium electrode upon potential is shown in Figure 34. The potential scale is referred to the flat band potential of the germanium, that is, where q and $\phi_{s\text{-}e}$ are identically zero. The minimum of the measured capacity at this point is entirely due to a minimum in the space charge capacity which is given by the formula for the capacity of a parallel plate condenser of thickness L_S and filled with a dielectric of permittivity ε_S:

$$(C_S)_{\min} = \varepsilon_S/4\pi L_S$$

The rise of the capacitance on each side of this point reflects the compression of the space charge layer as the increasing charge on the solid (negative on the cathodic side and positive on the anodic side) is more strongly attracted by the counter charge in the solution. Extreme anodic polarization causes the valence band to bend up to the Fermi level, and extreme cathodic polarization bends the conduction band down to the Fermi level. Further incremental additions of charges appear as ideal surface charges on the semiconductor, in this condition, and do not increase the potential drop across the solid ϕ_S but influence only ϕ_H and ϕ_{GC}.

6.3. Polarization of Ohmic Contacts

A comparison of current–voltage curves of redox systems for semiconducting and metallic electrodes reveals four striking differences which are explainable in terms of the physical properties of the solid. These are as follows

1. Rectification often occurs for the semiconductor, although symmetric anodic and cathodic behavior is the rule at a metal electrode. Redox systems having rather positive Nernst potentials often exhibit a cathodic branch due to reduction of the oxidized species, but the anodic current, due to oxidation of the corresponding reduced species, is blocked by the electrode. Similarly, redox systems having quite negative Nernst potentials exhibit only the anodic branch of the curve. For wide-band-gap intrinsic semiconductors, redox systems having low Nernst potentials, which therefore make neutral contacts to the solid, show neither oxidation nor reduction currents, that is, the electrode blocks the current in both directions.

2. The voltage which must be applied to the semiconductor electrode in order to cause a certain magnitude of the current to flow depends upon the nature of the redox system as it does at a metal, but

it also depends upon some power (≥ 1) of the thickness of the semi-conductor sample. Also, the current is normally proportional to some power of the voltage, rather than the exponential function of the voltage (Tafel's law) observed for a metal (Figure 35).

3. A potential-independent limit of the current is often seen at both semiconducting and metallic electrodes, due to control of the overall rate of the current generating process by some step which is not accelerated by potential. In most cases this step is associated with diffusion of the electroactive species through the solution to the electrode surface, when the current limit is identical on both types of solid. In other cases the current limit for the semiconductor is much less than for a metal, and is uninfluenced by stirring of the solution, although it is proportional in magnitude to the concentration of electroactive species in solution.

4. Impressive amplification of the current by irradiation of the electrode with visible or near ultraviolet light is observable for semiconductor electrodes but not for metallic electrodes. Photocurrents are usually measurable even from redox systems which make neutral contacts to semiconductors, and which therefore fail to show conduction "in the dark." The photocurrent is proportional to light intensity, but does not always depend linearly upon the concentration of the electroactive species in solution. It depends upon the wavelength of the light; the photocurrent may follow the absorption spectrum of the solid.

An explanation of the shape of the current–voltage characteristic of the semiconductor electrolyte junction can be reached upon consideration of the boundary value problem for the concentration of current carriers in the solid. This is based upon the equation which describes the continuity of the carriers, together with the kinetics of the heterogeneous "charge separation" reaction

$$D + S \rightleftharpoons A + C \tag{46}$$

which generates the carrier C in the surface of the solid. In this general equation, D represents the donor and A the acceptor, both dissolved in the solutions, and S represents an unoccupied carrier site in the solid.

In its most general form the continuity equation accounts for the rate of change of the concentration in any volume element due to transport of particles into or out of the volume by means of diffusion or migration, or due to recombination of electrons and holes. The electric field intensity at each point required for calculation of the flux of charged particles is a solution of Poisson's equation. This most general problem is formulated mathematically in terms of nonlinear partial differential equations which have no closed form solution, but in particular cases reasonably approximate equations can be written which may be solved analytically. For example, the steady-state concentration of electrons or holes and the corresponding current–voltage characteristic can be determined[58] for an *insulator** upon making the following assumptions:

(a) Diffusion can be neglected in comparison with migration as a mode of transport of carriers in the solid.

(b) The concentration of thermally generated carriers can be neglected in comparison with the concentration of injected carriers.

(c) It can be assumed that changes of the applied potential only affect the potential distribution within the solid, and leave uninfluenced the potential drop across the inner Helmholtz layer and the Gouy–Chapman diffuse layer. This assumption is particularly important in the context of this chapter, since it contrasts strikingly with the situation at a metal electrode, where changes of the applied potential only affect the potential distribution within the Helmholtz and Gouy–Chapman layers. The inspiration for the assumption comes, of course, from the comparison of the characteristic length for potential variations in the solid L_s with the thickness of the Helmholtz layer (~ 5 Å) and of the Gouy–Chapman layer (~ 1 Å) in concentrated electrolyte solutions; L_s can reach and even exceed 1 cm in a typical insulator.

The results reported below were obtained from a one-dimensional steady state model chosen to correspond to the experimental conditions used[59] in the measurement of Figure 35. These were obtained from a cell containing a thin, platelike single crystal of anthracene (the insulator) sandwiched between two electrolyte solutions, one consisting of 0.01 m Ce(SO$_4$)$_2$ in 0.5 m H$_2$SO$_4$, and the other containing 0.2 N NaCl. These solutions, in turn, were contacted by platinum metal electrodes which were connected to the voltage source and current measuring apparatus. The sodium chloride solution functioned merely as a neutral "hole collecting contact" and played no essential role in the experiment.

---

* A wide-band-gap intrinsic semiconductor provides insulation only if neutral or blocking contacts are applied to it. An Ohmic contact permits the injection of current carriers and the flow of continuous currents even through those materials normally classified as insulators.

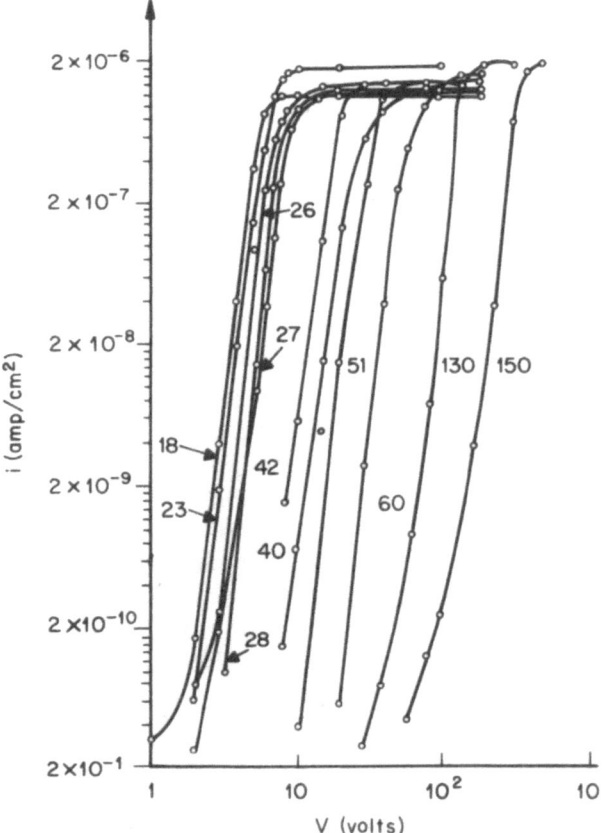

Fig. 35. Cathodic current–voltage curves at various thicknesses (in microns) of anthracene crystals. Electrolyte: 0.01 m Ce(SO₄)₂ in 0.5 m H₂SO₄ [55] (by permission of the Deutsche Bunsen Gesellschaft).

The current–voltage curve according to this approximation is given by[55]

$$V = -\tfrac{2}{3}E(0)X[(1 + d/X)^{3/2} - 1]$$

where $E(0)$ and X are the characteristic field and length within the solid near the electrolyte surface,

$$E(0) = i/un(0)e_0, \qquad X = \varepsilon_s i/8\pi u n^2(0)e_0^2$$

and d is the thickness of the insulator crystal. V represents the applied voltage, i the observed current, u the mobility of current carriers, and $n(0)$ the concentration of carriers at the injecting contact.

It is usually assumed that $n(0)$ is very large near an ohmic contact, so that $E(0)$ and X are small quantities. Then the dependence of current upon voltage becomes expressible in the form

$$i = (9\varepsilon_s/32\pi d^3)V^2$$

which is known as Child's law. It will be seen that this equation provides an interpretation of experimental fact 2 mentioned at the beginning of this subsection, concerning the dependence of the current upon voltage for a particular crystal, and the dependence of voltage upon crystal thickness at a particular current density. In this region (the region of variable current in Figure 35) the current is controlled only by properties of the solid, being determined by the rate at which the applied voltage can drain charge carriers out of an ample reservoir at the contact against the repulsion generated by the carriers already in transit across the crystal.

For an explanation of the current limit at high voltage, a consideration of the kinetics of the carrier generation reaction at the crystal surface is introduced. The rate of the charge separation is written in the form

$$i = i_0[(C_D/C_D^0) - n(0)C_A/n^0(0)C_A^0] \tag{47}$$

where F is the Faraday constant; i_0 is the exchange current; C_D and C_A are the concentrations of donor and acceptor species at the crystal surface under polarization; C_D^0 and C_A^0 are the corresponding concentrations at equilibrium; and $n(0)$ and $n^0(0)$ are the concentrations of carriers in the surface under polarization and at equilibrium, respectively. C_D and C_A are solutions of a diffusion problem in the solution. Comparison of this equation for the surface reaction rate with the corresponding one for metal electrodes [Eq. (30)] reveals two differences: First, the potential dependent coefficients of the concentration terms of Eq. (30) are missing from Eq. (47) due to the assumption that the potential difference across the Helmholtz layer, and hence the driving force for the electron transfer reaction, is unaffected by applied potential [assumption (c)]; second, a new term appears in Eq. (47) to account for the variation of the carrier concentration in the electrode surface. This concentration is practically constant in a metal.

In the steady state we can introduce

$$C_D/C_D^0 = 1 - i/i_D, \qquad C_A/C_A^0 = 1 + i/i_A \tag{48}$$

where $i_D = FD_D C_D^0/\delta_D$ is the characteristic diffusion rate of the donor species to a plane electrode, D_D is the diffusion coefficient, and δ_D is the thickness of the steady-state diffusion layer. $i_A = FD_A C_A{}^0/\delta_A$ is similarly defined. It is then clear that the current i tends to i_D as a positive limit as it does at a metal electrode, *provided that* $i_0 > i_D$. In the case $i_0 \ll i_D$, however, we have

$$i/i_0 \simeq 1 - [n(0)/n^0](1 + i/i_A)$$

whence it follows that the current tends to i_0 as a limit when the surface concentration of carriers is exhausted, $n(0) \to 0$. This situation, which is without parallel for a metal electrode, arises because the action of the applied potential is merely to drain off carriers from the insulator/electrolyte interface, thus reducing the current carried by the reverse of the charge separation reaction to zero. This leaves a potential-independent forward current equal to the exchange current as the maximum rate of charge transfer through the interface. This, therefore, is an explanation for the third experimental observation noted previously.

In particular, we gather that redox systems having *very* small exchange currents at wide-band-gap electrodes might not give rise to an observable injection current through the interface, for experimental problems complicate the measurement of currents of less than a certain magnitude. This is part of the cause of rectification at wide-band-gap semiconductor electrodes, for, as will become clear in the next section, redox systems which have a large exchange current for the injection of electrons into the conduction band of the solid do not at the same time have a large exchange current for the injection of the holes into the valence band. Hence only electron currents can be observed in such a system, and only hole currents can be observed, for the same reason, in solutions of oxidizing agents.

6.4. The Magnitude of the Exchange Current for a Redox System at a Semiconductor Electrode

The distribution of energy levels in a metal, a semiconductor, and a redox electrolyte solution are shown schematically in Figure 36.[60] A detailed theory of the magnitude of exchange currents[61,62] has been formulated under the following assumptions:

(1) The distributions of energy levels E in the oxidized and reduced species in the electrolyte are given by two Gaussian functions:

$$D_{ox}(E) = \tfrac{1}{2}(\lambda \pi k T)^{-1/2} \exp\{-(E - {}^\circ E_{ox})^2/4\lambda k T\} \qquad (49a)$$

$$D_{red}(E) = \tfrac{1}{2}(\lambda \pi k T)^{-1/2} \exp\{-(E - {}^\circ E_{red})^2/4\lambda k T\} \qquad (49b)$$

where $°E_{ox}$ and $°E_{red}$ are the most probable energies for the oxidized and reduced species respectively; the energy of reorganization λ is a characteristic parameter for the rearrangement of the solvation spheres of the reactants during electron exchange with the electrode. The following relation exists:

$$°E_{ox} - \lambda = °E_{red} + \lambda \equiv °E_{F,redox} \tag{50}$$

where $°E_{F,redox}$ plays the role in the electrolyte which the Fermi level E_F plays in the solid.

(2) For a semiconductor electrode electron exchange can occur either with the conduction band or the valence band. The probability of finding an electron at an energy level E in one of these bands is

$$W_{sc}(E) = n(E)D_{sc}(E) = \left[1 + \exp\left(\frac{E - E_F}{kT}\right)\right]^{-1} D_{sc}(E) \tag{51}$$

where $n(E)$ is the Fermi distribution, and $D_{sc}(E)$ is the density of electronic states in the electrode; $D_{sc}(E)$ vanishes in the band gap between the lower edge E_C of the conduction band and the upper edge E_V of the valence band. For an intrinsic semiconductor the Fermi level E_F lies halfway between E_V and E_C in the bulk of the electrode; near the electrode surface, however, the variation of the inner potential ϕ leads to a bending of the energy bands (see Figure 33). At the surface E_V and E_C are given by

$$\begin{aligned} E_C &= E_F + \Delta - e\phi_s \\ E_V &= E_F - \Delta - e\phi_s \end{aligned} \tag{52}$$

for an intrinsic semiconductor, where Δ is twice the band gap, and ϕ_s the potential at the electrode surface (in the bulk of the electrode ϕ has been set equal to zero).

(3) Electron transfer occurs between equal energy levels in the electrode and the solution with a certain probability κ which depends on the electronic properties of the electrode and the redox couple.

(4) The total exchange current is a sum of contributions from all energy levels of the following form:

$$i_0 = Fc_{ox}^s \int \kappa D_{ox}(E)D_{sc}(E)n(E) \, dE \tag{53}$$

where c_{ox}^s is the concentration of the oxidized species in the o.H.p. For semiconductors this integral splits into two parts:

one over the valence and the other over the conduction band. The corresponding contributions i_0^V and i_0^C to the exchange current density are then

$$i_0^C \sim \exp\{-(\lambda + E_C - E_F + kT \ln c_{red}^s/c_{ox}^s)^2/4\lambda kT\} \quad (54a)$$

$$i_0^V \sim \exp\{-(\lambda + E_F - E_V + kT \ln c_{red}^s/c_{ox}^s)^2/4\lambda kT\} \quad (54b)$$

It is clear from these equations that the band which (at the surface) is nearer to the Fermi level E_F gives the larger contribution. For wide-gap intrinsic semiconductors a large exchange current can only be observed if the electrolyte is either very oxidizing or very reducing, so that at the surface one of the bands is bent close to the Fermi level (see Figure 33b,c).

6.5. Photoeffects at Semiconductor Electrodes

We consider a system in which light is normally incident upon a semiconductor electrolyte interface which is polarized as before. Absorption of a photon by a solid can lead to photoconductivity by either of two mechanisms:

1. The intrinsic mechanism. Ionization can occur, followed by rapid thermalization of the positive and negative charges and drift of the carriers under the influence of the applied field and their mutual electrostatic attraction until they recombine or become discharged at one or the other of the contacts.

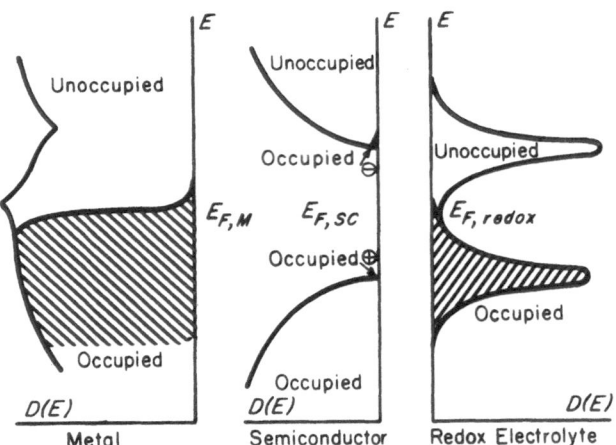

Fig. 36. Comparison of energy level distribution in a metal, a semiconductor, and a redox electrolyte under equilibrium conditions[52] (by permission of Interscience Publishers).

2. The extrinsic mechanism. A neutral exciton X can be created which moves through the crystal by energy exchange between neighboring molecules until it reaches the electrolyte interface, where it takes part as a reactant in a photochemical reaction:

$$X + D \rightarrow A + C \tag{55}$$

which results in the release of a carrier into one of the bands of the solid. Only the extrinsic mechanism will be dealt with further.[63]

A cross section through the potential energy surface of the photoinjection is shown in Figure 37; for a fuller treatment see ref. 63. Its rate can be expressed in a manner which is analogous to Eq. (47):

$$i_p = e_0[k_f^* C_X C_D - k_b^* n(0) C_A] = e_0 D_X (dC_X/dx)_{X=0} \tag{56}$$

where the rate constants k_f^* and k_b^* can be assumed to be potential independent, as before, and the concentrations refer to the plane of the interface. The concentration of excitons in the solid obeys a continuity equation

$$D_X(d^2/dx^2)C_X - (C_X/\tau_X) - \varepsilon I_0 \exp(-\varepsilon x) = 0$$

where D_X represents a diffusion coefficient for the excitons, τ_X is their lifetime, ε is the extinction coefficient of the solid, and I_0 is the light intensity. x is the distance from the electrolyte interface. This equation equates the net rate of diffusion of excitons into or out of a volume

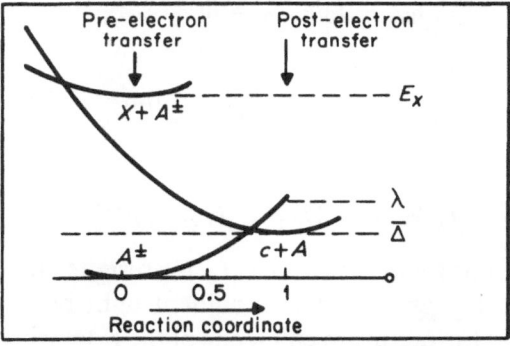

Fig. 37. Cross section through the potential energy surface for dark injection and photoinjection into a molecular crystal. $Z = 10^7 \, \text{cm}^4/\text{mole/sec}$; $D = 1.25 \times 10^3 \, \text{cm}^2/\text{sec}$; $\tau = 2 \times 10^{-8} \, \text{sec}$; $\lambda = 2 \, \text{eV}$[63] (by permission of *Electrochimica Acta*).

327

element to the difference between the rates of disappearance and creation inside the volume element. As before, the concentrations of carriers, donors, and acceptors are also solutions of appropriate diffusion boundary value problems.

The current–voltage behavior and the current limit imposed by diffusion of the donor species in solution i_D are not essentially different from the dark injection case, and are not considered here. However, we treat the case of another current limit which can arise when $i \ll i_D$ ($C_D \simeq C_D{}^0$), again making the approximations for an insulator which were detailed previously.

We find the following expression for the total current,[63] comprising a dark injection contribution i_0 and a photocurrent:

$$i/i_{lim} = 1 - n(0)/n^0(0) \qquad (57)$$

where

$$i_{lim} = i_0 + e_0 I_0 \left\{ \left[\frac{1+1}{\varepsilon(D_X\tau_X)^{1/2}} \right] \left[\frac{1+1}{k_f{}^*C_D{}^0(\tau_X/D_X)^{1/2}} \right] \right\}^{-1} \qquad (58)$$

i_{lim} is a potential-independent current limit which is reached when the applied electric field becomes so strong that the surface concentration of carriers tends to zero, $n(0) \rightarrow 0$. Energy carriers formed by the forward direction of reaction (55) are pulled through the crystal at this limit, and none recombine with the acceptor species according to the reverse of reaction (46).

Equation (58) can be rewritten

$$i_{lim} - i_0/e_0 I_0 = \Phi = f_1 f_2 \qquad (59)$$

where

$$f_1 = [1 + 1/\varepsilon(D_X\tau_X)^{1/2}]^{-1}$$

and

$$f_2 = [1 + 1/k_f{}^*C_D(\tau_X/D_X)^{1/2}]^{-1}$$

$i_{lim} - i_0$ is the photocurrent and $e_0 I_0$ represents the current which would flow if every photon in the incident light beam gave rise to a carrier; thus Φ is the quantum efficiency of the extrinsic photoinjection process. This has been factorized into f_1, the efficiency of collection of excitons by the surface, and f_2, the efficiency of the charge separation reaction. We note first that the photocurrent is proportional to the light intensity. Second, we observe that the photocurrent depends

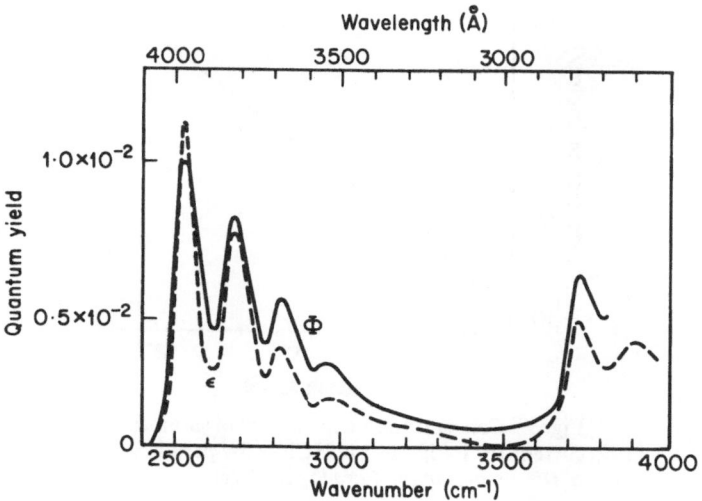

Fig. 38. Comparison of photoconduction spectrum with absorption spectrum of anthracene. Light polarized parallel to b axis. Electrolytic contact: 1 N NaOH (after Mulder[64]; by permission of *Philips Research Reports*).

upon the wavelength of the light through the influence of this parameter upon the extinction coefficient ε. In an anthracene crystal, for example, the mean free path of excitons $(D_X\tau_X)^{1/2}$ is much less than the penetration depth of the light ε^{-1}, so that $f_1 \sim \varepsilon(D_X\tau_X)^{1/2}$ and the plot of the photocurrent at various wavelengths parallels the absorption spectrum. See Figure 38.[64] Third, it follows from Eq. (59) that the photocurrent depends upon the concentration of the donor species in solution, provided that the rate constant of the photoinjection reaction k_f^* and the concentration of the donor C_D are not so high that $f_2 \approx 1$. Figure 39 shows that in the particular case of the photoinjection into anthracene from an iridium hexachloride solution[65] there is proportionality between f_2 and the concentration of the donor.

6.6. The Photovoltage

In the absence of irradiation a redox system establishes an interfacial equilibrium potential difference V_d which does not depend upon the nature of the solid, as previously stated. When extrinsic photogeneration of carriers occurs, however, the resulting accumulation layer of charges created adjacent to the electrolyte interface causes a

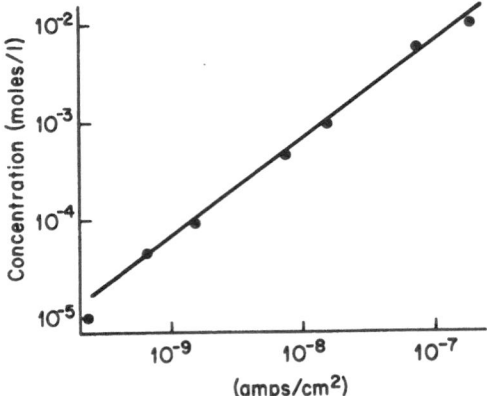

Fig. 39. Concentration dependence of saturation current for $IrCl_6^{2-}$ in 1 N HCl[62] (by permission of *Discussions of the Faraday Society*).

shift of the interfacial potential by the amount[66]

$$V_p - V_d = \frac{kT}{ze_0} \ln \frac{n_p{}^0(0)}{n_d{}^0(0)} = \frac{kT}{ze_0} \ln \frac{i_{lim}}{i_0} \tag{60}$$

where $n_p{}^0(0)$ is the equilibrium concentration of carriers at the surface in the presence of irradiation, and $n_d{}^0(0)$ is the concentration in the dark. Figure 40 presents the results of an experimental verification[56] of this relationship for the thallium and iridium hexachloride redox systems and the anthracene crystal electrode. These systems have values of z of 2 and 1, respectively, where z is the valence difference between oxidized and reduced species, and in confirmation of Eq.

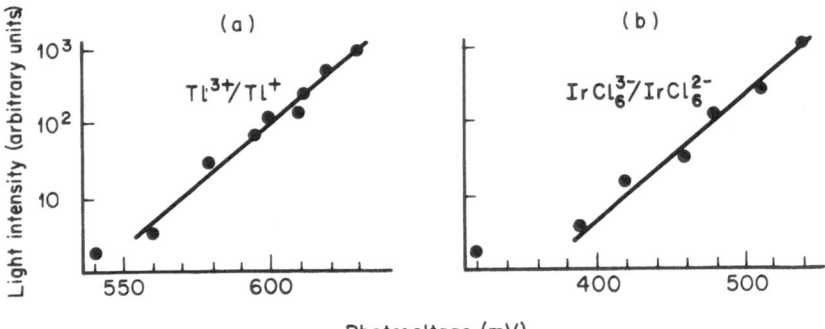

Fig. 40. (a) Dependence of photovoltage upon light intensity at interface between anthracene and Tl^{3+}/Tl^+ solution[62] (by permission of *Electrochimica Acta*). (b) Dependence of photovoltage upon light intensity at interface between anthracene and $IrCl_6^{2-}/IrCl_6^{3-}$ solution[62] (by permission of *Electrochimica Acta*).

(60), it can be seen that the gradients of the straight line are 29 and 58 mV, respectively,

$$\frac{d(V_p - V_d)}{d(\log i_{\text{lim}})} = \frac{kT}{ze_0} = \frac{58 \text{ mV}}{z} \qquad \text{at } 25°C$$

7. Some Aspects of Electrochemical Technology

In fundamental research on electrochemical kinetics we attempt to explain why a certain process occurs and what may be a suitable electrocatalyst for it, as well as to characterize the different steps of the reaction mechanism. In electrochemical technology some additional problems have to be solved. The process should occur at a reasonable rate, i.e., with a reasonable current density. This can be achieved by a high exchange current density and/or a *high effective surface area*. In addition we must ensure that the desired processes occurring at both electrodes are unaffected by unwanted side reactions. This reaction control is effected mainly by choosing a suitable substance as electrocatalyst.

In summary we can state that the nature and the detailed structure of the solid, i.e., of the electrode, are of essential importance in electrochemical technology. This will be demonstrated below by consideration of some typical examples.

Before doing so we have to explain the term *overvoltage*, a standard expression which is still widely used in the literature. Experience shows that in many electrode processes the exchange current density j_0 is of the order of 1 μA/cm^2 or less at the thermodynamic equilibrium potential ϕ_0. A measurable current (e.g., the beginning of bubble formation in gas evolution reactions) can only be obtained in these cases by the application of an overpotential of several hundred millivolts. This potential lag is called overvoltage. The phenomenon has been studied extensively in connection with hydrogen evolution at various metals (Table 4).

7.1. Overvoltage and the Rechargeable Lead–Acid Battery

During the discharge of the lead–lead dioxide cell with an aqueous sulfuric acid electrolyte, layers of PbSO$_4$ crystals are formed on the surface of the porous active electrode material at both battery

TABLE 4
Overvoltage for Hydrogen and Oxygen Evolution in Acid and Alkaline Electrolytes at Various Metals

Metal	Electrolyte	Overvoltage, mV		Electrolyte	Overvoltage, mV	
		Hydrogen	Oxygen		Hydrogen	Oxygen
Platinum	$1\,N\,H_2SO_4$	0	290	$1\,N$ KOH	0	360
Palladium	$1\,N\,H_2SO_4$	0	300	$1\,N$ KOH	0	310
Gold	$1\,N\,H_2SO_4$	80	450	$1\,N$ KOH	200	230
Nickel	$1\,N$ HCl	300	—	$1\,N$ KOH	250	—
Iron	$1\,N$ HCl	400	—	$1\,N$ KOH	100	—
Silver	$1\,N\,H_2SO_4$	250	—	$1\,N$ KOH	300	320
Graphite	—	—	—	Sat. NaCl	—	500
Lead	$1\,N\,H_2SO_4$	830	640	$1\,N$ NaOH	600	—
Mercury	$1\,N\,H_2SO_4$	1000	—	Sat. NaCl	1500	—

terminals (see Section 5):

$$(-)\quad Pb + SO_4^{2-} \rightarrow PbSO_4 + 2e^-$$

$$(+)\quad PbO_2 + 4H^+ + SO_4^{2-} + 2e^- \rightarrow PbSO_4 + 2H_2O$$

The open-circuit voltage E_0 is 2.06 V for a $4\,M\,H_2SO_4$ electrolyte. At technical current densities (10–$100\,mA/cm^2$) the terminal voltage E_{term} is still well above 1.8 V as a result of the high surface area of the active material and the fairly high exchange current density. From Table 4 we see that a measurable rate of water decomposition starts at 2.7 V, i.e., at 1.47 V more than would be expected from the thermodynamic value of 1.23 V. But during the charging process we observe that above 2.1 V oxygen evolution sets in, followed by the formation of hydrogen toward the end of the charging time at ~ 2.6 V. Thus the operation of the most important electrochemical power source is only possible because of the high overvoltages of hydrogen and oxygen at the electrocatalysts lead and lead dioxide, respectively.

7.2. The Production of Chlorine and Alkali via the Amalgam Process

In the chlor-alkali electrolysis via the amalgam process (cathodic deposition of alkali ions on mercury) the situation is similar to that in the lead–acid cell. Again, only the selection of a suitable electrocatalyst for the negative terminal with high hydrogen overvoltage results in the formation of alkali amalgam, according to the equation

$$Na^+ + e^- + xHg \rightarrow NaHg_x$$

For the deposition of sodium ions on sodium a standard potential of -2.7 V vs. s.h.e. would be required. In this case a preceding hydrogen evolution could not be avoided. Using mercury as the cathode material the deposition potential depends on the sodium concentration in the amalgam (0.2% under industrial conditions), in accordance with the Nernst equation. In practice one has about -1.78 V (see Figure 41). At a clean mercury surface the hydrogen evolution needs a more negative potential (-1.94 V, see also Table 4). But the catalytic properties of mercury are very sensitive to impurities already in the ppm range. Therefore the brine has to be purified carefully, especially with respect to traces of heavy metals.

For the positive terminal the selection of a catalyst which favors chlorine formation (see Figure 41a) is not difficult. Almost every metal (including graphite) shows higher exchange current densities for chlorine than for oxygen, to such an extent that chlorine evolution begins first (Figure 41b). The reaction

$$2Cl^- \rightleftharpoons Cl_2 + 2e^-$$

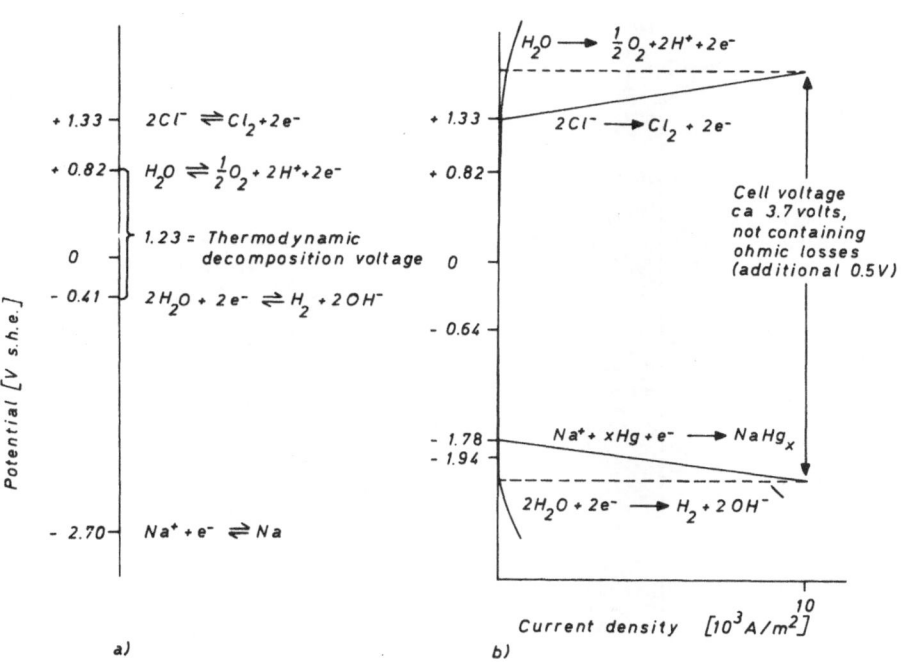

Fig. 41. Current–potential relationship in the chlor-alkali electrolysis: (a) thermodynamic deposition potentials for saturated NaCl solution ($pH \sim 7$); (b) chlorine and alkali amalgam formation in competition with water decomposition.

is therefore said to be reversible. In practical cells today graphite and titanium are used as anode material because of their excellent corrosion resistivity against chlorine.

7.3. Reactions at a Three-Phase Boundary; the Hydrogen–Oxygen Fuel Cell[67]

By a three-phase electrode we mean a porous body (cylinder, disk, or plate) which simultaneously acts as catalyst for the charge transfer reaction and as electronic conductor, in the pores of which a gas phase (fuel gas or oxygen) and a solution phase (electrolyte) are both present. This does not imply, however, that in every case a three-phase boundary, gas–liquid–solid, is formed in the pores of the electrode; frequently the electrolyte—if only in the form of a very thin film—wets the whole of the pore walls in the gas-filled part of the electrode. Such an electrode is called a *gas diffusion electrode*.

In Figure 42 possible reaction paths are indicated by the arrows: (1) diffusion of gas molecules through the film, adsorption and charge

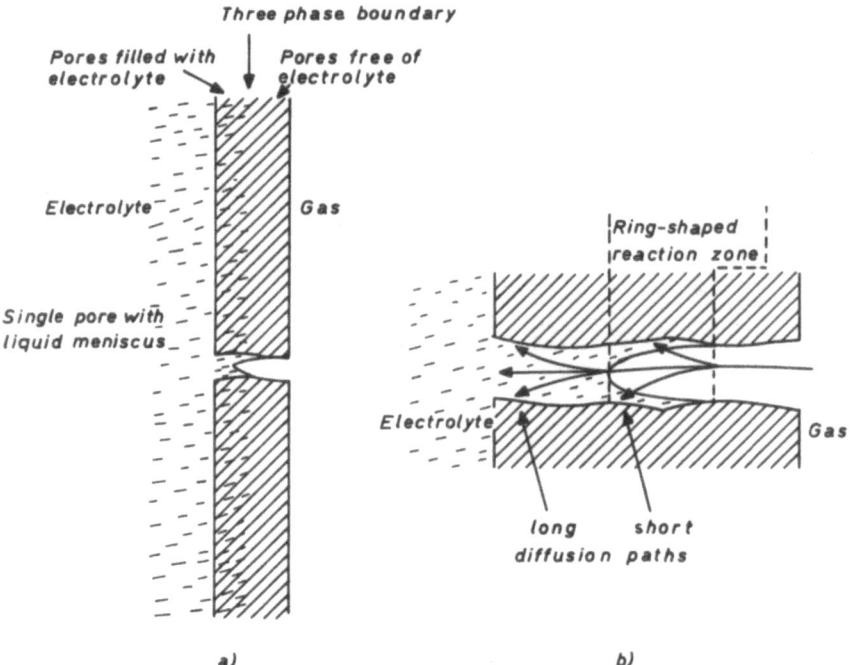

Fig. 42. Porous gas diffusion electrode: (a) section through the electrode; (b) idealized single pore with liquid meniscus.

transfer at the pore wall; (2) diffusion of gas molecules into the electrolyte-filled pores; (3) a small part of the reactant is chemisorbed on the bare pore wall and after surface diffusion the electrochemical reaction takes place at the three-phase boundary.

The contributions of the various reaction routes to the total current depend on pore structure, electrolyte, gas pressure, and the working temperature.

When plates of gas diffusion electrodes are arranged, as in Figure 43, they simultaneously serve as electrocatalyst, current

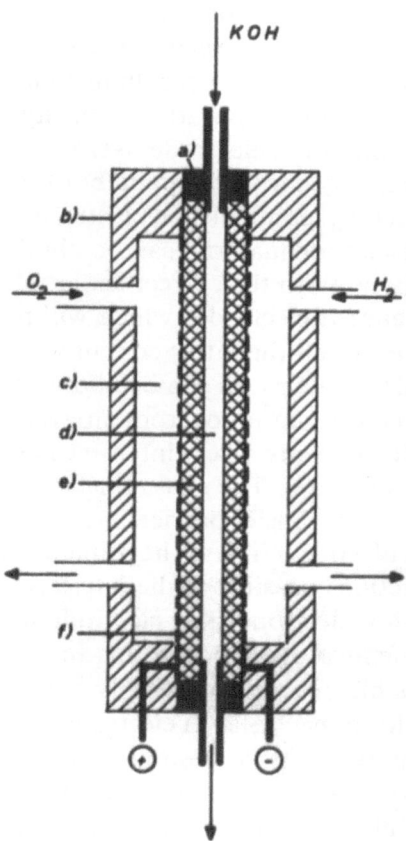

Fig. 43. Section of an individual hydrogen–oxygen cell with gas and electrolyte circulation: a, gasket; b, frame; c, gas chamber; d, electrolyte; e, porous metal plates; f, supporting grids as current collectors.

conductor, and distributor of gas behind the electrodes. Gases and electrolyte pass around a circulatory system.

The flooding of the pore system can be discouraged by the following precautions:

1. Use of a gas pressure adequate to maintain continuous free passage of gas through the pores.

2. Use of an electrode constructed of two layers differing in porosity—on the gas side, for example, with a pore size of 30–60 μm and on the electrolyte side, 10–20 μm. Because of surface tension in fine capillaries, the electrode layer on the electrolyte side is completely filled with solution, and menisci are established in the region between the layers of finer and coarser pores.

3. Use of hydrophobic treatment. Parts of the surface of the larger pores are covered with water-repelling molecules. This involves in most practical cases the incorporation of the hydrophobic material during the preparation of the electrode itself. Thus, by mixing active carbon with, e.g., polyethylene powder and hot-pressing at a temperature above the softening point of the plastic, a satisfactory hydrophobic and active electrode material can be obtained, with adequate electronic conductivity. With the correct choice of the plastic/carbon ratio, this method affords electrodes which will not become flooded with electrolyte (even when the latter contains alcohol as fuel).

Remarkable current densities can be obtained from such H_2–O_2 cells: several hundred mA/cm^2 at overpotentials of only 100–200 mV. The circulating systems ensure a very intensive mass transfer onto the surfaces of the porous plates. The cause for the observed overpotentials is mainly the charge transfer barrier.

The electrodes of a fuel cell have the principal function of making electrochemical reactions possible at the interface between phases by donating or accepting electrons. The electrode material is not itself concerned in the chemical reaction. In this sense, every electrode at which any appreciable reaction occurs is a catalyst electrode. In practice, however, the name "catalyst electrode" is used only for those electrodes that can promote a desired reaction on a technically interesting scale. For low-temperature fuel cells ($< 100°C$), adequate catalytic activity of electrode materials is an essential consideration.

The catalytic activity of an electrode depends on a number of properties of the electrode material. The three most important functions are:

1. Chemisorption of the reactants at the electrode surface.

2. Facilitation of the interface reaction by the dissociation of adsorbed molecules into atoms or by the cleavage of reactive groups.
3. Lowering of the activation energy of charge transfer.

In addition, the affinity of the electrode for the components of the electrolyte has a considerable influence.

Since a large heat of adsorption of reactants facilitates interface reactions, it is the metals that strongly adsorb hydrogen, such as palladium, platinum, and finely divided nickel (skeleton nickel) that are good catalysts both for hydrogen deposition and hydrogen oxidation. The metal structure determines catalytic properties mainly through the relation between the energy levels of the metal electrons and the "exchange levels" of the absorbate in the chemisorbed layer. As in heterogeneous catalysis, not only the composition of the catalyst, but also its state of subdivision and its surface structure are important. The greatest possible disorder and distortion of crystalline structure are desirable in the surface.

7.4. The Electrochemical Fluidized Bed Reactor; Organic Electrosynthesis

It was discovered by Müller and Schwabe[68] that an inactive metal wire dipping into a solution of formic acid containing a suspension of finely divided platinum, palladium, or other noble metals would show a potential. With sufficient movement of the wire or the electrolyte, this potential obtains a value characteristic of the solution and the suspended catalyst, caused by the contact between the wire and the catalyst particles. This involves the transfer of electrons from the particles to the electrode. The formic acid dissolved in the electrolyte serves to renew the "charge" of the catalyst particles. Current densities of more than $100 \, \text{mA/cm}^2$ (with respect to the current collector) can be obtained by the use of such a *catalyst suspension* device.

By continuous saturation of the electrolyte, gaseous reactants (e.g., dissolved hydrogen) can be converted. Enlarging the cell described above leads to a fluidized bed reactor (Figure 44), well known from modern technology in the field of heterogeneous catalysis. The electrochemical reactor is of increasing interest in organic electrosynthesis,[69] providing a reaction technique which offers the advantage of a good selectivity (in choice of electrode material and working potential) and hence a relatively high yield. In addition to the Kolbe

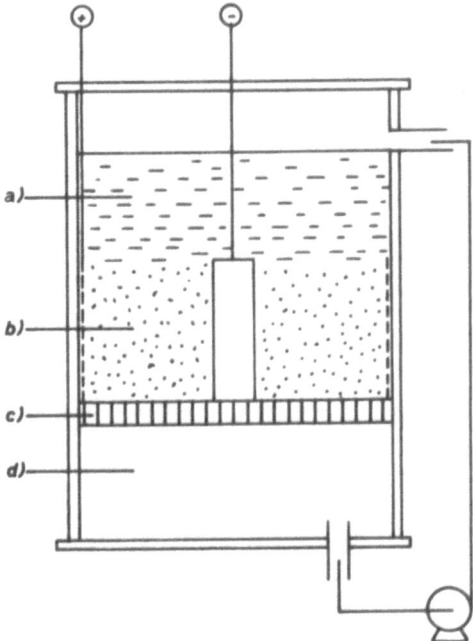

Fig. 44. Electromechanical fluidized bed reactor: a, electrolyte; b, catalyst particles; c, distribution plate; d, recirculated electrolyte.

reactions, the syntheses of adiponitrile, dimethylsebacate, propylene oxide, etc. have been developed in industry in recent years.

References

1. J. O'M. Bockris and A. K. N. Reddy, *Modern Electrochemistry*, Plenum Press, New York (1970).
2. H. H. Bauer, *Electrodics*, Thieme-Verlag, Stuttgart (1972).
3. J. Barlow, in *Physical Chemistry, an Advanced Treatise* (H. Eyring, W. Jost, and D. Henderson, eds.), Vol. IXa, pp. 162–246, Academic Press, New York (1970).
4. P. Delahay, *Double Layer and Electrode Kinetics*, Interscience, New York, London, Sydney (1965).
5. R. Parsons, in *Advances in Electrochemistry and Electrochemical Engineering* (P. Delahay and C. W. Tobias, eds.), Vol. I, pp. 1–64, Interscience, New York, London (1961).
6. R. S. Perkins and T. N. Anderson, in *Modern Aspects of Electrochemistry* (B. E. Conway and J. O'M. Bockris, eds.), Vol. 5, pp. 203–245, Plenum Press, New York (1969).

7. O. Stern, Zur Theorie der elektrolytischen Doppelschicht, *Z. Elektrochem.* **30**, 508–516 (1924).
8. A. N. Frumkin, Wasserstoffüberspannung und Struktur der Doppelschicht, *Z. Phys. Chem.* **164**, 121–133 (1933).
9. D. C. Grahame, Discreteness-of-charge-effects in the inner region of the electrical double layer, *Z. Elektrochem.* **62**, 264 (1958).
10. H. L. Friedman, *Ionic Solution Theory*, Interscience, New York, London (1962).
11. H. U. Harten, in *Festkörperprobleme* (F. Sauter, ed.), Vol. III, pp. 81–125, F. Vieweg & Sohn, Braunschweig (1964).
12. H. Gerischer, in *Physical Chemistry, an Advanced Treatise* (H. Eyring, W. Jost, and D. Henderson, eds.), Vol. IXa, pp. 463–542, Academic Press, New York (1970).
13. S. Trasatti, Work function, electronegativity, and electrochemical behaviour of metals. Pt. II: Potentials of zero charge and electrochemical work functions, *J. Electroanal. Chem.* **33**, 351–378 (1971).
14. K. J. Vetter, *Electrochemical Kinetics*, p. 114, Academic Press, New York (1967).
15. V. G. Levich, in *Physical Chemistry* (H. Eyring, W. Jost, and D. Henderson, eds.), Vol. IXb, pp. 986–1074, Academic Press, New York, London (1970).
16. R. A. Marcus, On the theory of electron-transfer reactions, *J. Chem. Phys.* **43**, 679–701 (1965), and references therein.
17. J. O'M. Bockris, R. J. Mannan, and A. Damjanovic, Dependence of the rate of Electrodic redox reactions on the substrate, *J. Chem. Phys.* **48**, 1898–1904 (1968).
18. H. Gerischer, Über den Zusammenhang zwischen dem Mechanismus der elektrolytischen Wasserstoffabscheidung und der Adsorptionsenergie des atomaren Wasserstoffs an verschiedenen Metallen, *Z. Phys. Chem. NF*, **8**, 137–153 (1956).
19. R. Parsons, The rate of electrolytic hydrogen evolution and the heat of adsorption of hydrogen, *Trans. Faraday Soc.* **54**, 1053–1063 (1958).
20. E. Schwarzer und W. Vielstich, Zum Einfluß von auf Platin adsorbiertem Blei auf den Mechnismus der Ameisensäure-Oxidation, *Chemie Ing. Techn.* **45**, 201–203 (1973).
21. H. Binder, A. Köhling, and G. Sandstede, in *From Electrocatalysis to Fuel Cells* (G. Sandstede, ed.), pp. 59–80, Battelle Seattle Research Center and the University of Washington Press, Seattle and London (1972).
22. H. H. Bauer, *Electrodics*, pp. 72–98, Georg Thieme Verlag, Stuttgart (1972).
23. N. A. Balashova and V. E. Kazarinov, in *Electroanalytical Chemistry* (A. J. Bard, ed.), Vol. 3, pp. 135–197, Marcel Dekker, New York (1969).
24. F. G. Will and C. A. Knorr, Untersuchungen des Auf- und Abbaus von Wasserstoff- und Sauerstoffbelegungen an Platin mit einer neuen instationären Methode, und Untersuchungen von Adsorptionserscheinungen an Rhodium, Iridium, Palladium und Gold mit der potentiostatischen Dreieckmethode, *Z. Elektrochem.* **64**, 258–275 (1960).
25. S. Bruckenstein and J. Comeau, Electrochemical mass spectroscopy, *Disc. Faraday Soc.* **56**, 285–292 (1973).
26. J. Heitbaum, Discussion remark, *Disc. Faraday Soc.* **56**, 305–307 (1973).
27. F. G. Will, Hydrogen adsorption on platinum single crystal electrodes, *J. Electrochem. Soc.* **112**, 451–455 (1965).
28. T. Kuwana, Optically coupled electrochemical studies, *Ber. Bunsenges.* **77**, 858–871 (1973).

29. J. O'M. Bockris and B. E. Conway, Determination of the Faradaic impedance at solid electrodes and the electrodeposition of copper, *J. Chem. Phys.* **28**, 707–716 (1958).

30. J. O'M. Bockris and A. R. Despić, in *Physical Chemistry, an Advanced Treatise* (H. Eyring, D. Henderson, and W. Jost, eds.), Vol. 9b, pp. 611–730, Academic Press, New York (1970).

31. J. O'M. Bockris and H. Kita, The dependence of charge transfer and surface diffusion rates on the structure and stability of an electrode surface: copper, *J. Electrochem. Soc.* **109**, 928–939 (1962).

32. W. W. Lossew, *Doklady Akad. Nauk SSSR* **100**, 111 (1955).

33. J. O'M. Bockris, D. M. Drazić, and A. R. Despić, The electrode kinetics of deposition and dissolution of iron, *Electrochimica Acta* **4**, 325–361 (1961).

34. W. Lorenz, Oszillographische Überspannungsmessungen, *Z. Elektrochemie* **58**, 912–918 (1954).

35. W. Mehl and J. O'M. Bockris, Mechanism of electrolytic silver deposition and dissolution, *J. Chem. Phys.* **27**, 812 (1957).

36. H. Gerischer, Zum Mechanismus der elektrolytischen Abscheidung und Auflösung fester Metalle, *Z. Elektrochemie* **62**, 256–264 (1958).

37. W. Lorenz, Impedanzmessungen zum Mechnismus der Elektrokristallisation von Silber, *Z. Phys. Chem. (Frankfurt)* **17**, 136–140 (1958).

38. A. R. Despić and J. O'M. Bockris, Kinetics of the deposition and dissolution of silver, *J. Chem. Phys.* **32**, 389–402 (1960).

39. G. Wranglén, Dendrites and growth layers in the electrocrystallization of metals, *Electrochimica Acta* **2**, 130–144 (1960).

40. P. B. Price, D. A. Vermilyca, and M. B. Webb, *Acta Met.* **6**, 524 (1958).

41. O. Kardos and D. G. Foulke, in *Advances in Electrochemistry and Electrochemical Engineering* (P. Delahay and C. W. Tobias, eds.), Vol. 2, pp. 145–233, Interscience Publishers, New York (1962).

42. M. A. Brimi and J. R. Luck, *Electrofinishing*, p. 72, Elsevier, New York (1965).

43. H. Seiter and H. Fischer, Über den Zusammenhang zwischen Abscheidungsbedingungen und Wachstumsform bei der Elektrokristallisation von Kupfer, *Z. Elektrochemie* **63**, 249–257 (1959).

44. M. Fleischmann and H. R. Thirsk, Anodic Electrocrystallization, *Electrochimica Acta* **2**, 22–49 (1960).

45. H. Fischer, *Elektrolytische Abscheidung und Elektrokristallisation von Metallen*, p. 393, Springer Verlag, Berlin, Göttingen, Heidelberg (1954).

46. C. W. Bunn and H. Emmett, Layer formation and crystal faces, *Disc. Faraday Soc.* **5**, 119–132 (1949).

47. F. C. Frank, The influence of dislocation on crystal growth, *Disc. Faraday Soc.* **5**, 48–54, 67 (1949).

48. E. Budevski, W. Bostanov, T. Vitanov, Z. Stojnov, A. Kotzeva, and R. Kaishev, Keimbildungserscheinungen an versetzungsfreien (1, 0, 0)-Flächen von Silbereinkristallen, *Electrochimica Acta* **11**, 1697–1707 (1966).

49. D. A. Vermilyea, in *Advances in Electrochemistry and Electrochemical Engineering* (P. Delahay and C. W. Tobias, eds.), Vol. 3, pp. 211–286, Interscience Publishers New York (1963).

50. D. A. Vermilyea, *J. Electrochem. Soc.* **101**, 388 (1954).

51. M. Fleischmann and H. R. Thirsk, The Behaviour of the lead dioxide electrode, *Trans. Faraday Soc.* **51**, 71–95 (1955).

52a. H. Gerischer, Semiconductor electrode reactions, in *Advances in Electrochemistry and Electrochemical Engineering* (P. Delahay, ed.), Vol. 1, Chapter 4, p. 139, Interscience, New York (1961).

52b. U. A. Myamlin and Yu. V. Pleskev, *Electrochemistry of Semiconductors*, Plenum Press, New York (1967).

53. M. J. Sparnaay, *The Electrical Double Layer*, Chapter 6, p. 271, Pergamon Press, Oxford (1972).

54. P. J. Boddy, The structure of the semiconductor electrolyte interface, *J. Electroanal. Chem.* **10**, 199–244 (1965).

55. W. Mehl and J. M. Hale, Insulator electrode reactions, in *Advances in Electrochemistry and Electrochemical Engineering* (P. Delahay, ed.), Vol. 6, Chapter 5, p. 399, Interscience, New York (1967).

56. W. Mehl, Reactions at organic semiconductor electrodes, in *Reaction of Molecules at Electrodes* (N. S. Hush, ed.), Chapter 7, p. 305, Wiley–Interscience, London (1971).

57. J. F. Dewald, *Bell System Tech. J.* **39**, 615 (1960).

58. J. M. Hale and W. Mehl, *Surface Sci.* **4**, 221 (1966).

59. W. Mehl, *Ber. Bunsenges. Physik. Chem.* **69**, 583 (1965).

60. H. Gerischer, *Z. Physik. Chem. N.F.*, **26**, 223, 325 (1960); **27**, 48 (1961).

61. V. G. Levich, in *Electrochemistry and Electrochemical Engineering* (P. Delahay, ed.), Vol. 4, Chapter 5, p. 249, Interscience, New York (1966).

62. R. R. Dogonadze, in *Reactions of Molecules at Electrodes* (N. S. Hush, ed.), Chapter 3, Wiley–Interscience, London (1971).

63. J. M. Hale and W. Mehl, *Electrochimica Acta* **13**, 1483 (1968).

64. B. J. Mulder, *Philips Res. Rpts. Suppl.* No. 4 (1968).

65. W. Mehl and J. M. Hale, *Disc. Faraday Soc. Newcastle* **45**, 30 (1968).

66. J. M. Hale, *J. Electrochem. Soc.* **115**, 208 (1968).

67. W. Vielstich, *Fuel Cells*, Wiley–Interscience, London, New York (1970).

68. E. Müller and K. Schwabe, Die Aufnahmefähigkeit der Platin-Metalle für Wasserstoff (Beziehungen zur Katalyse der Ameisensäure), *Z. Elektrochem.* **35**, 165 (1929).

69. F. Beck, Entwicklungsstand der Elektrosynthese organischer Verbindungen, *Chem. Ing. Techn.* **42**, 153 (1970).

70. P. Delahay and C. W. Tobias (Eds.), *Advances in Electrochemistry and Electrochemical Engineering*, Vol. 9, Interscience, New York (1973).

71. R. H. Muller (Ed.), *Optical Techniques in Electrochemistry, Advances in Electrochemistry and Engineering*, Vol. 9, John Wiley, New York (1973).

72. Optical studies of adsorbed layers at interfaces, *Symp. Faraday Soc.* **4** (1970).

73. K. S. Kim, N. Winograd, and R. E. Davis, *J. Am. Chem. Soc.* **93** (23), 6296–7 (1971); K. S. Kim and R. E. Davis, *J. Electron Spectrosc. Relat. Phenomena* **1**(3), 251–8 (1973); K. S. Kim, T. J. O'Leary, and N. Winograd, *Anal. Chem.* **45**(13), 2214–18 (1973); K. S. Kim, A. F. Gossmann, and N. Winograd, *Anal. Chem.* **46**(2), 197–200 (1974).

74. G. C. Bond, *Catalysis by Metals*, Academic Press, London (1962), p. 82.

75. A. T. Kuhn, C. J. Mortimer, G. C. Bond, and J. Lindley, *J. Electroanalyt. Chem. and Interfacial Electrochem.* **34**, 1 (1972).

76. A. T. Kuhn, H. Wroblowa, and J. O'M. Bockris, *Trans. Faraday Soc.* **63**, 1466 (1967).

Polymer Surfaces

H. L. Frisch
Department of Chemistry
State University of New York at Albany

G. L. Gaines, Jr.
General Electric Corporate Research and Development
Schenectady, New York

and

H. Schonhorn
Bell Laboratories, Inc.
Murray Hill, New Jersey

1. Introduction—Scope and Nature of Polymer Surfaces

Surfaces of organic polymers play a large role in many areas of current technology. Surface tension of the polymer melt is a significant factor in processing of the thermoplastics or the application of polymer coatings. The nature of the polymeric surface is involved in the phenomena of adhesion, wettability, and spreading of a polymer in contact with another phase. Some applications of organic plastics depend on such surface-related properties as friction and static surface charges. Polymers may be used in a finely divided state, e.g., in polymer latices, powder coatings, or as elastomeric filler particles in impact-resistant plastics. Besides the external bounding surfaces of a bulk polymer sample, we must not forget internal polymer interfaces separating, e.g., microphase inhomogeneities in copolymers, or polymers from reinforcing filler particles.

There is a large body of experimental information on the wetting of polymers, polymer melt surface tension, interfacial tension between different pairs of polymer melts, and the modification of polymer surfaces by additives and chemical reactions. These results are summarized in this chapter. Excluded from consideration are studies of polymer molecules at interfaces neither of which is a bulk polymer. Those interested in such studies may find references in recent reviews or articles on polymer monolayers,[1,2] adsorption of polymers from solution on solids (Ref. 3, p. 69; Refs. 4–6), and the surface tension of polymer solutions (Ref. 3, p. 209; Refs. 7–9). An extensive recent review by Rosoff[5] discusses these topics, as well as many which we shall treat in this chapter.

Biopolymer surfaces, in particular their organization into membranes and organelles of living cells, is a huge subject that could not be reviewed even within the confines of all the pages used in this treatise. The scope of this chapter is thus restricted to deal solely with synthetic, organic polymers, including the polysiloxanes. The macromolecules we shall deal with consist of many chemical repeat units held together by covalent bonds. Each repeat unit in turn consists of one or more structural units which reflect closely the atomic composition of the monomers from which the polymer is formed. The architecture of the basic linear polymer chain can be modified by branching or cross-linking. Previous chapters in this treatise have dealt with the composition, reactivity, and structure of the bulk polymer. We expect that the reader is aware of some of the special problems associated with polymers arising from their molecular weight heterogeneity and the existence of many classes of polymer chain defects due to compositional, optical, geometric, and head-to-tail isomerism, as well as inadvertent branching. An immediate consequence of this architectural irregularity of the backbone chains of polymers is the possibility of two distinct modes of solidification of polymers. The first is crystallization with a definite melting point T_M. Chapters 6 and 7 of Volume 3 of this treatise deal extensively with the necessary prerequisites for this mode of solidification and the morphology of the resulting crystalline polymer. The second mode is vitrification (glass formation), a more or less gradual freezing out of the conformational motions, with a well-defined glass transition temperature T_g separating the glassy from the rubbery "solid." This process, which also occurs in partially crystalline polymers, is discussed in Chapter 11 of Volume 5 and Chapter 7 of Volume 3 of this treatise. The second process is the only solidification process associated with the large class of amorphous polymers, which are devoid of any long-range X-ray order. The polymer chains in this class of polymers are either too irregular and/or their irregularities are too bulky to fit into a well-defined crystallographic space lattice.

Polymer single crystals are currently too thin for macroscopic measurements of their surface properties. The wetting of polycrystalline mats of single crystals has been studied. Otherwise, the measurements on polymeric solids reviewed in this chapter have been carried out on bulk samples of either partially crystalline polymers, in which the crystallites are imbedded in a matrix of amorphous material, or on wholly amorphous polymer material. In view of the fact that this is a treatise on the properties of solids, the question of what constitutes a polymeric "solid" is pertinent. A practical answer to this question is afforded by the results of mechanical measurements on bulk polymer samples.

Polymers exhibit time-dependent mechanical properties including a remarkable degree of viscoelasticity. Dynamic mechanical experiments yield both an apparent elastic modulus (in tension or shear) E and the mechanical damping tan δ, where δ is the loss angle. At a convenient frequency of, say, 100 Hz, a typical linear amorphous polymer in the glassy region below T_g has a modulus of about 10^{10} dynes/cm^2. As the temperature is increased the next region of mechanical response is the transition region which includes T_g, which can be taken to be, in first approximation, the point of inflection of the modulus curve or the maximum in the damping curve. In this region the modulus drops by a factor of one thousand. The third region is called the rubbery plateau, where the modulus remains roughly constant. As the temperature is further increased this is followed by a region of elastic or rubbery flow, and finally liquid flow with very little elastic recovery. In these regions the damping curve again begins to increase and the modulus may drop below 10^5 dynes/cm^2. In a cross-linked (amorphous) polymer the last two regions are absent, since there is no flow, the cross-links prevent the chains from slipping past each other. The presence of crystallinity increases the modulus significantly above T_g, with a sharp drop in the modulus and increase in mechanical damping curve at the melting point T_M. Thus in crystalline polymers the demarcation point between the elastomeric (rubberlike) "solid" and rubbery "melt" can be taken to be given by T_M, while for amorphous solids it lies in the vicinity of the end of the rubbery plateau region.

The fact that the demarcation between the "melt" and the "solid" is far from sharp justifies in part the inclusion in our review in this treatise of an extended discussion of the surface and interfacial tensions of polymer melts. The complicated dynamical response of polymers to applied stresses precludes the reliable estimation of surface tensions or surface free energies of polymers by such methods as mechanical necking experiments, which have been applied to other classes of solids. Ultimately it is the many internal degrees of freedom implied by the macromolecular nature of polymers which is responsible for this extensive and varied time dependence of their mechanical and thermal properties.

In preparing this chapter, we asked many of the eminent contributors to our knowledge of polymer surfaces to indicate in what ways polymer surfaces are (or are not) distinctive. It is the consensus of their replies and our own opinion that synthetic organic polymers are really van der Waals or molecular solids in terms of their local surface properties, even though they may remain solid at much higher temperatures than other members of that class by virtue of their large molecular weight or the presence of cross-linking. The fact that even the bulk polymer is nonuniform in composition, molecular weight, and structure leads to many new ways in which the interfacial region can differ from the bulk. One must deal with "contamination" from within as well as without by surface-active substances. With some notable exceptions, most polymers which we consider are hydrophobic and polar sites in such a hydrophobic matrix present special problems. Last but not least, polymer surface samples are usually not pretreated extensively, and often cannot be studied under conditions that prevent their contamination. (A solvent swabbing of the surface of a polymer is not like the extensive care in pretreating a semiconductor single-crystal surface for a high-vacuum LEED study.) The disadvantage of this is that polymer surfaces as used in various studies by different investigators may not be identical. On the other hand, the surfaces studied in this fashion may not differ as markedly from the routinely cleaned surfaces employed in many technological applications.

2. Some Fundamental Concepts

The basic thermodynamics of surfaces[10-12] has already been covered elsewhere in this treatise.[13] We merely group together for convenient reference a number of concepts and definitions together with some quantitative relations between certain thermodynamic entities. Let A be the total Helmholtz free energy of a system composed of two (indefinitely large) bulk phases α and β, separated by an interfacial region near or within which is contained a suitably defined flat dividing surface of total area a. Let A^{α} and A^{β} be the Helmholtz free energies per unit volume, multiplied by the volume of the homogeneous portion of each phase. Denote by n_i, μ_i, and Γ_i the number of moles in the system, the chemical potential, and the surface excess of the component i. Two, in general distinct, entities can be defined: The first is the *specific Helmholtz surface free energy*, which is defined by

$$\bar{a}_{\alpha\beta} = (A - A^\alpha - A^\beta)/a = \bar{a}_{\beta\alpha} \qquad (1)$$

The second is the surface tension $\gamma_{\alpha\beta} = \gamma_{\beta\alpha}$, representing the reversible work of stretching the suitably defined dividing surface

$$\gamma = \left(\frac{\partial A}{\partial a}\right)_{T,V,n_i} = \left(\frac{\partial G}{\partial a}\right)_{T,P,n_i} \qquad (2)$$

with G the total Gibbs free energy of the system. For a sensible choice of the dividing surface one can show that $\gamma_{\alpha\beta}$ and $\bar{a}_{\alpha\beta}$ are related by

$$\gamma_{\alpha\beta} = \bar{a}_{\alpha\beta} - \sum_i \Gamma_i \mu_i \qquad (3)$$

where the summation extends over all components in the system. Only for a pure, single-component liquid ($\alpha = L$) and its vapor ($\beta = V$) can we expect $\gamma_{LV} = \bar{a}_{LV}$, since for multicomponent systems some adsorption will occur and $\sum_i \Gamma_i \mu_i$ will not strictly vanish. The existence of strains in solids may make Eq. (3) a more attractive alternative to (2) as a defining relation for the surface tension, since it involves only integral quantities. We will refer to two surface tensions associated with systems with a solid phase. The first is γ_{SV} of a solid in equilibrium with a vapor phase, while the second is γ_S, a surface tension of the bare solid in contact with vacuum.

The surface tension of liquids or interfacial tension of liquid pairs can be reliably measured experimentally. This is true also of the viscous polymer melts. On the other hand, *there exists currently no direct, reproach-free method of measuring the surface tension or specific surface free energy of a macroscopic surface of a polymeric solid*. The surface free energies which are discussed, and possibly even measured, of a crystalline nucleus of a polymer within an amorphous matrix have little relation to the surface tensions or specific surface free energies of a macroscopic polymer surface in contact with vapor or a simple non-polar liquid. In any case these matters have already been discussed in Chapter 7 of Volume 3 of this treatise.

Thermodynamics of highly curved interfaces becomes very complicated and we shall refer our readers to specialized treatises and reviews[11,12,14] on this subject. Caution must be exercised whenever a phase domain becomes small, e.g., less than 10^{-5} cm in lineal extent. Polymers introduce further problems since the definition of a polymeric substance allows for possibilities of isomerism and molecular weight heterogeneity. Thus the surface tension of an isotactic polymer melt should (and can) differ from that of the syndiotactic polymer melt. Even in the absence of isomerism a heterodisperse polymer melt will not possess precisely the same molecular weight distribution in the immediate vicinity of a phase boundary as in the bulk (e.g., because of a difference of configurational entropy in the interface region of shorter chains as compared to longer chains). Furthermore, polymer solids can be quenched or otherwise prepared in metastable states which can exhibit hysteresis in surface as well as bulk properties. In short, the large size, the extensive range of mechanical viscoelastic relaxation times, and the molecular inhomogeneity of polymer molecules aggravate characteristic difficulties in applying continuum, phenomenological theory (hydrodynamics or thermodynamics) to the description of experimental surface phenomena. When a liquid spreads on a solid surface the advancing edge may be quite irregular and not reproducible and may merge into a thin film whose profile shows many steps in height differing by many orders of magnitude in scale down to the molecular.[15-19] What is the physical significance of the disjoining pressure when an apparent steady state in spreading has occurred? Can one extend the concept of a contact angle to the edge of the film? Ordinarily, macroscopic observations using a low-power $(10 \times)$ microscope are used in determining the contact angle at the apparent intersection of the three phase boundaries of vapor and a liquid and a solid. Even the temporal attainment of a situation in which the contact angle may possess a simple thermodynamic significance is fraught with ambiguity. Current studies of the hydrodynamic motion of fluids in the immediate neighborhood of a moving contact line show that the Stokes equations (with plausible boundary conditions) are insufficient to describe the motion.[20,21]

In the remainder of this chapter we review almost exclusively information derived concerning polymer surfaces from suitably selected macroscopic measurements where the experimenter expects no serious difficulties with the application of a continuum theory within the precision of the experiment. Even then, further simplifying approximations may be made in interpreting measurements. For example, effects on the advancing contact angle due to the deformation of a solid surface in the process of wetting[22] will be ubiquitously neglected, the perturbation being presumably small.[5] We take this opportunity though to warn our readers that the price of a simplified exposition is sometimes scientific error.

The importance of the specific surface free energies or surface tensions arises because many measurable equilibrium surface quantities of macroscopic surfaces can be expressed in terms of their differences, e.g., differences in surface tensions. Simple instances are afforded by the measurable spreading pressures and contact angle θ. Thus the spreading pressure of a liquid on a solid π_{SL} and that of a vapor on a solid π_e can be written[11]

$$\pi_{SL} = \gamma_S - \gamma_{SL}, \qquad \pi_e = \gamma_S - \gamma_{SV} \tag{4}$$

The contact angle where a vapor, liquid, and a nondeformable solid meet can also be expressed in this fashion. Thus

$$\cos \theta = (\pi_{SL} - \pi_e)/\gamma_{LV} \tag{5}$$

in terms of measurable entities, or using (4), it can be expressed in terms of the three relevant surface tensions, via the so-called Young equation:

$$\cos \theta = (\gamma_{SV} - \gamma_{SL})/\gamma_{LV} \tag{6}$$

For stable equilibrium $\gamma_{LV} > 0$ and $\pi_e > 0$, but π_{SL} may have either sign. If $\cos \theta \geq 1$, the solid is wetted completely by the liquid, which can spread freely over the surface; if $\cos \theta < -1$, the solid is not wetted. For intermediate values of $\cos \theta$ between ± 1 the cosine of the equilibrium contact angle is a direct measure of wettability. Equation (6) shows that the equilibrium contact angle and γ_{LV} determine the reversible work in replacing a unit area of solid–vapor interface by a solid–liquid interface, the adhesion tension $\gamma_{SV} - \gamma_{SL}$. This entity is involved in the reversible work of adhesion W_A between a solid and a liquid in the presence of adsorbed vapor:

$$W_A = \gamma_{SV} - \gamma_{SL} + \gamma_{LV} = \gamma_{LV}(1 + \cos \theta) \tag{7}$$

This equation is referred to as the Harkins-modified Dupré relation.

Most measurements in the literature of the contact angle as a thermodynamic equilibrium property are actually measurements of the advanced (or advancing) contact angles within a minute or so of contact line displacement. The reported average deviations are commonly $\pm 2°$, though sometimes only $\pm 1°$. Often on causing a rapid displacement of the contact line, one finds that the new static contact angle depends on the direction of the movement of the contact line.[20] The difference between the advancing and the receding contact angles $\theta_a - \theta_r$ is called the contact angle hysteresis.[23-25] Contact angle hysteresis can depend on the time interval between movement and measurement, on contamination, and on many aspects of the state of the substrate surface. In the case of solid surfaces these are known to include roughness, heterogeneity, presence of viscoelastic stress, and the content of liquid dissolved or dispersed in it.[24a,b]

There are many regimes of contact angle hysteresis, depending on the velocity of contact line displacement. Here we are concerned solely with the regime in which the contact line is caused to move sufficiently slowly (velocities of less than 1 mm/min) so that the difference between advancing and receding contact angles is sensibly constant and reproducible. In this quasistatic regime the difference between advancing and receding contact angles of test liquids on a solid may provide a rough insight into the distribution of surface heterogeneity and effects due to increasing surface rugosity. This technique is based on theoretical studies of a quasistatic advance of a liquid drop on model patchy surfaces.[26-28] With heterogeneous or rough surfaces a conflict arises in the minimization of the total free energy of the system due to the differing quasistatic constraints at the vapor–liquid interface, where Laplace's equation of capillarity has to be satisfied, and the Young equation (6) at the three-phase boundary. This provides a mechanism of quasistatic hysteresis for patchlike surfaces (patches larger than 0.1 μm of different surface free energy) and surfaces with rugosities of the order of 0.5 μm or larger.

This work of Good,[25] Johnson and Dettre,[26,27] and Neumann and Good[28] is of considerable significance to us since it provides the basis of a novel method of surface characterization of polymers. Johnson and Dettre's simple model[26] consists of concentric circular rings of alternating homogeneous surface with contact angles θ_1 and θ_2, $\theta_1 > \theta_2$. The result of their analysis is illustrated in Figure 1, which is a schematic plot of the advancing and receding angles versus the increasing proportion of the high-surface-energy surface. The receding contact angle decreases sharply toward θ_2 as soon as a small proportion of high-energy surface is introduced. The advancing contact angle remains essentially constant until there is a very high proportion of high-energy surface. It follows that: (1) The advancing contact angle is a characteristic of the low-energy portion of the surface and the receding angle that of the high-energy portion. (2) Absence of hysteresis indicates a homogeneous surface. (3) A predominantly low-energy surface will show poor reproducibility of the receding contact angle, while a scatter in the values for the ad-

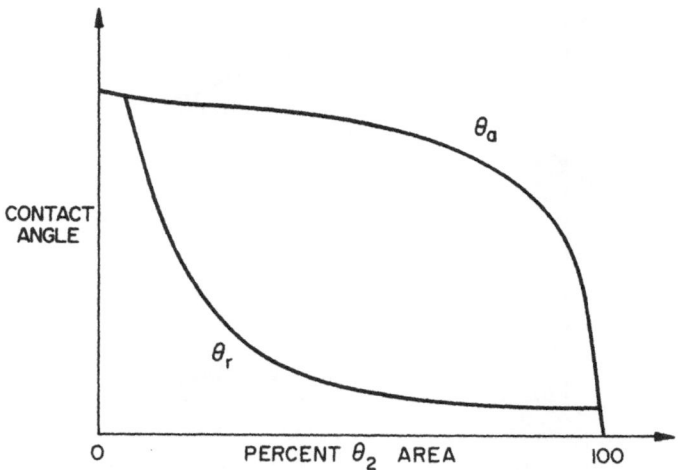

Fig. 1. Predicted advancing (θ_a) and receding (θ_r) contact angles on a model surface with concentric rings of two different wettabilities, $\theta_1 > \theta_2$. (After Johnson and Dettre.[26])

vancing contact angles implies a predominantly high-energy surface. The predictions of this model are in agreement with the more general model of Neumann and Good[28] and many observations.

Johnson and Dettre's model[27] of a rough surface consists of a surface composed of concentric grooves of an energetically homogeneous substrate. The variation of the contact angles with increasing roughness is shown schematically in Figure 2. The advancing contact angle increases gradually, while the receding contact angle decreases sharply as roughness increases, until a composite surface of solid and trapped vapor is attained. Subsequently there is a gradual decrease in the advancing contact angle and a more rapid increase of the receding contact angle with further increase in rugosity. The predictions of these simple models appear reasonable providing the assumptions concerning the geometry of the surface and the dynamics of spreading can be met by the experiment.

Zisman and co-workers have studied the wetting behavior of many solid surfaces by a variety of pure liquids, often members of chemically homologous series such as the *n*-alkanes. If the cosines of the advancing contact angles θ_a for several liquids on a given solid surface (such as a solid polymer) are plotted against the liquid surface tensions γ_{LV}, it is observed that the points lie near a straight line or in a narrow rectilinear band (cf. Figure 3). The intercept of the cos θ vs. γ_{LV} plot of this band or straight line at cos $\theta = 1$ (zero contact

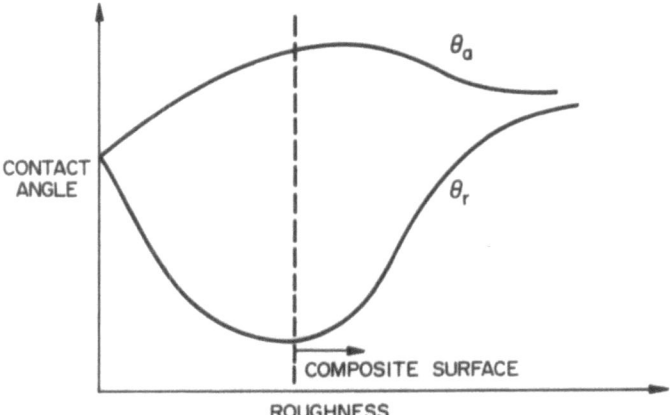

Fig. 2. Predicted advancing (θ_a) and receding (θ_r) contact angles on a model rough surface. In the region denoted "composite surface" vapor is trapped in the surface irregularities beneath the liquid layer. (After Johnson and Dettre.[27])

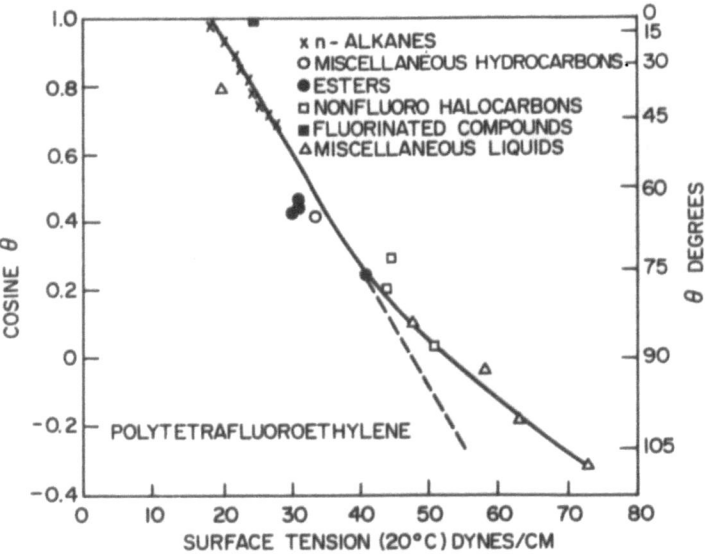

Fig. 3. Contact angles of various liquids on polytetrafluoroethylene, plotted to estimate the critical surface tension for wetting. (Reproduced from Zisman,[23] by permission.)

angle) represents the liquid surface tension below which complete wetting is expected (at least for the class of liquids investigated) on the given solid surface. This limiting surface tension is called the critical surface tension (for wetting) γ_c. Mathematically, then, in the vicinity of $\gamma_{LV} = \gamma_c$ one has empirically

$$\cos \theta = 1 + K(\gamma_c - \gamma_{LV}) \tag{8}$$

with K as the limiting slope of the rectilinear band. As Fox and Zisman have already emphasized in one of their earliest papers on the subject,[29] "... the concept of γ_c must be used with caution, since ... γ_c varies between liquid types (used to determine it). In particular, it is not valid to construe γ_c as a measure γ_s." Clearly γ_c is only an approximate characteristic of the surface, unlike γ_s or $\bar{a}_s$. Rosoff[5] has summarized the arguments on these points in great detail. Experimentally Johnson and Dettre[30] in an elegant study of the wettability of low-energy liquid surfaces by a series of alkanes have shown that an analog of Eq. (8) can be applied to liquid substrates and the difference between γ_c and γ_s (here referring to the liquid substrate) ranged from 5 to 50 dynes/cm.

Even though γ_c does not have an absolute thermodynamic significance, it is an extremely useful figure of merit for classifying surfaces of organic polymers, among other materials, in regards to their wetting behavior. The observed values of γ_c (at 20–25°C) for organic polymers vary from about 50 to 60 dynes/cm for polar polymers as urea–formaldehyde and polyamide–epichlorhydrin resins,[31] to 10 dynes/cm for certain acrylic polymers bearing perfluorinated side chains.[32] The limiting slope K also varies only over a narrow interval, roughly from 0.03 to 0.04.[10] The status of the contact angle as a thermodynamic entity is clearly dependent on the fact that it involves a difference in surface tensions, the adhesion tension. Any inferences from contact angle data concerning the surface characteristic entities γ_s or $\bar{a}_s$ must be based on modelistic, extrathermodynamic assumptions.[33]

Solid surfaces are often classified into so-called low-energy and high-energy surfaces, where the terms in principle refer to the specific surface free energy but are usually interpreted experimentally to mean low or high wettability. (The specific surface free energies of many solids are not accessible to direct measurement, but γ_c can be measured easily.) Most solid organic polymers are presumed to have specific surface free energies which are less than 100 ergs/cm^2, and so are classed as low-energy surfaces.[5]

3. Characterization and Structure of Polymer Surfaces

To characterize fully an interface, one is interested in the values of various physical and chemical properties of the matter composing it, first, as a function of the depth into the interface and, second, as a function of two directions normal to the depth. This requires combinations of many analytical tools and techniques. Each experimental technique provides data on a specific property (or properties) which is an average over a characteristic depth of sampling its range. In the case of polymer surfaces one is interested in: (1) the elemental and molecular composition; (2) the distribution in molecular weight, cross-link density, extent of branching, and chain isomers; (3) variation in density and percent of crystallinity; (4) changes in morphology or extent of ordering; (5) the nature and extent of preferential orientation of the molecular segments; (6) the distribution of regions of varying surface free energy (surface energetics), patches of high polarity, etc; (7) the superficial surface area (particularly of finely divided polymer samples); and (8) the microtopography of the surface.

Hagstrum and McRae have reviewed in the second chapter of Volume 6A the experimental methods employed to study the structure of surfaces. Some of these techniques, e.g., LEED or Auger spectroscopy, have not yet been successfully applied to polymer surfaces for various reasons. Thus, the high-energy electron beams used to excite Auger emission may cause radiation damage to polymer surfaces. This has also limited the application of electron and ion microprobe techniques. The nature of the nuclei composing organic polymers prevents the use of Mössbauer spectroscopy except in certain polymers such as the polysiloxanes. Newer techniques appear very promising but have not actually been developed for applications to polymer surfaces; examples are secondary ion mass spectroscopy or ion neutralization spectroscopy. Certain techniques such as ellipsometry[5,34] have been very successfully applied to the study of monolayers of adsorbed polymers, but are outside the scope of this review since they have not been applied directly to a bulk polymer surface. Finally we must not forget that in studying, say, a polymer–metal interface, the presence of metal or metal oxide can be detected by a variety of techniques [X-ray fluorescence or electron microprobe (particularly as attachments to a scanning electron microscope), atomic absorption spectroscopy, etc.]. Barring effects due to voids or nonpolymeric impurities, etc., such studies provide some insight into the distribution of the polymer in the interface by the argument that regions containing high concentrations of metal or metal oxide are largely devoid of polymer. Some useful techniques for the characterization of adherend surfaces (to polymers) have been recently reviewed by Hamilton.[35]

In what follows we shall present short descriptions of experimental techniques that respond directly to the ponderable matter composing the organic polymer interface. We will intersperse a few examples of how a combination of a number of such techniques provides at least a partial characterization of some selected polymer surfaces. A complete characterization of polymer surfaces along the lines of the eight or so categories which we listed above is still largely a matter for the future.

3.1. Surface Characterization Techniques

The best-known technique to determine surface composition and some information on orientation and/or degree of crystallinity is attenuated total reflectance (ATR) infrared spectroscopy.[5] This technique is often augmented by comparison of its results with results of standard and modified infrared transmission studies on bulk films (~ 1 mil or more in thickness). Following Harrick,[36] in ATR the penetration of the incident beam into the surface region of a polymer film adhering to, say, a KRS-5 crystal is approximately one-fifth of the wavelength or about $2\,\mu m$ for radiation of frequency $720\,cm^{-1}$. It thus has a range of about $1\,\mu m$ and measures a relatively thick surface layer. Good contact is required between the sample and the spectrometer prism, which may be a source of difficulty with rough or inflexible films.

An example of the application of this technique is Luongo and Schonhorn's[37] infrared study of the surface region of polyethylene (PE) nucleated on gold or aluminum (high-energy surfaces) and polytetrafluoroethylene (PTFE, a low-energy surface). The chemical composition of the surface region and that of the bulk appear spectroscopically to have the same chemical nature but there are characteristic differences in the intensities of certain infrared bands which are known from studies of bulk samples[38] to be suitable for following relative changes in crystallinity. Luongo and Schonhorn followed, in particular, the absorbance ratio R of the strong $730\text{-}cm^{-1}$ and $720\text{-}cm^{-1}$ components of the methylene rocking doublet shown in Figure 4. The $730\text{-}cm^{-1}$ component of the doublet is associated with CH_2 groups in the crystalline regions. In the PE unit cell the in-phase rocking modes

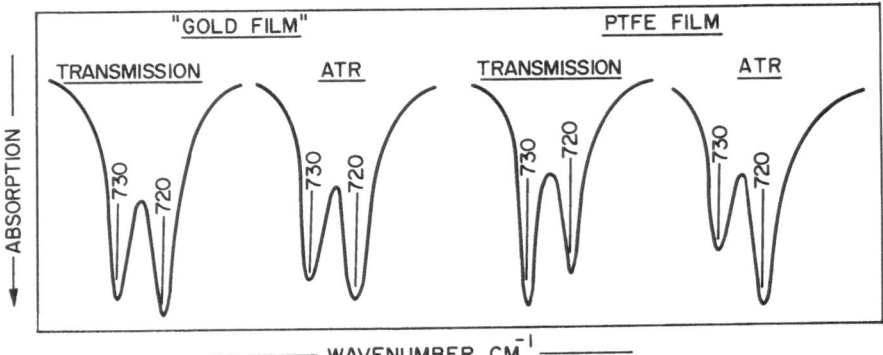

Fig. 4. Transmission and ATR infrared spectra of thin polyethylene films nucleated against gold and PTFE. (From Luongo and Schonhorn.[37])

have a dipole moment change parallel to the crystallographic *a* axis represented by the 730-cm^{-1} component. The 720-cm^{-1} component is due to the CH$_2$ groups in the amorphous phase with a contribution from the crystalline phase due to the out-of-phase modes with moments parallel to the *b* axis. A crystallinity standard for comparison of the absorbance ratio $R_{720/730}$ was provided by measurements on single crystals of C$_{36}$H$_{74}$ and checked against measurements of single-crystal PE aggregates. To study the orientation of the crystallites, $R_{720/730}$ was studied by transmission and ATR as a function of the angle of incidence of the incident spectrometer radiation on the film surface. The conclusions reached by the authors of this study were that while both PE films nucleated against gold and PTFE possess some crystallinity, the surface region ($\gtrsim 2~\mu$m) of the film nucleated against PTFE is considerably less crystalline as seen with the ATR technique. The film nucleated against PTFE had disk-shaped spherulites whose average diameter was about 20 times larger than the spherulites of the film nucleated against gold. The size of the spherulites was determined from optical polarization microscopy. This accounted for the anisotropy with respect to the angle of incidence. These studies provide strong evidence for difference in the surface crystallinity in the films nucleated against various surfaces. Incidentally, no residue of gold was detected by electron probe, and atomic absorption spectroscopy indicated from 0 to 0.01 equivalent monolayer of gold remaining on the PE film. The chemical purity of the film generated against PTFE could be checked only by the absence of effects in the infrared spectrum. The wetting behavior[39] of the films nucleated against gold and PTFE is different. Thus, for example, the contact angle (20°C) of glycerol was 53° with the film nucleated against gold and 81° with the film nucleated against PTFE. This is not inconsistent with the spectro-scopically determined differences in crystallinity of the surface region.

The Henke-design X-ray tube[40] generates long-wavelength or "soft" X rays and thereby extends the usefulness of X-ray fluorescence analysis down to atomic number five, by what is termed *soft X-ray spectroscopy*. The effective depth of analysis is about 1 μm into the surface.[41,42] It is a valuable technique in its own right as well as a useful complement of the infrared ATR technique.

X-ray photoelectron spectroscopy (ESCA), developed by Siegbahn and co-workers, allows chemical structure analysis even in polymers in a range one thousand times less deep than either ATR or soft X-ray spectroscopy. Basically, this is because photoelectrons ejected from atoms deep in the sample are either self-absorbed by the sample or lose part of their kinetic energy and contribute only to the spectral background. The effective escape depth varies with the electron kinetic energy and surface structure but is of the order of a few tens of angstroms.[41-43] Besides providing a semiquantitative elemental analysis, ESCA spectra exhibit small shifts (chemical shifts) due to the environment of an atom in a molecule. Thus, for example, in the carbon spectrum from poly(vinyl fluoride) there is a doublet corresponding to the carbon atoms in CHF and CH_2 groups. The electron density is less around the fluorine-substituted carbon than around the carbon with only hydrogen substituents, and the binding energy is correspond-ingly increased.

Certain spectroscopic techniques can only be applied to the much higher area internal surfaces attained at the interface between polymer and dispersed, small filler particles. A variety of nuclear magnetic resonance studies have been carried out on filled polymers; usually the filler employed was carbon black.[44] Proton spin resonance studies support the view that a portion of polymer in the vicinity of the filler is immobilized. Kaufman et al.[45] report the observation of two spin–spin relaxation times T_2 in the bound rubber phases of polybutadiene and ethylene–propylene rubber, each reinforced with 50 parts per hundred parts rubber of an SAF black (155 m^2/g surface area). The amount of fully immobilized polymer was only 4% of the total, but its shorter T_2 indicated the loss in mobility. Inorganic fillers may have larger effects on local segmental mobility than carbon black. Lipatov and Fabuliak[46] had earlier reported large shifts with tem-perature in the minima in the NMR signal of poly(methyl methacry-late) and polystyrene filled with silica. Somewhat smaller effects were observed with these polymers when filled with finely dispersed PTFE. These authors also interpret their results as due to a decrease of segmental motion, mainly determined by a decrease of the number of conformations in the interface rather than due to significant changes of interaction energy at the surface.

Kulividze *et al.*[47] studied the width of the proton resonance line in the system (crystallizing) polybutadiene at low temperatures and hence decreased the intensity of the broad component of the resonance signal derived from the crystalline phase. Only small effects were observed which could be attributed to a loss of mobility of the amorphous polymer. The spin–lattice relaxation time T_1 and T_2 decreased slightly on addition of carbon black. The position of the glass transition (as judged from the narrowing of the NMR signal) was almost unaffected by the carbon black. Droste *et al.*[48] report two spin–lattice relaxation times, one for the bulk polymer and one for polymer "bound" at the interface, for clay-filled phenoxy resin. They estimate the thickness of the polymer surface layer at 30 Å, or about 25% of the polymer is "associated" with the surface. This agreed with the estimates previously obtained of the thickness of the polymer surface layer of styrene–butadiene rubber/HAF carbon black of Kraus and Gruver.[49] Kraus and Gruver obtained their estimate from the dilatometrically determined shift of T_g on filling the rubber. The 4% immobilized polymer of Kaufman *et al.*[45] would correspond to a layer of about 5 Å. Clearly the range of the NMR technique, under favorable circumstances, can be quite small. In passing, we have already referred to estimates of the depth of the surface layer of polymer bound to filler particles obtained from dilatometric measurements of T_g.[44] Similar inferences can be drawn from "mechanical spectroscopy," i.e., linear viscoelastic creep or relaxation measurements[44] and sorption measurements of low-molecular-weight vapors by filled polymer.[50] Further discussion of the interpretation of these measurements (which have led to consistent estimates of the surface layer) would take us too far afield into the study of polymer–filler interactions and away from the subject matter of this chapter.

The surface areas of finely divided polymer samples have been obtained from standard gas adsorption techniques whose theoretical and experimental basis has been reviewed in Chapter 5 of Volume 6A. The results of such measurements will be separately discussed somewhat later in this section.

The principal tools which are available for studying the micro-topography of polymer surfaces are various forms of both optical and electron microscopy. Below we shall list and briefly describe the most useful microscopic techniques.

Electron microscopy is a powerful tool for studying surface roughness and inhomogeneity. Scanning electron microscopy (SEM) analyzes the electrons back-scattered from a polymer surface that has been coated with a thin, vacuum-evaporated metallic or carbon film. This technique has a great depth of field, but lateral resolution is limited to a few hundred angstroms. Transmission electron microscopy requires replication techniques to study solid surfaces, but has the advantage of at least an order of magnitude higher resolution.[41,42] The effectiveness of transmission electron microscopy depends largely on the success in suitably preparing a specimen for investigation, i.e., microtoming or staining a section. Besides providing information about the lateral surface microtopography, this technique allows one to prepare sections penetrating the surface layer or region of the sample. New frontiers in polymer ultrastructure investigations may be opened by special techniques of sample preparation such as freeze-cleaving of the specimen (e.g., to minimize deformation of the sample) borrowed from biological and medical electron microscopy. The potential for usefully applying these techniques of freeze-cleaving or etching to the study of the surfaces of polymeric membranes or polymeric lattices[51] is large.

Optical microscopy has also been an extremely useful tool when there exists a special surface region whose extent is large enough to be resolved. An example is provided by the transcrystalline regions sometimes observed in crystalline polymers, which we shall discuss in greater detail elsewhere in this section. We have already mentioned the use of polarization microscopy in surface studies of crystallinity. The use of optical microscopy and interferometry to study the interfaces of polymers dissolving in solvents has been reviewed and summarized by Ueberreiter.[52] The concentration gradient of the solvent can be followed by the number of interference lines produced by a beam of monochromatic light. The dissolution of a typical glassy polymer such as polystyrene at 25°C in toluene produces a multilayered interface. The composition changes continuously in going from the solvent to the pure polymer. First, there is a liquid layer surrounding the solid, followed by a gel layer containing swollen polymer in a rubberlike state. The solvent concentration does not fall to zero beyond the gel layer, because a polymer which is glassy at room temperature has to take up a large amount of solvent to be plasticized in the so-called solid-swollen layer. Even this is not the last one containing solvent molecules. An infiltration layer exists in which solvent enters preexisting channels and holes in the solid polymer surface without creating new holes. The total depth of all these layers is about 5×10^{-2} cm and, as expected from a diffusion-controlled process, increases according to an Arrhenius law with increasing temperature, and decreases somewhat with the Reynolds number for flowing solvents until a limiting value is achieved.[52] The dissolution of crystalline polymers can be regarded as a "melting" in a liquid of foreign material; the rate of dissolution is proportional to the velocity of crystallization.[52]

Besides these spectroscopic and microscopic techniques, which in one way or another involve an external source of electromagnetic radiation as a probe for the polymer molecules in the surface layer or region, there exist characterization techniques which do not involve external sources of electromagnetic radiation. These fall into a variety of different categories depending on the physical nature of the probe involved. First, valuable information about the thickness and mechanical properties such as the frequency-dependent elasticity, viscosity, loss angle, etc., can be obtained from propagation and attenuation measurements of ultrasonic waves. Special techniques have been devised to separate the surface sound waves from the bulk modes.[53] The range of these measurements extends from a fraction to tens of microns. Second, one can study sufficiently thin (~ 1 μm) microtomed sections of the surface region or films of a polymer by various chemical and physical means. Thus the chemical reactivity of the sorption and diffusion of small molecules in and through such thin sections or films can be investigated. Third, stylus instruments (Talysurf, etc.) can be used to mechanically probe the microtopography of polymer surfaces. Finally, wetting studies can themselves be employed to characterize surfaces. The range of critical surface tension, contact angle, and contact angle hysteresis (see Section 2) measurements is quite short and measured at most in tens of angstroms since these would be affected most by the outermost molecules or layers of molecules. It has been found that stretch-oriented films of polypropylene or Teflon FEP exhibit anisotropic contact angles, being higher for the liquid front advancing or retreating perpendicular to the direction of stretch than in the parallel direction.[54] The observed contact angle anisotropy is probably due to the anisotropic molecular force field and not secondary effects such as directional variations in surface roughness.

Among the most extensively studied polymeric surfaces have been those of fluoropolymers, both control surfaces and those subjected to special surface treatments. These have been studied by combinations of infrared spectroscopy (ATR and transmission), ESCA, soft X-ray spectroscopy, contact angle, contact angle hysteresis, and electron and optical microscopy.[41,43,55–58] The polymers most studied are untreated commercial polytetrafluoroethylene (Teflon TFE) and tetrafluoroethylene–hexafluoropropylene copolymer (Teflon FEP) films and the same films subjected to a variety of surface etchings intended to increase the adhesiveness of these to other polymers, e.g., polyurethane, epoxy, or phenolic resin adhesives. These include treatment with (1) sodium in naphthalene/tetrahydrofuran, (2) sodium in liquid ammonia, or (3) molten potassium acetate at 325°C. The color of the fluorocarbon surface is changed to dark brown to black by these treatments. Microscopic measurements indicate the color change is localized within a surface region of a few tenths of a micron.[57] The advancing contact angle of water (25°C) on PTFE changes from 108° for the unmodified film to 62–66° for the etched films.[57] Similarly Teflon FEP exhibits a decrease of the advancing water contact angle from 109° to 52°.[41] The infrared spectra of the original PTFE (either of a composite layer built up from seven 5-μm films or the differential spectrum) is greatly changed in the region of 1600 cm^{-1} as a result of the three etching treatments (cf. Figure 5). The presence of the intense absorption band at 1600–1700 cm^{-1} has been ascribed to the presence of conjugated double bonds and/or the valence vibrations of C=O groups.[57] Surface films subjected to Na in naphthalene/tetrahydrofuran or to potassium acetate

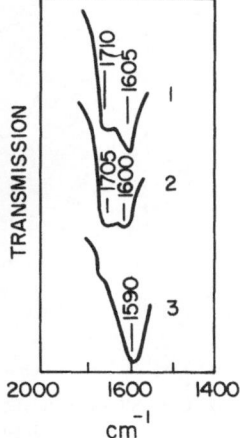

Fig. 5. Differential infrared spectra of polytetrafluoroethylene modified by (1) the sodium–naphthalene complex, (2) solution of sodium in liquid ammonia, and (3) molten potassium acetate. (From Borisova *et al.*[57])

treatment exhibit spectra indicating the presence of the hydrophilic OH and CO groups as well as unsaturation. Films treated with Na in anhydrous ammonia exhibit indications of surface NH groups as well.[57] These results are consistent with the work of Dwight and Riggs[41] and Collins *et al.*[43] using, among other techniques, ESCA and soft X-ray spectroscopy. The presence of nitrogen in the sodium/ammonia-etched films of both Teflon TFE and FEP is confirmed. The soft X-ray data indicate that the depth of etch after 60 sec of exposure is about 0.3 μm in PTFE and 0.07 μm in FEP. A comparison of the electron microtopography of FEP subjected to this treatment and the original, shown in Figure 6, indicates that etching creates a spongelike top layer that has partly come off, exposing less porous substrate. The ESCA spectra of the Teflon FEP original and the etched specimens, together with advancing and receding water contact angles, are shown in Figure 7. The original polymer spectrum shows two sharp peaks: fluorine (695 eV) and carbon in a highly oxidized state (295 eV) characteristic of fluorocarbons. The high contact angles and small hysteresis show that FEP has a very low and nearly homogeneous surface energy. After sodium-etching, the fluorine peak is absent and an intense oxygen peak is observed. Resolution of the carbon peak now centered at 285 eV indicates that it could arise from three contributions—hydrocarbon, carbonyl, and carboxyl-type carbon in an intensity ratio 4.8/1.7/1.0, respectively. This is consistent with the ATR infrared spectra of heavily etched

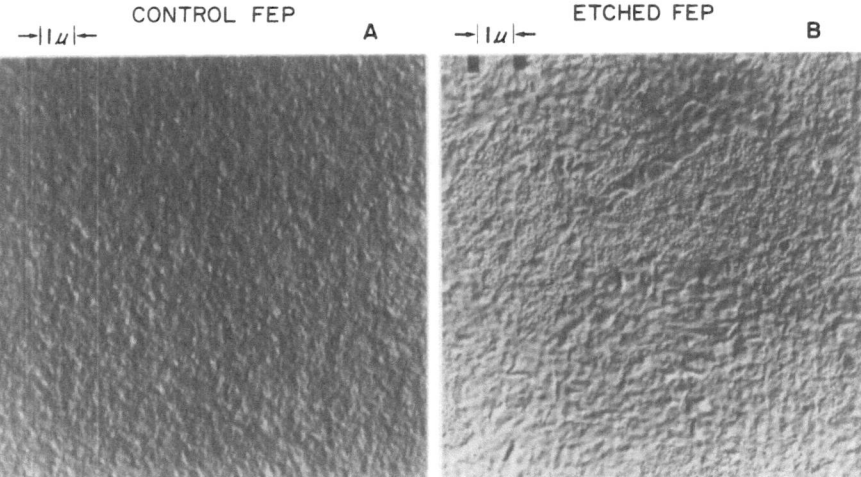

CONTROL FEP   A   ETCHED FEP   B
→|1μ|←   →|1μ|←

Fig. 6. Electron micrographs of FEP surfaces before and after exposure to sodium in liquid ammonia. (From Dwight and Riggs.[41])

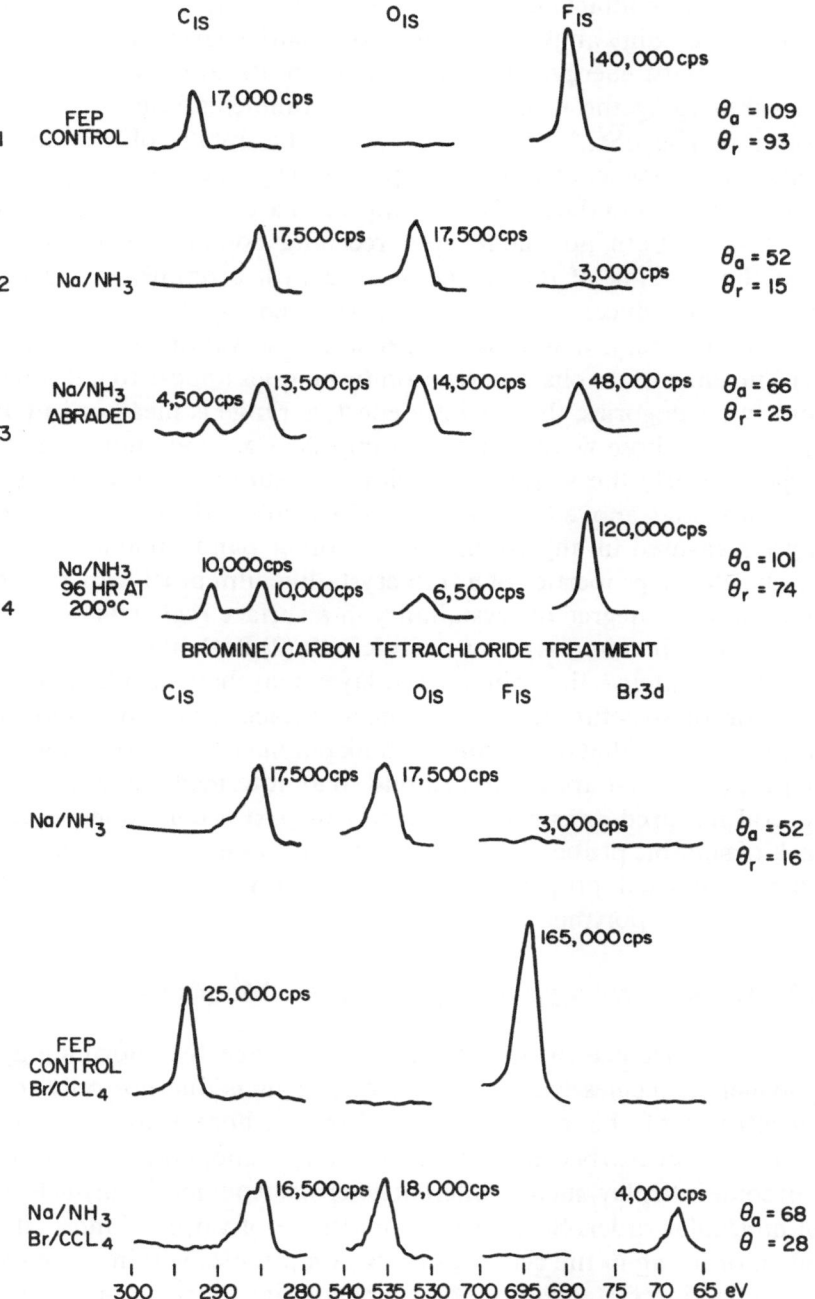

Fig. 7. ESCA spectra and water contact angles on FEP surfaces after various etching treatments. (From Dwight and Riggs.[41])

films.[47] The contact angle results show[41,58] that the surface is polar. The low receding angle may reflect some surface roughness as well as higher surface energy. The presence of unsaturation was confirmed by submerging the etched films in a bromine/carbon tetrachloride solution. The ESCA spectrum confirms the uptake of bromine not only by the presence of the bromine peak, but by the changed appearance of the carbon and the unchanged appearance of the oxygen peak. The increase in both advancing and receding contact angles suggests that the reaction of the etched surface with bromine/carbon tetrachloride introduces patches of low surface energy.[41]

In summary, a number of general observations can be made: (1) Current surface characterization techniques appear to fall broadly into two categories, those whose effective range is measured in angstroms and those whose effective range is about one thousand times larger. Clearly the wetting behavior of a surface such as the polyfluorocarbons above follows the ESCA characterization of a *surface layer* measured in angstroms. On the other hand, significant effects on the elastic properties of a thin crystalline film of PE can arise due to a modified degree of crystallinity in a surface *region* whose extent is measured in microns or tens of microns. (2) Both have to be investigated separately—the thin surface layer may have a different composition or structure from the thicker surface region and both may have properties that differ from the bulk polymer.[41,59] (3) The absence of properties that appear to respond to an intermediate range, say of several hundred angstroms, is clearly in part a consequence of the lack of suitable probes with that range. This intermediate region may possess unusual properties since the average diameter (radius of gyration) of a polymer coil has this dimension.

3.2. Transcrystallinity and Polymer Surface Morphology

The existence of an interface can change the morphology of polymer molecules lying in its vicinity. This is due not only to the anisotropy of the intermolecular forces, whose range cannot be excessive, but also because new, longer-ranged cooperative phenomena can come into play, such as different heterogeneous modes of nucleation of molecular order. Such order can also be of various kinds—it can be an ordering in the center of mass of molecules within some space lattice, as in crystallization, or a preferential ordering of the orientation of molecules which may not involve a first-order phase transition. In principle, an amorphous polymer which is on the verge of being able to crystallize could exhibit some degree of crystallinity when it is allowed to cool on a suitable substrate. This has not yet been reported, but the modification of the crystallinity of crystalline polymer on a variety of substrates has been investigated.

While the bulk (or interior) of most semicrystalline polymer samples is spherulitic, there are several possibilities for the surface region, among them: (1) spherulites smaller or equal in diameter to those in the interior, (2) elongated spherulites originating at the surface and propagating normal to the surface (this is transcrystallinity), and (3) oriented lamellar crystals, originating at the surface and extending into the interior. The first two categories can sometimes be observed under certain conditions in injection-molded samples where molecular orientation exists in the melt.[59] Apparently, in transcrystalline specimens the cooling of the surface of a polymer melt causes spherulites to nucleate very closely together on the surface. The lamellae which grow from these nuclei can propagate only in one principal direction normal to the surface since growth in the lateral directions is restricted by neighboring spherulites. These conditions lead to the formation of a surface region, as shown in Figure 8, in which only

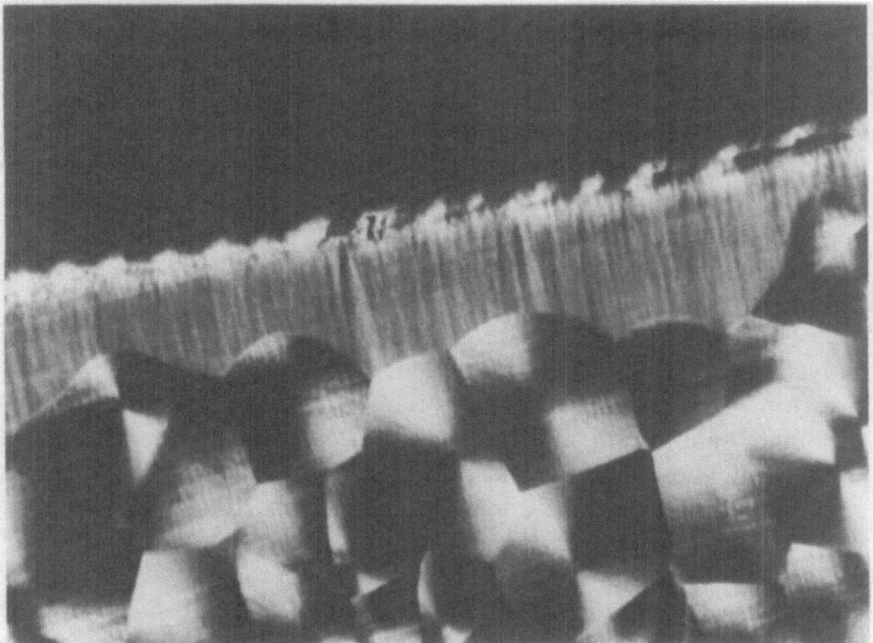

Fig. 8. Micrographs showing transcrystalline growth near the polyethylene/aluminum interface. (From Kwei *et al.*[60])

very narrowly divergent spherulitic sectors develop and give an overall rodlike appearance.[60]

These transcrystalline regions as observed in polyamides,[61,62] polyurethanes,[61] and polyolefins[59,60,63,64] have thicknesses estimated from photomicrographs[59,60,63,64] which are often in excess of 10 μm. The exact value of the thickness of the transcrystalline region (if any) is dependent on many factors such as: (a) thermal properties and history of the polymer film[59-65] (e.g., its melt temperature, crystallization temperature, and to some extent cooling rate) and (b) the nature of the surface and the polymer in contact with it.[59-65] It appears plausible that the occurrence of transcrystallinity can be attributed to massive heterogeneous nucleation induced at the mold surface.[59] To be effective, the mold surface should have a nucleating efficiency equal to or greater than that of adventitious nuclei present in the polymer. For polyolefins,[59] as the crystallization temperature approaches the melting point, the activity of the mold surfaces is found to increase, leading invariably to transcrystalline formation. The degree of activity of various mold surfaces correlates with the known activity of specific dispersed nucleating agents having similar chemical structures.

Such transcrystalline layers of 10 μm or more in thickness are effectively macroscopic. One expects, therefore, that in thin films (such as in adhesive layers) where the surface zone is a substantial portion of the total volume, the contribution of the transcrystalline region to the overall physical properties of the film may become significant. Experiments have shown that these transcrystalline regions have different mechanical properties, diffusion and solubility coefficients, and surface properties, from the bulk interior of the polymer.

The dynamic linear viscoelastic properties of transcrystalline polyethylene (PE) and polypropylene (PP) films prepared by compression molding between two flamed sheets of copper, sulfo-chromated aluminum, and Mylar have been studied[60] as a function of their thickness. The PE and PP films, independently of their thickness, had densities of 0.955 and 0.911 g/cm^3, respectively. The dynamic Young's modulus E and its real and imaginary components E' and E'' were measured at 23°C by the use of the Vibron dynamic viscoelastometer at a frequency of 110 Hz. Figures 9 and 10 show plots of the measured values of E' and E'', respectively, vs. thickness of the polyolefin film t. One sees that both E' and E'' decrease with increasing thickness t of the film.

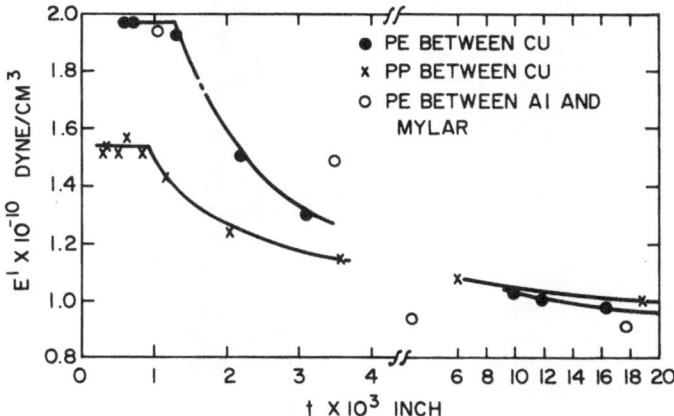

Fig. 9. The dependence of E' on thickness for molded polyolefin films. (From Kwei *et al.*[60])

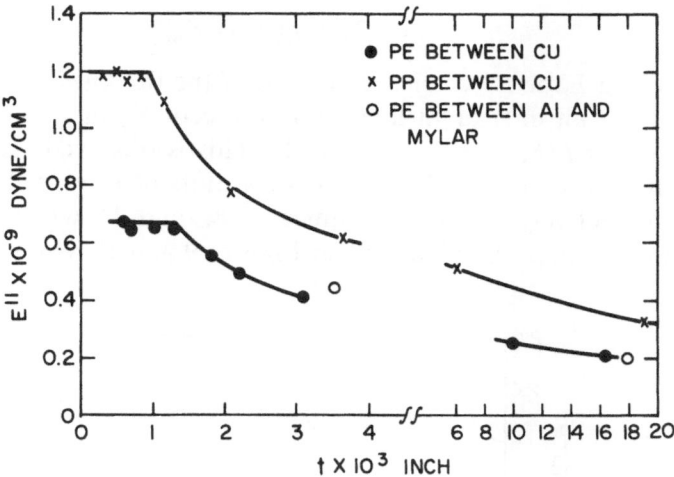

Fig. 10. The dependence of E'' on thickness for molded polyolefin films. (From Kwei *et al.*[60])

The observed variation of E with t can be easily rationalized by the fact that the transcrystalline surface regions and the bulk regions comprise a "parallel" combination in response to the applied oscillation load indicated by the arrows F shown in Figure 11. If $E_s = E_s' + iE_s''$ is the transcrystalline surface region Young's modulus, of total thickness $\frac{1}{2}t_s$, and $E_B = E_B' + iE_B''$, is the corresponding Young's modulus of the interior bulk region of thickness $t - t_s$, then a parallel

367

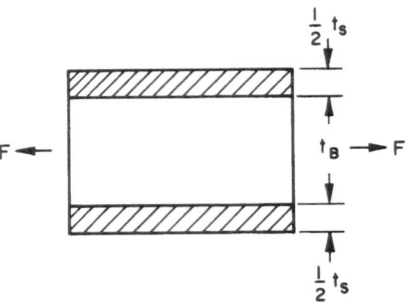

Fig. 11. The "parallel" combination of the surface and bulk region in response to the applied force *F*. (From Kwei *et al.*[60])

combination of these should result in a mechanical response given by

$$E = E' + iE'' = [1 - (t_s/t)]E_B + (t_s/t)E_s$$
$$= E_B + t_s(E_s - E_B)t^{-1} \tag{9}$$

Thus plotting E' or E'' vs. the reciprocal of the thickness of the film t^{-1} should result in straight lines with intercept E_B' or E_B'' and slope $t_s(E_s' - E_B')$ or $t_s(E_s'' - E_B'')$, respectively. This is observed, as can be seen from Figures 12 and 13, which show plots of E' vs. t^{-1} and E'' vs. t^{-1}, respectively. Using the values of E_s' and E_s'' which can be read from the curves of E' and E'' in Figures 9 and 10 as $t \rightarrow 0$, one

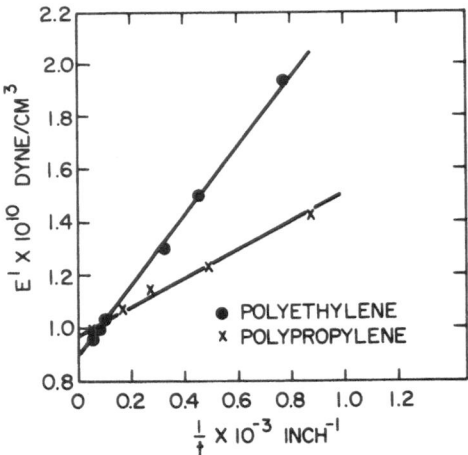

Fig. 12. Plot of E' vs. t^{-1} according to Eq. (9). (From Kwei *et al.*[60])

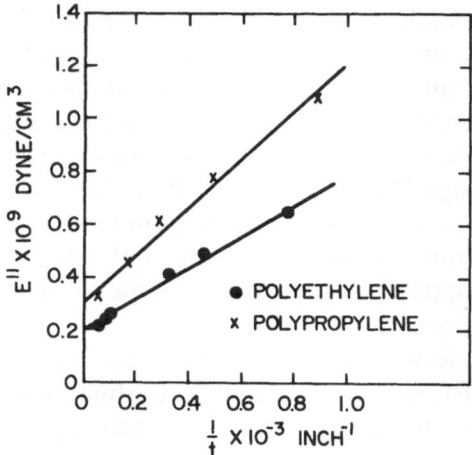

Fig. 13. Plot of E'' vs. t^{-1} according to Eq. (9).
(From Kwei *et al.*[60])

can obtain values of t_s from the slopes and intercepts of Figures 12 and 13. The values of E_s', E_B', E_B'', and t_s obtained in this way are shown in Table 1. The ratios of the real and imaginary parts of the Young's modulus of the surface to interior regions are 2 and 3–4, respectively. This indicates rather marked differences in the mechanical properties. The thickness t_s obtained from mechanical measurements is in good agreement with the values of t_s estimated from photomicrographs. Similar results have recently been obtained for polyamides.[65]

These results have immediate possible implications for the cohesive failure of composites in which a semicrystalline polymer is used as an adhesive. Criteria of failure derived from simplified versions of linear theory such as Griffiths' for brittle fracture and the critical stress in rupture relate directly the tensile stress in brittle

TABLE 1

Mechanical Properties of the Surface and the Bulk Regions

	PE	PP
E_B', dynes/cm^2	9×10^9	9.7×10^9
E_s', dynes/cm^2	1.97×10^{10}	1.53×10^{10}
E_B'', dynes/cm^2	2×10^8	3×10^8
E_s'', dynes/cm^2	6.7×10^8	1.20×10^8
t_s (from $E_s' - E_B'$), mils	1.22	0.97
t_s (from $E_s'' - E_B''$), mils	1.27	1.05

failure to $(E')^{1/2}$ and the critical stress to E', respectively. Unfortunately, these criteria are not applicable to real polymers, without extensive modification or reinterpretation, since ultimate properties of polymers involve highly nonlinear and history-dependent response. If indeed the transcrystalline regions prove on direct experimental test to be somewhat "stronger" and "tougher" than the interior regions, this should be used in designing adhesive combinations. Clearly, more direct determinations of the ultimate strength properties of the transcrystalline as opposed to the interior regions of polymer samples would be very desirable.

Sorption balance measurements of the uptake of ethane in transcrystalline PE films and on films with the transcrystalline region ground away have been carried out.[63] Small-angle X-ray diffraction confirmed differences in orientation in the two regions and that these were not essentially affected by the grinding. In a quenched transcrystalline PE film sample the transcrystalline region was sufficiently thick that when the interior region was removed, a direct measurement of ethane uptake into the transcrystalline region film along could be undertaken. From the ethane uptake (measured at 23°C) diffusion D and solubility coefficients k could be determined for the whole film as well as the corresponding entities D_B and k_B for the interior region and in the case of quenched film D_s and k_s of the transcrystalline surface region. The values of these coefficients for the quenched sample were $D = 3.4 \times 10^{-8}$, $D_B = 1.7 \times 10^{-8}$, and $D_s = 3.6 \times 10^{-8}$ cm²/sec, and $k = 0.635$, $k_B = 0.63$, and $k_s = 0.555$ mg/cm³ atm. The ratio of D/D_B falls from 2.1 to 1.2 when the polymer is allowed to cool very slowly as opposed to being quenched, the smallest depth of surface region being formed on slowest cooling.

X-ray diffraction studies[63] support the view that the differences in diffusion behavior between the surface and interior regions are due to differences in orientation of the lamellar boundaries. In the transcrystalline region the lamellar boundaries, along which diffusion is strongly preferred, are oriented predominantly normal to the surface, as opposed to the interior region. Diffusion should be more rapid because the lamella boundary diffusion paths are relatively long and direct in the surface region.[63] These differences would have some very specific implications for those adhesive composites involving, e.g., a transcrystalline polymer film and a tacky liquid or another polymer solution to which the Voyutskii–Vasenin diffusion theory of adhesion has particular bearing. If the rate of interdiffusion of the adhesive and adherend, on initial wetting, is the key factor in producing an adhesive (or autoadhesive) joint as is claimed in certain cases, then differences in diffusion (and solubility) coefficients become directly applicable, e.g., the specific work of peeling is predicted to be directly proportional to the square root of the diffusion coefficient in the surface region. Also, environmental or aging effects in adhesive failure may involve mass transport of moisture, oxygen, plasticizers, etc. Whether structural change, accumulation producing a weak boundary layer, or chemical reaction deteriorating or enhancing the adhesive joint is involved is immaterial as long as the kinetics of the process is to some extent diffusion limited.

3.3. Adsorption on Solid Polymers

Studies of adsorption, either from the gas or liquid phase, have contributed much to understanding the properties of many solid surfaces. Adsorption processes may also be directly relevant to practical problems in such areas as lubrication, detergency, and textile manufacture. However, in the case of polymers it has generally proven difficult to decide whether conventional sorption measurements reflect true adsorption, i.e., a process limited to the polymer surface, or some combination of surface plus bulk diffusional or swelling (absorption) effects. In some cases, of course, sorption clearly is a bulk phase process, e.g., in ion exchange with conventional polyelectrolyte resins. Such matters are excluded from discussion here.

Low-temperature adsorption of inert gases (nitrogen and the rare gases) has been widely used to estimate surface area. For finely divided polymer samples or shavings, apparently satisfactory measurements have been obtained with polyethylene,[67-70] polypropylene,[69-71] nylon,[67] and poly(methyl methacrylate).[72] Hysteresis, presumably due to gas permeation into the polymer, was observed in nitrogen and argon sorption by polystyrene at 65–75°K.[73] With polytetrafluoroethylene, the low energy of the surface apparently leads to difficulty in surface area measurement.[74]

The low surface energy of polytetrafluoroethylene has provoked study of sorption of a variety of vapors by this polymer. Water adsorption suggested the presence of a small fraction of polar sites on the surface.[75] Sorption of hydrocarbons appeared to include significant premeation in Graham's experiments,[76] but Whalen[77] could not confirm this result. He pointed out, however, that capillary condensation at points of particle contact produces most of the measured vapor uptake at high relative pressures. The low-pressure adsorption measurements suggested that in his sample $\sim 3\%$ of the surface consisted of energetic sites, while the rest of the surface adsorbed hexane or octane as a low-density two-dimensional gas. Other gases whose sorption by polytetrafluoroethylene have been reported are H_2[68] and CF_4.[78]

Solution sorption studies on polymers have been numerous, usually involving specific technological goals. It is impossible to mention here more than a few of the types of studies which have been carried out. Perhaps the greatest amount of attention has been paid to the possibility of using such measurements to determine the particle size in polymer latices. The soap titration technique[79] involves determining the excess amount of soap required to reach the critical micelle concentration in solution when the latex is present. This provides a surface area, and hence particle size, estimate if an area per adsorbed soap molecule can be assigned. Gerrens[80] has compiled a list of such area values, as well as references to the technique. Paxton's results, however, suggest that such areas must be used with caution, since he finds that different latices yield different areas.[81]

The sorption of surfactants, both ionic and nonionic, on textile fibers has also been widely studied. With the more polar polymers, such as cellulosics and nylon, solubility in the polymer is generally important.[82]

A variety of organic compounds are sorbed from both aqueous and nonaqueous solutions by styrene- or methacrylate-based macroporous resins of high surface area.[83]

The fact that finely divided polyethylene, polypropylene, and polytetrafluoroethylene catalyze the oxidation of hydrocarbons in solution by molecular oxygen may be related to the sorption of these materials by the polymers.[84]

In all of the types of studies so far mentioned, the polymer is examined in a finely divided form, either as fiber, latex, or powder (which may be obtained by drying a latex prepared by emulsion polymerization). The surfaces of such samples are difficult to characterize; in view of what is noted elsewhere in this review, it seems likely that their properties are highly dependent on the methods of preparation. Especially in the case of materials derived from emulsion polymerization processes, where surfactants are known to be present from the preparation, the actual surface composition may be quite unrelated to the bulk polymer.

There is little quantitative information on adsorption on macroscopic plane surfaces of solid polymers. That preferential adsorption from certain solutions may occur on such surfaces may be inferred, however, from a variety of observations. Ellison and Zisman[85] observed that the wettability of nylon by *n*-decane was markedly reduced (contact angle increased from ~ 0 to 40°), after the polymer has been exposed to a solution of perfluorolauric acid; apparently the fluorinated acid was adsorbed. Similar results have been obtained with poly(methyl methacrylate).[86] Polyethylene terephthalate[85] and bisphenol-A polycarbonate[86] do not show this effect, remaining completely wettable by decane after treatment with the perfluorinated acid solution. Some of the observations on boundary lubrication of polymers (Section 6) are also consistent with adsorption phenomena.

These observations must be considered as very preliminary, especially in view of the dependence of surface properties on sample history, as noted in our previous discussion. Frozen-in strains in the surface may have important effects on adsorption; it has been reported[87] that the bending or stretching of polyethylene increases adsorption of sodium stearate or sodium dodecylsulfate from aqueous solutions severalfold.

4. Surface and Interfacial Tension of Polymer Melts

We have already remarked on the relevance to this review of measurements of the surface and interfacial tensions of polymer melts, as well as the difficulties which the viscous or viscoelastic character of these materials introduce. It seems appropriate therefore, to consider both the experimental methods available and the results of these measurements. (A more detailed account has also been given in a recent review.[88])

4.1. Methods of Measurement

There are, of course, a very large number of methods for the measurement of liquid surface and interfacial tensions. For viscous polymer liquids only the static methods seem to be completely satisfactory; it has not been possible to demonstrate that pull, detachment, or bubble pressure measurements can always be made slowly enough to yield accurate results with highly viscous liquids. For example, the ring method (used manually) exhibits rate effects with liquids of viscosity greater than ~ 5 P. While capillary rise is one of the static methods, the very slow attainment of equilibrium (because of the resistance to flow in a narrow capillary) makes it unsatisfactory for highly viscous materials.[89,90] The methods which have been developed to the stage where their results may be considered most reliable are the Wilhelmy plate method, as described by Dettre and Johnson,[91] the pendant drop method as modified by Roe,[92,93] the sessile bubble method, used by Sakai[94] and Hata,[95] and the rotating drop or bubble technique, proposed by Vonnegut[96] and refined by Princen *et al.*[97] In the case of the last, however, the experimental difficulties are considerable and the precision of measurement so far attained has not equalled that of the first three.[98] (An early attempt[99] to investigate polymer interfacial tensions by the breakup of a molten thread has never been fully developed.)

4.1.1. The Wilhelmy Plate Method[91]

When a solid object is partly immersed in a liquid which wets it, there is a force acting downward on it due to the liquid surface tension. If the object is a flat plate which is completely wetted (contact angle zero), and its position is adjusted so that the lower edge is in the plane of the free liquid surface, this force is simply equal to $\gamma_{LV}P$, where P is the plate perimeter. Hence a simple measurement of the difference in the apparent weight of a plate (of glass, platinum, etc.) hanging in air and then in contact with the liquid surface permits calculation of the surface tension.

Assuring complete wetting equilibrium with viscous polymer liquids requires considerable care. The technique developed by Dettre and Johnson consists in immersing the plate to different measured depths, then withdrawing to the zero-immersion level and allowing the liquid on the plate to drain until the force becomes constant. When the withdrawal step of each immersion–withdrawal cycle is begun at the same measured force on the plate, it is found that the weight difference ΔW at equilibrium is given by

$$\Delta W = \gamma_{LV}P + xA \tag{10}$$

where x is the depth of immersion and A is a constant which measures the weight of the thin, tenacious film of liquid adhering to the plate above the meniscus. Plotting $\Delta W/p$ vs. x and extrapolating to $x = 0$ then gives γ_{LV}. The measurements are conveniently carried out with a recording force-measuring device such as an Instron tensile tester.

An important advantage of the plate method used in this way is that knowledge of the liquid density is not required. On the other hand, for some systems the achievement of zero contact angle may be difficult. This restriction becomes especially severe for measurements of interfacial tensions between liquids, and while such measurements have been reported for small-molecule liquids of low viscosity,[100] none has been attempted in polymer systems.

4.1.2. The Pendant Drop Method [92,93]

The shape of a pendant drop under the influence of gravity and surface tension is governed by a known differential equation. Andreas *et al.*[101] first proposed that measurements of two diameters of the drop could be used to determine the surface tension through this relationship. Their procedure involved determination of the maximum diameter D_e and a second diameter D_s located at a distance D_e above the drop vertex (Figure 14). The ratio of these diameters is a shape parameter S which has a known relationship to the factor H in the equation

$$\gamma = g\rho D_e^2/H \tag{11}$$

(g is the gravitational acceleration and ρ is the liquid density, or density difference for interfacial tension measurement). Tables of H for measured values of S are available.

An important advantage of the method for viscous liquids resides in the fact that measurements on a drop photograph can provide a

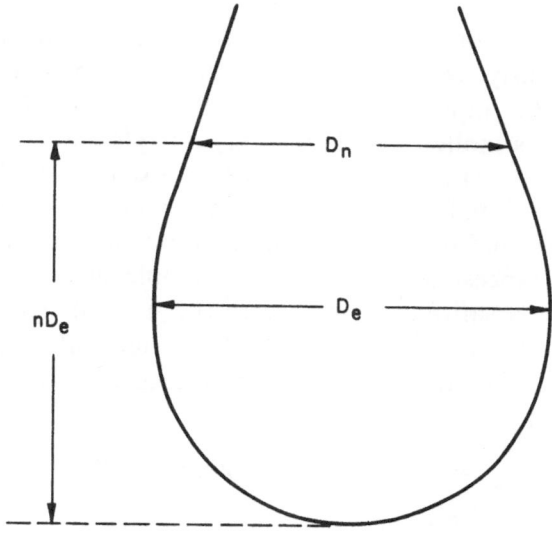

Fig. 14. Shape of a pendant drop. If $n = 1$, $D_n = D_s$.
(After Roe *et al.*[92])

375

criterion for hydrodynamic equilibrium of the drop, by verifying conformity to the differential equation of the profile. Roe and his co-workers pointed out that diameters at several heights can be measured, and the corresponding shape factors must be self-consistent if the drop is in hydrodynamic equilibrium.[92] They provided tables* of $1/H$ vs. S_n, where the S_n are the ratios D_n/D_e taken at various positions nD_e above the drop vertex ($n = 0.8, 0.9, 1.0, 1.1, 1.2$). This verification is especially important if the surface tension itself is changing with time (due to a slow adsorption process, for example), since it permits separation of such phenomena from viscous flow effects.

Measurements of both surface and interfacial tension of polymers by the pendant drop method have now been reported from several laboratories. The technique does not require any particular solid–liquid angle (except that the contact angle be constant over the surface from which the drop is suspended, so that the drop shape will constitute a figure of revolution), and there is minimal solid–liquid contact to impede fluid flow. As in all techniques except the Wilhelmy plate method, accurate knowledge of the liquid density is required. One experimental difficulty which does arise when the surface tension changes with time is that an initially stable drop may detach if a sufficient decrease of surface tension occurs.

4.1.3. The Sessile Bubble Method[94,95]

Several alternate methods based on drop, bubble, or meniscus profiles obviously exist, all based on the fundamental differential equation for the liquid surface profile. For liquids of high surface tension, such as molten metals or inorganic glasses, the dimensions of a sessile drop resting on a plane surface are often used. Liquids of low surface tension, however, will tend to exhibit low contact angles on most solids, and the sessile drop method is not convenient under these circumstances. Instead, a sessile bubble under a flat, wetted plate provides a satisfactory geometry (Figure 15). The dimensions usually measured are x_{90} and y_{90}, the equatorial radius and the distance from equator to vertex; the surface tension may then be calculated from

$$\gamma = g\rho(x_{90})^2/BF^2 = g\rho(y_{90})^2/BG^2 \tag{12}$$

---

* These tables are available from the Library of Congress, ADI Auxiliary Publications Project.

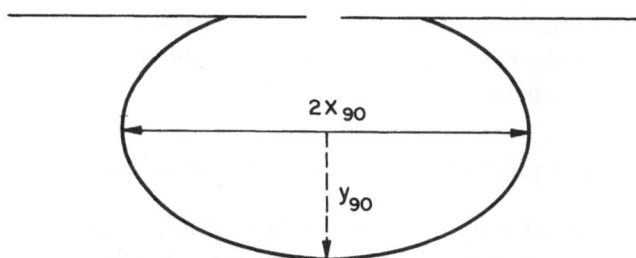

Fig. 15. Shape of a sessile bubble.

where the parameters B, F, and G, which are functions of the ratio x_{90}/y_{90}, have been computed and tabulated by Staicopolus.[102,103]

The method was first applied to polymer surface tension measurement by Sakai, and has been extensively used by Hata and his coworkers for both surface and interfacial tension measurements. [For interfacial tensions, of course, the "bubble" is a drop of the second liquid, and the difference in density is required in Eq. (12).] The advantages and limitations are quite similar to those of the pendant drop technique, except that drop detachment is eliminated as a difficulty. To date, no analysis has been given to assess hydrodynamic equilibrium from measurements of several dimensions, but this extension doubtless could be accomplished.

4.1.4. The Rotating Bubble Method[96,97,104]

In this technique, a bubble (or drop of a second liquid) contained in a liquid in a horizontal cylindrical tube is subjected to a centrifugal field. If the tube is rotated rapidly about its long axis, the drop or bubble elongates; the force due to centrifugal acceleration is balanced by the surface or interfacial tension. Measurements of the volume and length of the bubble, the density difference, and the speed of rotation are required to calculate the surface tension. The method has been applied to several polymer systems by Patterson *et al.*[98] They found that equilibrium was attained very slowly in very viscous ($10^3 P$) liquids, and used an extrapolation procedure to estimate equilibrium values; the results they obtained were in reasonable agreement ($\pm \sim 1$ dyne/cm) with values derived from other methods. However, neither the cause of the slow equilibration nor the validity of the extrapolation has been critically examined.

Because of the experimental complexity and need for accurate measurement of an additional parameter (rotation speed), this method would seem to have few advantages except in special cases. It has been suggested that it may be particularly adaptable to systems with very low interfacial tension.[104]

4.2. Measured Values of Surface and Interfacial Tension

The range of polymer melt surface tensions so far reported is not large, by comparison with a variety of other liquids. Thus, molten metals and fused salts exhibit surface tensions of 100–1000 dynes/cm, while the values for liquified inert gases are of the order of 10 dynes/cm or less. As with relatively nonpolar organic liquids—their small-molecule analogs—polymer liquid surface tensions are generally in the range 10–50 dynes/cm. Table 2 indicates typical values of both surface and interfacial tensions for some polymer melts. (These values must be compared only in a qualitative way, however, since each refers to a particular polymer sample, and these may differ considerably in molecular weight, distribution, or branching.) It is only recently that the influence of such details of molecular architecture has been examined.

TABLE 2
Surface and Interfacial Tension (dynes/cm) of Molten
Polymers[88]

Surface tension, 150°C	
Polydimethyl siloxane (10^6 cs)	13.6
Polypropylene (melt index 1000, 5% crystalline)	22.1
Polyethylene: branched (Bakelite DYLT)	26.5
linear (Alathon 7050)	28.1
Polystyrene ($M_n = 9300$)	30.8
Polyethylene glycol (Carbowax 6000)	33.0
Interfacial tension, 150°C	
Polyethylene/polypropylene	1.1
Polyethylene/polystyrene	5.0
Polyethylene/poly(vinyl acetate)	10.5
Polydimethyl siloxane/polypropylene	3.0
Polydimethyl siloxane/polystyrene	6.0
Polydimethyl siloxane/poly(vinyl acetate)	8.2
Polydimethyl siloxane/polyethylene glycol	9.8

4.2.1. Dependence of Surface and Interfacial Tensions on Molecular Weight and Temperature

A considerable amount of experimental evidence has now accumulated that the surface tensions of homologous series of liquids, including polymeric fluids, of increasing molecular weight M vary linearly with $M^{-2/3}$.[105–107] If this observation is generally valid, it suggests that polymer melt surface tensions should be discussed in terms of the two terms in the expression

$$\gamma = \gamma^\infty - (k/M^{2/3}) \tag{13}$$

where γ^∞ is the surface tension extrapolated to infinite molecular weight—presumably an intrinsic characteristic of the polymer repeat unit—and the slope k defines the extent of the molecular weight dependence. Comparisons of homologous series of substituted alkanes,[105] as well as some polymers with different chain-terminating groups,[108] indicate that the end groups determine k for a given repeating unit backbone.

The surface tensions of polymeric fluids decrease with increasing temperature in a roughly linear fashion, as is the case for other liquids; the temperature coefficients for polymers, however, are rather small, typical values being 0.05–0.08 dyne/cm per °C. To date, precise data on temperature dependence of surface tension of homologous series of chain molecules are very limited, and interpretation is accordingly uncertain. If the available information on the n-alkanes, alkenes,[107] and anionically polymerized polystyrenes[106] is considered in the context of Eq. (13), however, it appears that the slope k increases slightly with temperature, but the change is small enough that $-d\gamma^\infty/dT$ also lies in the range 0.05–0.08 dyne/cm per degree.

For interfacial tensions between dissimilar polymer melts, only very limited information is available. A preliminary study[109] suggests that a similar molecular weight dependence exists, which may be at least approximately expressed as

$$\gamma_{12} = \gamma_{12}^\infty - (k_1/M_1^{2/3}) - (k_2/M_2^{2/3})$$

Such a relation, of course, implies a criterion for miscibility between the two polymer fluids, since the interfacial tension becomes zero for molecular weights smaller than those that satisfy the equation

$$(k_1/M_1^{2/3}) + (k_2/M_2^{2/3}) = \gamma_{12}^\infty \tag{14}$$

In all cases so far examined the temperature coefficients of interfacial tension are small; a typical value is -0.01 dyne/cm per degree.

4.2.2. Effect of Additives on Surface and Interfacial Tension

It has long been known that surface properties of bulk polymers can be modified by additives. Friction and wettability are altered, for example, when fatty amides are incorporated in polyethylene,[110] or various partially fluorinated compounds are added to poly(methyl methacrylate) or poly(vinyl chloride).[111a,111b] Poly(dimethyl siloxane)–organic block copolymers produce marked surface modification at very low concentrations.[112a,112b]

The study of surface activity by measurement of surface tension lowering in small-molecule systems is commonplace, of course. A few measurements have been reported in polymer systems. In these systems diffusion is often slow, and incompatibility between high-polymer species is the rule. For these reasons, the effects which can occur are more complex, and have not yet been mapped in any detail.

Because of practical interest in polyurethane foams, the reduction of the surface tension of commercial polypropylene glycol by siloxane–polyether block copolymers has been studied.[113a–113c] At concentrations as low as a few tenths of a percent, some of these copolymers reduce the (room temperature) surface tension of the polyglycol from 32 to ~ 21 dynes/cm, i.e., to a value characteristic of pure poly(dimethyl siloxanes). Similar results have been obtained for a (dimethyl siloxane)–(styrene) block copolymer in polystyrene at elevated temperatures; in this case, the rate of surface tension reduction appears to be diffusion controlled.[114]

4.3. Theoretical Approaches to Polymer Surface and Interfacial Tension

Attempts have been made to apply to polymers most of the correlations and theories originally directed to understanding the surface tension of small-molecule liquids. While qualitative agreement with experimental values has been achieved, it appears that high-polymer liquids do not conform quantitatively to these theoretical predictions any more than do most nonpolymeric liquids. More complete and precise data, for both surface and bulk properties of polymer liquids, is needed to assess the significance of the discrepancies between theory and experiment.

Roe[115] suggested that the parachor for a polymer segment, calculated by summing atomic parachor values, could be used to estimate the surface tension from the equation which defines the parachor P,

$$\gamma = (P\rho)^4/M \tag{15}$$

where M is the molecular weight (taken by Roe to refer also to the repeating segment) and ρ is the density. This, of course, implies that the molecular weight and temperature dependences of γ arise only from the corresponding variation of ρ. Surface tensions estimated in this way appear to agree with experiment within about $\pm 15\%$ for a number of polymers.[89,115,116]

Schonhorn[117] has used the scaled particle theory of rigid spheres to develop an equation for the surface tension of polymers in terms of the cohesive energy density; the result resembles Hildebrand's empirical relation between surface tension and solubility parameter.[118] Schonhorn also noted a correlation between the temperature dependence of surface tension and viscosity.[119] Both Roe[120] and Patterson and Rastogi[121] have discussed the application of the principle of corresponding states and cell theories of liquids to polymer surface tensions. The latter authors have suggested that a direct test of the corresponding states principle for surface tension is provided if the quantity $\gamma\beta^{2/3}\alpha^{1/3}/k^{1/3}$ is a universal function of αT (β is the isothermal compressibility, α the thermal expansion coefficient, k the Boltzmann constant). They found that data for several polymers, including polymethylenes, polyisobutylene, poly-(dimethylsiloxanes), polyethylene oxide, and polypropylene oxide, as well as other polyatomic liquids, all fitted a single curve when plotted in this way. Data for polystyrene also fits this correlation within about $\pm 10\%$ if the compressibility data of Matsuoka and Maxwell[122] are used. On the other hand, the reduction parameters used by Roe to obtain a corresponding states plot (based on the Prigogine–Saraga cell theory) overestimate the surface tension of polystyrene by 20–30%.[106] Patterson and Rastogi also found that cell model calculations agreed reasonably, but not quantitatively, with their empirical corresponding states plot.

Siow and Patterson[123] have compared the solubility parameter, parachor, and corresponding states approaches in regard to their ability to predict $d\gamma/dT$ and the $M^{-2/3}$ molecular weight dependence. They concluded that these concepts are closely related, in each case yielding temperature and molecular weight dependences through changes in free volume. The solubility parameter relation does not predict either $d\gamma/dT$ or the molecular weight relation well. The parachor and corresponding states correlations, however, do yield fairly good predictions for the slopes of γ vs. T and $M^{-2/3}$, even though the absolute magnitudes of the calculated surface tensions are only approximately correct.

LeGrand and Gaines[105] also attempted to correlate the molecular weight dependence of surface tension with bulk properties through a crude free volume argument. Essentially, this assumed that the molecular weight variation of the surface tension and the glass transition temperature T_g both reflected free volume changes, and hence should be related. Agreement with experiment was poor, reflecting, at least in part, difficulties with the T_g–molecular weight relationships assumed.[106]

Stewart and von Frankenberg[124] have applied the significant structure theory to the surface tension of polyethylene; again, agreement is only qualitative.

Mixtures of smaller oligomeric chain molecules seem to be capable of treatment by lattice-model equations,[125,126] but the significance of this observation is not yet clear. Rastogi and St. Pierre[127] applied several regular solution equations to their data on the surface tensions of mixtures of polyethylene oxide, polypropylene oxide, and poly(epichlorohydrin), without success. No attempt has yet been made to develop a theory for such mixtures.

In summary, it would appear that the surface tension of polymer fluids can be treated approximately within the context of existing theories of the liquid state, and qualitative estimates of surface tension values can be made, at least in some cases, from measurements of other physical properties. Probably the discrepancies so far observed between theory and experiment reflect at least as much deficiencies in our understanding of bulk polymer liquids as in the extension of that understanding to surface phenomena.

With regard to interfacial tensions between molten polymers, there are again some approximations originally developed by analogy to theories for low-molecular-weight fluids. Much analysis has so far attempted to apply the theories of Girifalco and Good[128] and Fowkes,[129] or modifications of them,[130] which state essentially that the interfacial tension is related to the surface tensions of the two separate liquids by expressions of the form $\gamma_{12} = \gamma_1 + \gamma_2 - D$. The deviation terms D have different significance in the several variants of these theories, but they are in practice evaluated by comparisons between different systems. (We give a brief summary of Fowkes' argument, with reference to solid–liquid interfacial forces and contact angles, in Section 5.2.) Wu[130a] has compared directly measured interfacial tension between polar polymer pairs with those which could be calculated on the basis of these theories from measurements of the interfacial tensions of the respective polymers against polyethylene. While qualitative correlation was achieved, the quantitative discrepancies were in some cases greater than 50%, and the theoretical values were as often too low as too high.

Recently Helfand and Tagami[131] formulated a statistical mechanical theory of the interface between immiscible polymers A and B. The other theories which we have discussed above are outgrowths or adaptations of free volume, cell, hole, or significant structures theories of low-molecular-weight fluids which are based on some lattice model of the liquid state augmented by some form of random mixing hypothesis of the connected statistical segments of the polymers. The scaled particle theory approach of Schonhorn hardly takes into account the connectedness of the polymer segments. The approach of Helfand and Tagami is more fundamental, being based on a self-consistent field which determines the configurational statistics of the polymer molecules in the interfacial region. At the interface energetic forces (determined essentially by the polymer A–polymer B interaction parameter χ) tend to drive the A and B molecules apart, but this separation must be achieved in such a way as to prevent a gap from opening between the polymer phases. The energetic force on, say, an A molecule must be balanced by an entropic force describing the tendency of A molecules to penetrate into the B phase because of the numerous configurations of the A molecule which do so.

Currently one of the weaknesses of the theory is that it has been developed only for the case that the A and B polymer molecules are so similar that they possess identical degree of polymerization Z, effective length of the monomer units b (thus the mean end-to-end distance Zb^2), density ρ_0, and compressibility κ. The authors recommend the use of the geometric mean when these properties are not actually identical. Under the circumstances, and with the considerable difficulty of assigning values to the parameters required, the practical utility of the theory at its present state of development is very limited.

In the Helfand–Tagami formulation the effective field $W_A(\mathbf{r})$ on a segment of A polymer is the reversible work of adding the segment at the point $\mathbf{r}$ where the densities are $\rho_A(\mathbf{r})$ and $\rho_B(\mathbf{r})$, less the work of adding the segment to bulk A, and is given by

$$\frac{W_A(\mathbf{r})}{kT} = \chi\frac{\rho_B(\mathbf{r})}{\rho_0} + \zeta\left[\frac{\rho_A(\mathbf{r})}{\rho_0} + \frac{\rho_B(\mathbf{r})}{\rho_0} - 1\right] \tag{16}$$

$$\zeta = 1/\rho_0 kT\kappa$$

The first term describes the fact that A–B contacts are energetically less favorable than A–A contacts, and the second term represents the force of cohesion which prevents the opening of a gap between the phases. To obtain the self-consistent configurations of this system, one focuses on the quantity $q_A(\mathbf{r}, t)$, the ratio of the density at $\mathbf{r}$ of the initial segments of A chains of length Ztb, $0 < t \leq 1$, to this density in bulk A. Since the segment at Zt is the origin of two independent random walks of length Ztb and $Z(1 - t)b$, the overall segment density of A at $\mathbf{r}$ is

$$\rho_A(\mathbf{r}) = \rho_0 \int_0^1 dt\, q_A(\mathbf{r}, t)q_A(\mathbf{r}, 1 - t) \tag{17}$$

The quantity $q_A(\mathbf{r}, t)$ satisfies the Pouchly'–Di Marzio–Edwards diffusion equation, which, for the dividing surface at $x = 0$ and A-rich phase located at $x > 0$, can be written

$$\frac{1}{Z}\frac{\partial}{\partial t}q_A(x, t) = \frac{b^2}{6}\frac{\partial^2}{\partial x^2}\, q_A(x, t) - \frac{W_A(x)}{kT}q_A(x, t) \tag{18}$$

subject to

$$q_A(x, 0) = 1, \qquad q_A(\infty, t) = 1, \qquad q_A(-\infty, t) = 0$$

Equation (18) has been solved asymptotically for effectively infinite degree of polymerization of the chains and $\chi/\zeta \to 0$ (since $\chi/\zeta \leq 10^{-3}$), yielding the density profiles

$$\rho_A(x, \chi) = \rho_0\{1 + \exp[-2(6\chi)^{1/2}x/b]\}^{-1}$$
$$\rho_B(x, \chi) = \rho_0\{1 + \exp[2(6\chi)^{1/2}x/b]\}^{-1} \tag{19}$$

with an effective interface thickness $\rho_0[(d\rho_A/dx)_{x=0}]^{-1} = 2b/(6\chi)^{1/2}$. A straightforward computation of the surface part of the grand potential yields the interfacial tension γ:

$$\gamma = \frac{kT}{\rho_0}\int_0^1 d\lambda \int dx\, \chi\rho_A(x, \lambda\chi)\rho_B(x, \lambda\chi) = \left(\frac{\chi}{6}\right)^{1/2}\rho_0 bkT \tag{20}$$

TABLE 3
Comparison of Calculated and Measured Interfacial Tensions

Polymer pair	Interfacial tension, dynes/cm		χ	$b,^*$ Å	Specific volume,* cm^3/mole
	Calc.	Exptl.[162]			
PS/PMMA	1.0	1.5	0.01	6.5	96
PMMA/PnBMA	2.0	1.8	0.07	6.1	114
PnBMA/PNA	1.9	1.9	0.05	6.3	107

* Geometric mean.

In comparing this theory with experiment, one must carefully restrict oneself to cases where χ is not too large and the assumption of symmetry between A and B is tenable. Since χ is not directly available, it can be approximated from swelling data using the Hildebrand solubility parameter δ,

$$\chi = (\delta_A - \delta_B)^2/\rho_0 kT$$

so that (20) becomes, in terms of experimentally accessible parameters,

$$\gamma = |\delta_A - \delta_B| b(\rho_0 kT/6)^{1/2} \qquad (21)$$

Table 3 shows a comparison of theory to experiment. The interface thickness varies from about 20 to 50 Å in going from PMMA/PnBMA to PS/PMMA.

5. Wettability

5.1. Contact Angles

We have already considered (Section 2) the subject of wetting of a solid by a liquid in a general way. For completeness, we now describe briefly the methods commonly used for contact angle measurement, cite pertinent literature values for the contact angles on a variety of polymer surfaces, and finally explore the interpretation of these results.

It is not generally convenient to have available both the large sample of solid and the large volume of liquid needed for many other of the methods employed in contact angle measurements.[10,24,132–137] Most contact angle measurements on polymers have therefore been made by applying a liquid drop to a smooth surface. Zisman and co-workers[23] simply viewed a sessile drop through a microscope fitted with a goniometer scale, thus measuring the angle directly.

385

Leja and Poling[138] photographed sessile or clinging drops at a slight angle so that a portion of the drop profile was reflected from the surface, the angle of meeting of the direct and reflected images then being twice the contact angle. Ottewill (see Ref. 139) makes use of the captive bubble technique, wherein a bubble formed by the manipulation of a micrometer syringe is made to contact the solid surface which is immersed in the liquid, and the angle is measured from the surface. This has the advantage that one could swell or shrink the bubble to obtain advancing or receding contact angles.

The value of θ may also be obtained[140–142] by means of the reflection of light, as illustrated in Figure 16. It is clear that the beam emitted by the microscope tube is reflected back into the microscope only when the tube direction is perpendicular to the liquid surface. If the microscope is focused on this surface at the three-phase line, the angle which the tube makes with the vertical when the liquid appears bright is equal to the contact angle. Generally, measurement of θ should be taken at several points on the surface, heterogeneities contributing to mistaken values of θ. In addition, this method is only useful for values of θ below 90°.

Typical values of the contact angle of some simple liquids on polymeric surfaces are shown in Table 4.

It is also instructive to consider the temperature dependence of wetting at this point. As the temperature is increased, it is clear that both the liquid surface tension and the surface energy of the polymer should decrease. It is not clear that the value of θ will increase, decrease, or remain the same, since θ depends in a somewhat complicated

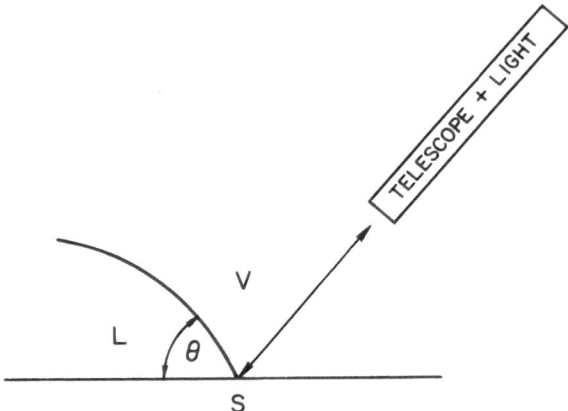

Fig. 16. Contact angle goniometer; reflection of light is observed in telescope.

TABLE 4
Typical Values of θ at 25°C for a Variety of Liquids on Polymer Surfaces

Polymer	Wetting liquid		
	Water	Glycerol	Formamide
Polyethylene	94	81	77
Nylon (66)	70	60	50
Polyethylene terephthalate	81	65	61
Polystyrene	91	80	74
Polytetrafluoroethylene	93	85	79
Polychlorotrifluoroethylene	90	82	82

way on the interplay between these phases when in contact. Barring increased compatibility, the meager data in the literature[136,143–150] would indicate that $d\theta/dT$ is essentially constant provided one is sufficiently far removed from the critical point of the wetting liquid. Near room temperature, the variation is small in all cases so far studied, being typically ± 0.05–0.15 deg/°C. However, Johnson and Dettre found a sharp drop in the contact angles for both octane and hexadecane on fluorinated ethylene–propylene copolymer as the temperature was raised to near the boiling point of the liquids. Petke and Ray observed a similar sharp decrease for superheated water contact angles (120–160°C).

Schonhorn has described an approach based on the Katayama–Guggenheim equation[151]

$$\gamma = \gamma_0[1 - (T/T_c)]^{11/9} \tag{22}$$

(where γ_0 is a constant and T_c is the critical temperature) which has been applied to the data of Johnson and Dettre.[24a] The treatment agrees reasonably well with their rapidly changing values of θ for low-molecular-weight hydrocarbons in contact with a fluoropolymer. The sharp decrease in θ as the temperature is raised can be easily visualized by examining Figure 17. Since T_c of the liquid is lower than that of the polymer, there will be a point where spreading will occur barring any degradation of the liquid or polymer. For *n*-octane on the fluoropolymer the intersection temperature is approximately 100°C. Since $d\theta/dT$ depends strongly on the difference between γ_L and γ_P, it is not expected to change significantly except when $\gamma_L - \gamma_P$ approaches zero.

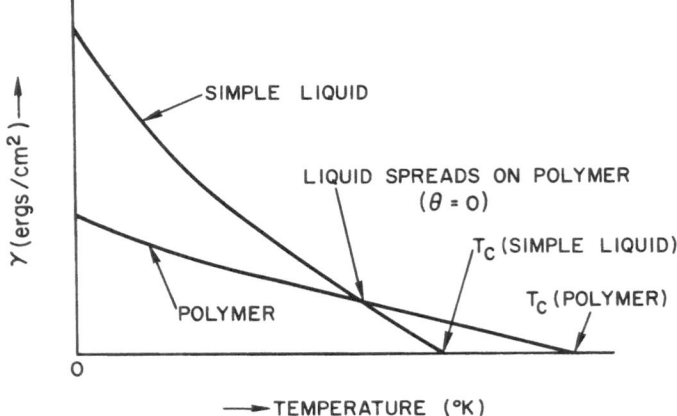

Fig. 17. Schematic variation of surface tension with temperature for a test liquid and a polymer. Where the lines intersect, spreading is expected.

5.2. Critical Surface Tension

We have discussed in Section 2 the concept of the critical surface tension for wetting. Wetting measurements with homologous series of organic liquids (e.g., *n*-hydrocarbons differing in chain length) often do yield quite accurately linear plots of $\cos\theta$ vs. γ_{LV}. For other liquids which are considerably more polar than the *n*-hydrocarbons it has been noted that a distinct curvature exists at higher γ_{LV} and lower $\cos\theta$.[23] A recent detailed discussion has been given by Dann.[152],* Only since the advent of the work of Fowkes[129,154] and Good[128,155] has any reasonable explanation been offered for this observation of curvature. Although when dealing with hydrogen-bonding liquids the intercept γ_c is less precisely defined, it still proves to be a useful parameter because it is characteristic of the solid surface. The general nature of the rectilinear band between $\cos\theta$ and γ_{LV} in the rapidly growing body of reliable experimental data led Zisman to use γ_c to characterize and compare the wettability of a variety of low-energy surfaces. Table 5 lists the critical surface tension of some halogenated polyethylenes.[156]

Within the last decade, an enormous body of literature has developed concerning the critical surface tension of wetting and its

---

* But also see Ref. 153.

TABLE 5
Critical Surface Tension of Some Halogenated Polyethylenes

Polymer	Structural formula	Critical surface tension, dynes/cm
Poly(vinylidene chloride)	H Cl H Cl —C—C—C—C— H Cl H Cl	40
Poly(vinyl chloride)	H H H H —C—C—C—C— H Cl H Cl	39
Polyethylene	H H H H —C—C—C—C— H H H H	31
Poly(vinyl fluoride)	H H H H —C—C—C—C— H F H F	28
Poly(vinylidene fluoride)	H F H F —C—C—C—C— H F H F	25
Polytrifluoroethylene	F F F F —C—C—C—C— H F H F	22
Polytetrafluoroethylene	F F F F —C—C—C—C— F F F F	18

interpretation. Table 6 itemizes the values of γ_c for a variety of polymers.[157]

From the work of Zisman and collaborators,[23] it is clear that the wettability of polymer surfaces is governed by the nature and packing of the outermost atomic species in the exposed groups of atoms in the surface and not by the nature and arrangements of atoms to a depth of ~ 20 Å below the polymer surface layer. The work of Schulman and Zisman[158] concerning close-packed adsorbed films of alkyl and perfluoroalkyl acids determined the effect of particular functionalities and their relative wettabilities. Table 7 attempts to itemize important functionalities of polymer surfaces. These findings exemplify the extreme localization of the attractive field of force around the solid surfaces of covalently bonded atoms in polymers.

The details of correlation of wettability with the molecular constitution of the surface presents some points of controversy, however.

TABLE 6
Critical Surface Tension γ_c of a Variety of Polymers

Polymer	γ_c, dynes/cm at 20°C
Polyethylene	31
Polypropylene	29
Poly(vinyl fluoride)	28
Poly(vinyl chloride)	39
Poly(vinylidene fluoride)	25
Polytrifluoroethylene	22
Polytrifluorochlorethylene	31
Polytetrafluoroethylene	18.5
Polyhexafluoropropylene	16–17
Polystyrene	33–43
Poly(vinyl alcohol)	37
Poly(vinyl acetate)	36
Polyethylene oxide	43
Polydimethylsiloxane	24
Nylon 6	46
Polyethylene terephthalate	41–47

TABLE 7
Wettability of Particular Functional Groups[23]

Surface constitution	γ_c, dynes/cm at 20°C
$-CF_3$	6
$-CF_2H$	15
$-CF_3$ and $-CF_2$	17
$-CF_2$	18
$-CH_2-CF_3$	20
$-CF_2CFH-$	22
$-CF_2-CH_2-$	25
$-CFH-CH_2-$	28
$-CH_3$ (crystal)	22
$-CH_3$ (monolayer)	24
$-CH_2-$	31
$-CClH-CH_2$	39
$-CCl_2-CH_2-$	40
$=CCl_2$	43

Thus Zisman[23] has emphasized the importance of "closest packing" of the exposed groups (e.g., $-CH_2-$, $-CH_3$, $-CF_2-$, or $-CF_3$ in hydrocarbon or fluorocarbon polymers) at the surface in leading to high contact angles, and there are many observations which support this correlation. On the other hand, Hoernschemeyer[159] has pointed out that an increased concentration of attractive centers should produce more interactions per unit area, and hence a lower contact angle. He has been able to show that the relative wettability of fluorocarbon and hydrocarbon surfaces is consistent with the calculated difference in packing density at the surface, assuming the intermolecular forces to be similar. Johnson and Dettre (Ref. 24a, p. 131) have suggested that Zisman's correlations of high contact angles with closest packing should instead be referred to a reduction in "interfacial diffuseness," i.e., to a reduced tendency for molecular interpenetration at the interface.

Observations on polymer surfaces have been invoked to support all quarters of this debate. Thus, the observation[39] that polyethylene single-crystal mats are more wettable than the melt-crystallized polymer has been interpreted as supporting Hoernschemeyer's thesis. On the other hand, Pittman and Ludwig[160] found that their fluoro-alkylacrylates became less wettable when crystallinity increased, either because of increase in fluoroalkyl chain length or from annealing. They suggested that the polyethylene single crystals constituted a special case, because of the possibility that the chain-fold surface of these crystals may have a higher energy or lower density groups. In view of other uncertainties about the relationship between bulk crystallinity and surface morphology, and such problems as the incomplete characterization of most polymer samples, it would appear that none of these experiments yet permits unambiguous interpretation.

For a quantitative analysis of wettability, it is instructive to consider the thermodynamic work of adhesion of a liquid on a solid, as given by Eq. (7); when $\cos \theta$ does vary linearly with γ_{LV} [i.e., when Eq. (8) applies], it is easily shown that the work of adhesion has a maximum value given by

$$W_{A(\text{max})} = (1/K) + \gamma_c + \tfrac{1}{4} K \gamma_c^2 \tag{23}$$

where K is the slope of the linear $\cos \theta$ vs. γ_{LV} plot.

From a wettability point of view it appears that liquid will interact strongly with polymer surfaces to yield a maximum in the work of adhesion which corresponds to the above relation. However, it must be borne in mind that this work of adhesion may bear little or no relation to the practical joint strength derived in adhesive bonding. Here one must include not only the wetting characteristics of the solid but also the mechanical properties of the surface of the substrate and the adhesive and any interplay resulting from applied stresses. If all that was required to do was to comply with the rules of wetting, the problems of adhesive bonding would all but vanish.

Recently, Girafalco and Good[128] and Fowkes[129] have stimulated considerable interest in analyzing the relationship between the work of adhesion and intermolecular forces. Although both groups have obtained essentially identical results, it appears to be more fruitful to consider the approach of Fowkes in analyzing new polymeric species.

Fowkes,[129] Kaelble,[161] Wu,[116,162] and Dann[152] consider that

$$W_A = \sum W_A{}^i \tag{24}$$

where the superscript i stands for distinguishable kinds of molecular interactions such as hydrogen bonding, dispersion interaction, acid-base interactions, π bonding, Keesom dipole–dipole interactions, Debye dipole–induced dipole interactions, etc.

Fowkes proposed that individual components which lead to the surface energy of the solid and surface tension of the liquid may or may not interact. For example we may consider water, phase 1, in contact with an oxide-free metal surface, phase 2. Fowkes has considered that the surface energy of the metal can have components due to electronic interactions, forming metal "bonds" and the normal dispersive forces; thus

$$\gamma_2 = \gamma_2{}^m + \gamma_2{}^d \tag{25}$$

Similarly, water may be considered to have two components to its surface energy due to dispersion forces and hydrogen bonding; thus

$$\gamma_1 = \gamma_1{}^h + \gamma_1{}^d \tag{26}$$

In Fowkes' scheme it appears unlikely that the metallic bonds interact with water molecules or that water molecules form hydrogen bonds with the metal. Consequently, only dispersion forces interact. Fowkes proposed that the interfacial energy between the metal and water is given, therefore, by

$$\gamma_{12} = \gamma_2 + \gamma_1 - 2(\gamma_2{}^d\gamma_1{}^d)^{1/2} \tag{27}$$

where

$$W_A{}^d = 2(\gamma_2{}^d\gamma_1{}^d)^{1/2}$$

It is clear that the same situation is true for polymer–liquid interactions. Only those interactions that are similar are important in determining the interfacial tension. For certain liquids, for example, the normal hydrocarbons, only dispersion forces are operative, $\gamma_1 = \gamma_1{}^d$. This allows for the determination of a variety of dispersion and hydrogen-bonding components to the surface energy by measuring the interfacial tension of these systems and computing $\gamma_2{}^d$ since, in general,

$$\gamma_2 = \gamma_2{}^d + \gamma_2{}^h \tag{28}$$

where γ_2 can be measured independently, and we can compute $\gamma_2{}^h$. Similarly, using mercury as a liquid in conjunction with an *n*-hydrocarbon, Fowkes has determined both $\gamma_1{}^d$ and $\gamma_1{}^m$. Hence, by a stepwise process of choosing one liquid in which only dispersion forces are operative and another in which other (e.g., hydrogen bonding) forces are operative, a table of molecular interaction components may be constructed. For example, the contributions to the surface tension of water (72.8 dynes/cm at 20°C) are 21.8 ± 0.7 ergs/cm² due to dispersion interactions and 51.0 ergs/cm² due to hydrogen bonding.

For pairs of liquids that interact mainly by dispersion forces, the cohesive energy density concept has been adapted by Girafalco and Good[128] to yield

$$\gamma_{12} = \gamma_{23} + \gamma_{13} - 2\Phi_{12}(\gamma_{23}\gamma_{13})^{1/2} \tag{29}$$

in which Φ_{12} is a correction term to account for disparity in the gram-molecular volumes V_1 and V_2

$$\Phi_{12} = \frac{4(V_1/V_2)^{1/3}}{(V_1^{1/3} + V_2^{1/3})^2}$$

Experiments have shown that Φ for the water aliphatic hydrocarbon systems are not in agreement with experiment. This notion led Fowkes to propose that

$$\gamma_{12} = \gamma_{23} + \gamma_{13} - 2(\gamma_{23}^d\gamma_{13}^d)^{1/2} \tag{30}$$

when only dispersion interactions are operative. Melrose[163] discusses in detail the significance of this approach when admittedly the surface entropy has been ignored in the geometric mean procedure. Table 8 lists the dispersion contributions to the surface tension for some commonly used liquids in contact angle measurements.

TABLE 8
Dispersion Components of the Surface Tension of Common
Liquids Used in Wettability Studies

Liquid	γ_{13}^d, dynes/cm	γ_{13}, dynes/cm
Dimethylsilicones	16.9 ± 0.5	19.0
Water	21.8 ± 0.7	72.8
Glycerol	37.2 ± 0.2	64.5
Formamide	39.5 ± 7	58.2
Mercury	200 ± 7	484

The Fowkes equation in conjunction with the Young equation leads to

$$\cos \theta = -1 + 2(\gamma_{23}^d \gamma_{13}^d)^{1/2}/\gamma_{13} \qquad (31)$$

assuming π_e, the spreading pressure, is zero. If the molecules of the liquid phase interact with the solid with only dispersion forces, $\gamma_{13}^d = \gamma_{13}$, then the contact angle of such a liquid on the surface establishes γ_{23}^d. Generally, γ_{13}^d is obtained from Table 8 and the contact angle measured, to enable a computation of γ_{23}^d values for a variety of low-surface-energy polymers. Since dispersion forces predominate on many polymer surfaces, the values of γ_{23}^d will be close to the surface energy of the material. Table 9 illustrates this with a compilation of γ_{23}^d values for common polymers.

To extend the Fowkes equation to systems including hydrogen bonding, Owens and Wendt,[164] Dann,[152] and Wu[116,165] considered that

$$\gamma_{12} = \gamma_{23} + \gamma_{13} - 2(\gamma_{23}^d \gamma_{13}^d)^{1/2} - 2(\gamma_{23}^h \gamma_{13}^h)^{1/2} \qquad (32)$$

where the work of adhesion is

$$W_A = 2(\gamma_{13}^d \gamma_{23}^d)^{1/2} + 2(\gamma_{13}^h \gamma_{23}^h)^{1/2}$$

By analogy to Eq. (31), we may write that

$$\cos \theta = 2 \left[\frac{(\gamma_{13}^d \gamma_{23}^d)^{1/2}}{\gamma_{13}} + \frac{(\gamma_{13}^h \gamma_{23}^h)^{1/2}}{\gamma_{13}} \right] - 1 \qquad (33)$$

By choosing pairs of liquids on a particular surface, Owens and Wendt[164] solved pairs of simultaneous equations to yield γ_{23}^d and γ_{23}^h values for a variety of polymers and compared these to the critical surface tensions of wetting (Table 10). Kaelble[161] has employed a

TABLE 9
The Dispersion Component of the Surface Energy of
Some Polymeric Materials

Solid	γ_{23}^d, ergs/cm^2
Polyhexafluoropropylene	18
Polytetrafluoroethylene	19.5
Paraffin wax	25.5
Polymonochlorotrifluorethylene	30.8
Polyethylene	35
Polystyrene	44

similar approach to that of Owens and Wendt[164] but with greater sophistication. The reader is referred to his book for these extensive tabulations.

Petke and Ray[150] have made extensive measurements of the temperature dependence of wetting and report the critical surface tension of a variety of polymers as a function of temperature. Table 11 itemizes several of the $d\gamma_c/dT$ values for polymeric species. The temperature range covered by Table 11 is from 25 to approximately 100°C.

In summary, it must be pointed out that the concepts of the critical surface tension, as well as the separability (on a simple additive basis) of the dispersion force contributions, are clearly simplified approaches which seek to introduce some order into a diversity of experimental observations. They are extremely useful for qualitative ranking of systems, but must be used with care because of the simplifications which are inherent in them. In addition to

TABLE 10
Components of Surface Energy for Various Solid Polymers

Polymer	γ_{23}^d	γ_{23}^h	γ_{23}	γ_c
Polyethylene	33.2	0.0	33.2	31
Poly(vinyl chloride)	40.0	1.5	41.5	39
Poly(vinylidene chloride)	42.0	3.0	45.0	40
Poly(vinyl fluoride)	31.3	5.4	36.7	28
Poly(vinylidene fluoride)	23.2	7.1	30.3	25
Polytrifluorethylene	19.9	4.0	23.9	22
Polytetrafluoroethylene	18.6	0.5	19.1	18.5
Polyethylene terephthalate	43.2	4.1	47.3	43
Poly(methyl methacrylate)	35.9	4.3	40.2	39
Nylon 66	40.8	6.2	47.0	46
Polystyrene	41.4	0.6	42.0	43

TABLE 11
Temperature Dependence of the Critical Surface Tension

Polymer	γ_c at 25°C, dynes/cm	$-d\gamma_c/dT$, dynes/cm °C
Polyethylene	36.0	0.05
Polystyrene	33.0	0.06
Polyacetal	38.0	0.04
Polycarbonate	34.5	0.04
Polyethylene terephthalate	40.0	0.06
Fluoropolymer (FEP Teflon)	19.5	0.05

several sources already cited, recent discussions by Wu[165] and Cortes[166] concerning alternative formulations of the Girifalco–Good–Fowkes theory continue to emphasize this point. While no rigorous general calculation of interaction energies across a phase boundary has yet been attempted, Israelachvili[167] has very recently shown how a direct calculation of van der Waals dispersion forces across an interface, using macroscopic theory, is related to the simple additivity theories. He also compares his results with calculations of Padday and Uffindell,[168] who used a more detailed microscopic model, including many-electron effects.

5.3. The Kinetics of Spreading of Polymer Melts on Solid Surfaces

Although the terms "wetting" and "spreading" are often used synonymously, it has been suggested that *wetting* should refer to the equilibrium state achieved by a liquid in contact with a solid surface, while *spreading* should denote the kinetic process by which this equilibrium could be achieved. We consider here the rate processes by which a mass of polymer melt spreads on a horizontal solid substrate, a process of considerable importance for adhesion. Early observations indicated that the apparent advancing contact angle $\theta(t)$ appears to approach reproducibly the apparent (advancing) final contact angle θ_∞ after time lags varying from minutes to hours.[169,170] The macroscopic rate of spreading can be followed by (1) the temporal evolution of the contact angle $\theta = \theta(t)$ to θ_∞, or (2) following the adherend area covered by the spreading drop. Low-power $(10 \times)$ microscope observations have been employed to follow the spreading process. Unfortunately, for technical reasons neither the earliest stages [usually $\theta(\pm) \sim 90°$; less than 30 sec] nor the final stages (in the case where $\theta_\infty = 0$, the last minute of angle) can be followed accurately.

Maintenance of a constant temperature is very important. The limitations of the time of observation and the temperature range over which the polymer–substrate systems are studied are important to keep in mind since there may well be several kinetic stages of spreading.[170,171] If the substrate is isotropic and flat, one would expect that the shape of the spreading droplet would approximate microscopically a surface of revolution; the advancing locus of the three-phase contact would be a circle of radius $R(t)$. A characteristic initial lineal dimension of the drop would be given by $l_0 = (m/\rho)^{1/3}$, where m is the mass of the drop and ρ its density. Providing l_0 is sufficiently small [so that the gravitational stress is negligible compared to the capillary stress in the melt, $l_0 < (2\gamma/\rho g)^{1/2}$, with g the gravitational acceleration], one can neglect effects due to gravity.[170]

Following Schonhorn et al.,[170] who first applied dimensional scaling to this problem, one would argue as follows: If bulk hydrodynamic viscous resistance balanced the spreading forces arising solely from effects due to conventional capillarity, the mathematical formulation of this problem would consist of (a) the Navier–Stokes and continuity equations for the velocity fields of creeping flow in the melt and vapor phase, (b) the kinematic equations relating the velocity fields to the temporal evolution of the geometric parameters describing the shape of the spreading droplet [in particular $F(t)$ and $\theta(t)$], and (c) initial and boundary conditions. Among the latter would be a condition that on the flat, rigid solid substrate at the three-phase contact locus the net viscous radial fluid stress is equated to the spreading pressure determined by the conventional instantaneous spreading tension (γ surface tension of the melt)

$$\sigma(t) = \gamma_{SV} - \gamma_{SL} - \gamma \cos \theta = \sigma - \gamma \cos \theta \qquad (34)$$

with σ the maximum value of $\sigma(t)$. Scaling distances by l_0; pressure, viscous, and capillary stresses by σ; velocities by η/σ, with η the melt viscosity; and neglecting the viscous resistance of the vapor phase, there must be dimensionless functions f and h of the indicated arguments so that

$$R(t)/l_0 = f(\sigma t/\eta l_0 ; \sigma/\gamma) \qquad (35)$$

and

$$[\cos \theta(t)]/\cos \theta_\infty = h(\sigma t/\eta l_0 ; \sigma/\gamma) \qquad (36)$$

Equations (35) and (36) would predict that spreading data at fixed η/σ, ρ, and σ/γ would be scaled by $m^{1/3}$ arising from l_0; i.e., there would be a mass effect. If the stationary contact angle θ_∞ is given by the equilibrium contact angle of Young's equation (6), then $\sigma/\gamma = \cos\theta$. The scaling results are of course predicated on the existence of a satisfactory, unique, and nonsingular solution of this Navier–Stokes boundary value problem, which in view of known difficulties[20,21] is not at all obviously true. It was also pointed out that if sufficient bulk viscous flow occurs, the droplet shape would depart measurably from a spherical segment but might appear as a characteristic "cap" about the center of the drop and an advancing, projecting "foot."[170] If a final equilibrium contact angle is achieved, one expects, if gravity is negligible, that the shape of the drop is that of a spherical segment.[24a,24b,27]

Approximate hydrodynamic solutions which do not satisfy exactly the complete boundary value problem have been obtained for fixed simple shapes, for a spherical segment[172–174] and a cone.[175] These solutions neglect either radial[174,175] or vertical flow[172,173] and exhibit the scaling shown in Eqs. (35) and (36). These solutions do not explain why different shapes of the spreading drop are to be observed.

The experimental observations are conflicting. Spreading of polyethylene,[170,172,176–179] ethylene–vinyl acetate copolymer,[170,172] polydimethylsiloxanes,[174,175] fluoroalkyl methacrylate polymer,[171] styrene–acrylate copolymer,[174] and polystyrene[180] has been studied, some on low- as well as high-energy surfaces and over a range of temperatures. Both spherical segment[170,172,174,176–178,180] and "cap" with projecting "foot"[171,175] shapes have been observed. The scaling of Eqs. (35) and (36) and the mass effect have been observed in only some cases.[172,174,175] In others[177,181] where the drops also approximated a spherical segment the spreading was described by a time-temperature superposition principle (35) given by

$$R(t)/l_0 = F(\gamma t(\cos\theta_\infty)/\eta L_w)$$
$$(\cos\theta)/\cos\theta_\infty = H(\gamma t(\cos\theta_\infty)/\eta L_w)$$
$$(37)$$

with L_w a temperature- and mass-independent "wetting length" characteristic of the polymer–substrate system. The order of magnitude of the mean radial spreading velocity in all of these observations is about 10^{-2} mm/min.

The approximate analytical solutions[172–175] give qualitatively the observed rates of spreading when a mass effect is observed. They do not account for the variety of shapes or the nature of L_w in (37).

It has been suggested that microirregularities of the substrate surface determine L_w.[177,181] Yet experimental observations of L_w do not correlate with surface roughness as determined by a Talysurf trace[170] and on the same surface glycerol exhibits an L_w about one thousand times larger than ethylene–vinyl acetate copolymer.[179] All these studies have indicated some factors which are important in spreading. (They all agree, for example, that the velocity γ/η is an important factor in spreading.) At present, however, it cannot be said that a clear understanding of the process has been achieved. Different phenomena may be important for different systems or different temperatures and surface preparations. Besides a rate of spreading determined by bulk viscous flow, there may be a regime of spreading which might, for example, be controlled by the rate of spreading of an extremely thin and hence invisible precursor, primary film.[181,182]

6. Friction, Lubrication, and Adhesion of Polymers

6.1. Friction and Lubrication

The friction of unlubricated polymers (against themselves or "hard" surfaces such as glass or steel) can be satisfactorily explained, in the large, in terms of a simple adhesion mechanism.[183–186] In surfaces which are in contact over a real area a, adhesion occurs and the adherend regions must be sheared if sliding is to take place. If the specific shear strength of the junctions is s, the frictional force F_f can be written

$$F_f = as \tag{38}$$

where s is very close to the bulk shear strength of the polymer. Actually, s increases slightly with load and is only approximately constant in the usual experimental conditions. Apart from the adhesion term, a smaller contribution may arise from a ploughing or deformation of the surface,[187] which can contribute to energy loss via elastic hysteresis. These contributions become increasingly important for very soft materials, or rubbery polymers with large hysteresis, but may often be neglected for semicrystalline or glassy polymers. For a hemispherical slider of radius R and a surface covered by "hemispherical" protuberances of radius r, n per unit area, a modified Howell–Lodge[188] multiasperity model of a for deformation modes intermediate between elastic and plastic deformation yields[183] (for fixed velocity of sliding)

$$F_f = ksn^{0.26}r^{0.52}(W/E)^{0.96}R^{0.13} \tag{39}$$

with W the load, E an effective modulus, and k a dimensionless proportionality constant. This expression accounts quantitatively for the frictional behavior of crystalline polar nylon 66 and nonpolar polyethylene and is in qualitative agreement with the variation of the frictional force of poly(vinyl chloride) with the measured elastic compliance. The deformation and shear processes at the regions of real contact depend in the expected way on the viscoelastic properties of the polymer[186]: Friction varies with sliding speed and temperature and the curves of friction against velocity show one or more peaks, the position of which shifts to higher velocities at higher temperatures.

The ratio of the frictional force to the normal load W is the coefficient of friction μ. The importance of relaxation processes is particularly evident in the more unusual behavior of polytetrafluoroethylene (PTFE). The friction at low velocities of sliding of PTFE is considerably lower ($\mu = 0.04$–0.09) compared to the other crystalline polymers. nylon, or low-density polyethylene ($\mu \approx 0.3$). With increasing velocity a limiting value of $\mu \approx 0.3$ is obtained also for PTFE. At low velocities there is transferred a thin (less than 100 Å) oriented film of PTFE to the countersurface; at higher velocities the transfer increases and takes the form of lumps and streaks. Thus the low friction observed when PTFE slides on clean surfaces is not the result of poor adhesion; rather it is due to the easy shear of relevant units of the PTFE crystal.[186] Similar behavior is observed with high-density polyethylene and may be a consequence of the smooth molecular profiles of these polymers.[189]

Lubrication of polymers is complex. Lubricants can diffuse both in and out of polymers and thus affect the bulk mechanical properties (e.g., by plasticization), changing a, as well as affect friction by more specific surface mechanisms. A case in point is the effect of water on the frictional behavior of nylon.[185,190,191] For nylon sliding on nylon one observes an immediate drop in friction on addition of a drop of water due to the reduction by plasticization of the shear strength of the surface layer of nylon. With prolonged exposure to water (when it can be absorbed into the bulk) the polymer is softened (hardness, elastic modulus, and shear strength fall), which leads to an increase in the area of contact and a consequent increase in friction. With larger molecules the penetration rate into the polymer and its plasticization are considerably smaller. The main effect of fatty acids or paraffins may be to provide a thin surface film of low shear strength. For nonpolar hydrocarbons the reduction in friction is small (as compared, e.g., to metals) probably due to the fact that these molecules are not able to form a well-condensed protective film on the polymer surface.[185,190]

Some instances are known where a monomolecular film of lubricant is effective; e.g., Fort[191] has shown that an applied mono-layer of octadecanol effectively lubricates polyethylene terephthalate, and Allan[110] has concluded that monolayers of fatty acid amides effectively lubricate polyethylene. On the other hand, when similar fatty acid amides were incorporated in vinylidene chloride acrylonitrile copolymers, Owens[192] found that the additive existed as bulk crystals on the surface. Similarly, much larger amounts than monolayer thicknesses, possibly solid crystalline condensed deposits, were found to be effective lubricants when polyethylene containing fatty acid amides was heated at low temperatures.[193] While initially the diffusion of the amide is rate limiting, it appears that the presence of the solid amide on the surface inhibits further diffusion, and that the amides diffuse preferentially into the hollows and depressions on the surface. The lubricating properties of the amides appear to be simple functions of their rheological characteristics and as such affect s in (38). The commercial application of self-lubricating and thus self-healing polymers has been pointed out by Bowers *et al.*[194] as well as others.[112]

The role of the "other" nonpolymeric surface in the lubrication of polymers must not be forgotten. In the case of glass and other hydrophilic surfaces water can form a film a few molecules thick which can survive if the local contact pressures are not too large as, e.g., for glass sliding on polyethylene or rubber but not glass on glass.[185]

6.2. Adhesion and Polymer Surfaces

Although adhesive-bonding technology has a rather long and varied history, the science of adhesion is relatively new. Many opposing viewpoints have been formulated to account for the large body of experimental data. The merit of any theory is to account for this existing body of data and, hopefully, to provide insight to guide the scientist to perform new experiments, leading to further understanding.

As one examines the science of adhesion as concerned with polymers, it soon becomes apparent that a wide variety of disciplines in both physics and chemistry is needed and that no one discipline is sufficient to enable the investigator to draw conclusions about the operations of a particular system. Surface chemistry, surface physics, polymer physics, rheology, fracture mechanics, etc., are but a few of the important fields of scientific endeavor with which the investigator should be familiar before beginning an excursion into the science of adhesion. Adhesion is concerned initially with the phenomenon of making an adhesive joint (i.e., wettability, relative surface energetics of both phases, and kinetics of wetting). These are purely surface considerations. Adhesive joint strength is concerned with the breaking strength of a bonded assembly. Once an adhesive joint has been formed, interfacial forces appear no longer to be of primary concern, since true interfacial separation probably never occurs under ordinary failure conditions, except at points where molecular contact did not exist prior to breaking the adhesive joint. What is of prime importance in considering adhesive joint strength is the mechanical response of the composite to an applied stress. Probably a more realistic description of the breaking strength of an adhesive joint would be based on a mechanical deformation theory of adhesive joint strength.

Several recent compendia[195–210] review in a detailed manner the current efforts in the science of adhesion. We cite these references to enable the reader to obtain as in-depth a study as he deems desirable. The literature on the surface treatment of polymers and the analysis of adhesive joint strength is beyond the intended scope of the present review.

7. Concluding Remarks

Any review of this field would be incomplete without at least broadly classifying some of the major areas of descriptive chemistry aimed at modifying the surfaces of polymers. We have already mentioned a variety of effects which can alter the surface properties of a formed polymer sample. These include the purely physical treatments which can alter morphology in the surface region, and adsorption of certain compounds, e.g., the perfluorinated fatty acids. The incorporation of surface-active additives was noted in connection with lubrication; such additives may also markedly alter wettability. Jarvis *et al.*[111a] found that small amounts of various partially fluorinated compounds reduced the wettability of poly(methyl methacrylate) and poly(acrylamide). Similar effects have been found for dimethylsiloxane block copolymer additives.[112] None of these processes has been extensively studied, but it is already apparent that considerable control of surface properties is possible for certain polymers.

A very large amount of work aimed at chemically modifying polymer surfaces has been reported. In an extensive review, Angier[211] classifies these as surface grafting by irradiation, chemical surface grafting, and chemical treatment of surfaces. The first two categories involve the production of a surface layer of a second polymer, grafted to the base polymer. There may be more or less interpenetration or copolymerization, depending largely on the permeability of the base polymer to the monomer of the second. Initiation of the polymerization may be by high-energy radiation (such as γ rays), lower-energy radiation (ultraviolet), or usual chemical initiating species such as peroxides (formed in situ by oxidation of the base polymer surface), or ozone.

Among the chemical treatments summarized by Angier are oxidation, especially of polyethylene, to enhance its wettability and adhesion. Acid and alkali treatments, halogenation, sulfonation (concentrated or fuming sulfuric acid or chlorosulfonic acid), and treatment with chlorosilanes have been applied to a number of different polymers. Fluorinated polymers may be modified by treatment with alkali metals dissolved in liquid ammonia.

More recently, the surface layer of polyethylene has been cross-linked by reaction with gaseous fluorine.[212] Olsen and Osteraas[213,214] have reported a variety of modifications of polyethylene surfaces by reactive sulfur compounds, carbenes, and nitrenes, as well as difluoro-carbene ($:CF_2$) modification of several other polymers.[215] Amino groups have been attached to several polymers with the aid of radio-frequency plasmas;[216] this, and several other chemical modification techniques, has been used to permit binding of the blood anticoagulant, heparin, to plastic surfaces.[217] Nitrogen-containing functional groups can also be incorporated in polycarbonates by treatment with diamines or polyamines, such as triethylenetetramine.[218]

While quantitative studies of these surface-modifying treatments have been limited, it seems generally that relatively thick layers of material are actually involved. In the preparation of surface-grafted materials, appreciable weight gains are observed. One important reason for chemical reaction modification is to improve receptivity to dyes; clearly, if easily detectable increases in color density are produced, the amount of dye involved (and hence presumably the depth of reaction) must correspond to more than a few molecular layers. Hence, the treatments are "surface modifications" only in the sense that the bulk of the sample far from the surface is unaffected.

It seems likely that in most, if not all, such processes what is observed is a homogeneous chemical reaction occurring in a thin layer into which the reactants have diffused.

The reader will have noted that despite many incisive insights by workers in this field, our understanding of the surfaces of bulk polymers on a molecular basis is very limited. We have seen up to now that the local surface properties of organic polymers can be accounted for almost exclusively by treating them as a subclass of van der Waals or molecular solids of high molecular weight. In this zeroth-order description of the surface phenomenology of organic polymers there are no qualitative differences that distinguish the polymers as such. Such effects presumably exist, but produce perturbations under ordinary conditions which are too small to observe readily within current experimental precision. Conceivably such effects might be most prominent in the vicinity of special points of the phase diagram associated with phase transitions. Here we might first distinguish effects on surface properties such as a small maximum in the contact angle of a liquid wetting a polymer surface due to a phase transition in the bulk polymer. It is therefore significant that Neumann and Tanner[219] have reported such an effect for a polyfluorocarbon, occurring at the glass transition of the polymer. This, of course, has immediate consequences for very precise temperature extrapolations of liquid contact angles, but in a larger sense may herald the possibility that additional characteristic transitions of bulk polymers may be reflected in unusual behavior of the local surface properties. The glass transition is of course not a special characteristic of polymeric substances alone, but is observed for many typical small-molecule van der Waals substances such as, e.g., glycerine. Second, we might expect to distinguish phase transitions occurring solely within the interface region. For example, there does not yet appear to be any data on the critical temperature below which a liquid will not spread on a polymer surface. Perturbations of such a critical temperature, if they could be detected might provide a test of interfacial phase transitions. Even more speculative would be any discussion of a definitely characterized phase transition between different surface phases of a polymer, although such transitions may well be observed in the future.

Among our aims in writing this chapter, besides providing a review of the state of knowledge of polymer surfaces up until December 1973, was to provide a simple pedagogical guide to some of the special methods, results, and theories of this field. Many technologically important topics such as electrical charging effects and the triboluminescence of polymer surfaces simply could not be encompassed within the confines of the space allotted to this chapter.

Acknowledgments

We are grateful for many valuable suggestions and written comments from our colleagues; in particular we wish to thank Drs. A. W. Adamson, R. H. Dettre, F. R. Eirich, D. R. Fitchmun, P. J. Flory, R. J. Good, J. R. Huntsberger, D. I. James, D. H. Kaelble, D. G. Le Grand, Yu S. Lipatov, A. W. Neumann, S. Newman, J. F. Padday, A. K. Rastogi, R. J. Roe, D. Tabor, W. C. Wake, S. Wu, and A. C. Zettlemoyer.

References

1. D. J. Crisp, in *Surface Phenomena in Chemistry and Biology* (J. F. Danielli, K. G. A. Pankhurst, and A. C. Riddiford, eds.), Pergamon Press, New York, (1958), p. 23.
2. G. L. Gaines, Jr., *Insoluble Monolayers at Liquid Gas Interfaces*, Wiley–Interscience, New York (1966), pp. 172ff, 264ff; *Surface Chemistry and Colloids*, MTP Int. Rev. Sci., Physical Chemistry Series One, Vol. 7, Butterworths, London (1972), pp. 1ff.
3. R. R. Stromberg, in *Treatise on Adhesion and Adhesives* (R. L. Patrick, ed.), Marcel Dekker, New York (1967), p. 69.
4. J. J. Kipling, *Adsorption from Solutions of Non-Electrolytes*, Academic Press, New York (1965), Chapter 8.
5. M. Rosoff, in *Physical Methods in Macromolecular Chemistry* (B. Carroll, ed.), Marcel Dekker, New York (1969), p. 1.
6. Y. S. Lipatov, *Adsorption of Polymers* (in Russian), State Publishing House, Kiev, 1972.
7. G. L. Gaines, Jr., *J. Phys. Chem.* **73**, 3143 (1969).
8. G. L. Gaines, Jr., *J. Polymer Sci. A-2* **7**, 1379 (1969); **9**, 1333 (1971); **10**, 1529 (1972).
9. Y. S. Lipatov, *Uspekhi Khimii*, in press.
10. A. W. Adamson, *The Physical Chemistry of Surfaces*, Wiley–Interscience, New York (1967).
11. J. W. Gibbs, *The Collected Works of J. Willard Gibbs*, Vol. I, *Thermodynamics*, Yale University Press, New Haven, Connecticut (1928).
12. R. Defay, I. Prigogine, A. Bellemans, and D. H. Everett, *Surface Tension and Adsorption*, Wiley, New York (1966).
13. G. A. Samorjai, this volume, Chapter 1.
14. F. B. Buff, The theory of capillarity, in *Handbuch der Physik*, Vol. X, Springer Verlag (1960), pp. 281–304.
15. W. Hardy, *Proc. Roy. Soc. A* **112**, (1926).
16. D. H. Bangham, *J. Chem. Phys.* **14**, 352 (1946).

17. B. V. Derjaguin and S. M. Zorin, in *Proc. 2nd Int. Congr. Surface Activity (London)* (1957), Vol. 2, p. 145.
18. A. Sheludko, *Adv. Colloid Interface Sci.* **1**, 391 (1967).
19. J. F. Padday, *Special Disc. Faraday Soc.* No. 1, 64 (1970).
20. C. Huh and L. E. Scriven, *J. Coll. Interface Sci.* **35**, 85 (1971); see also G. E. P. Elliot and A. C. Riddiford, *Rec. Progr. Surface Sci.* **2**, 111 (1965).
21. E. B. Dussan V., PhD Thesis, Johns Hopkins Univ. (1972).
22. G. R. Lester, *J. Colloid Sci.* **16**, 315 (1961); *Nature* **209**, 1126 (1966).
23. W. A. Zisman, in *Contact Angle, Wettability and Adhesion*, Advan. Chem. Series No. 43 (1964), p. 1.
24a. R. E. Johnson, Jr. and R. H. Dettre, in *Surface and Colloid Science* Vol. 2 (E. Matijevic, ed.), Wiley–Interscience, New York (1969), p. 85.
24b. T. D. Blake and J. M. Haynes, *Progress in Surface and Membrane Science*, Vol. 6, Academic Press, New York (1973).
25. R. J. Good, *J. Amer. Chem. Soc.* **74**, 504 (1952).
26. R. E. Johnson, Jr. and R. H. Dettre, in *Contact Angle, Wettability and Adhesion*, Adv. Chem. Series No. 43 (1964), p. 112.
27. R. E. Johnson, Jr. and R. H. Dettre, *J. Phys. Chem.* **68**, 1744 (1964).
28. A. W. Neumann and R. J. Good, *J. Coll. Interface Sci.* **38**, 341 (1972).
29. H. W. Fox and W. A. Zisman, *J. Colloid Sci.* **7**, 109 (1952); W. A. Zisman, in *Contact Angle, Wettability and Adhesion*, Advan. Chem. Series No. 43 (1964), p. 1.
30. R. E. Johnson, Jr. and R. H. Dettre, *J. Coll. Interface Sci.* **21**, 610 (1966).
31. H. D. Feltman and J. R. McPhee, *Textile Research J.* **34**, 634 (1964).
32. M. K. Bernett and W. A. Zisman, *J. Phys. Chem.* **66**, 1207 (1962).
33. A. W. Adamson and I. Ling, in *Contact Angle, Wettability and Adhesion*, Adv. Chem. Series No. 43 (1964), p. 57; A. W. Adamson, *J. Colloid. Interface Sci.* **44**, 273 (1972) and references cited therein.
34. R. Peyser, D. J. Tutas, and R. Stromberg, *J. Polymer Sci. (A-1)* **5**, 651 (1967).
35. W. C. Hamilton, in *Appl. Polymer Symp.* No. 19 (1972), p. 105, and references cited therein.
36. N. J. Harrick, *J. Phys. Chem.* **64**, 1110 (1964).
37. J. P. Luongo and H. Schonhorn, *J. Polymer Sci. (A-2)* **6**, 1649 (1968).
38. S. Krimm, C. Y. Liang, and G. B. B. M. Sutherland, *J. Chem. Phys.* **25**, 549 (1956).
39. H. Schonhorn and F. W. Ryan, *J. Phys. Chem.* **70**, 3811 (1966).
40. B. L. Henke, in *Advances in X-ray Analysis*, Vol. 13 (B. L. Henke, J. B. Newkirk, and G. R. Mallett, eds.), Plenum Press, New York (1970), p. 1.
41. D. W. Dwight and W. M. Riggs, *J. Colloid Interface Sci.* **47**, 650 (1974).
42. D. M. Hercules, *Anal. Chem.* **44** (5), 106 (1972); W. M. Riggs and R, P. Fedchenko, *Amer. Lab.* **4**(11), 65 (1972).
43. C. G. S. Collins, A. C. Lowe, and D. Nicholas, *Europ. Poly. J.* **9**, 1173 (1973).
44. G. Kraus, *Adv. Polymer Sci.* **8**, 155 (1971).
45. S. Kaufman, W. P. Slichter, and D. D. Davis, *J. Polymer Sci.* A2 **9**, 829 (1971).
46. Y. S. Lipatov and F. G. Fabuliak, *Vysokomolek. Soed.* **10**, 1605 (1968).
47. V. L. Kulividze, S. G. Klinov, and N. N. Lezhnev, *Vysokomolek. Soed.* **9**, 1924 (1967).

48. D. H. Droste, A. T. Di Benedetto, and E. O. Stejskal, *J. Polymer Sci. A-2* **9**, 187 (1971); D. H. Droste and A. T. Di Benedetto, *J. Polymer Sci.* **13**, 2149 (1969).

49. G. Kraus and J. T. Gruver, *J. Polymer Sci. A-2* **8**, 571 (1970).

50. M. Panar, H. H. Hoehn, and R. R. Hebert, *Macromolecules* **6**, 777 (1973).

51. M. Furuta, *Polymer Letters* **11**, 113 (1973); R. G. Hocker, Private communication.

52. K. Ueberreiter, in *Diffusion in Polymers* (J. Crank and G. S. Park, eds.), Academic Press, London (1968), p. 219.

53. P. K. Tsarev and Y. S. Lunatov, *Mechanika Polymerov* **1972**, 195 (1972); V. I. Mishko, Y. S. Lunatov, and R. A. Veselovsky, *Vysoko molek. Soed.* **13**, 114 (1971); Y. S. Lunatov and V. F. Babitch, *Vysokomolek. Soed.* **10b**, 848 (1968).

54. R. J. Good, J. A. Kvikstad, and W. O. Bailey, *J. Colloid Interface Sci.* **35**, 314 (1971).

55. H. Schonhorn and F. W. Ryan, *J. Adhesion* **1**, 43 (1969).

56. K. Hara and H. Schonhorn, *J. Adhesion* **2**, 100 (1970).

57. F. K. Borisova, G. A. Galkin, A. V. Kiselev, A. Y. Korolev, and V. I. Lygin, *Kolloid. Zh.* **27**, 320 (1965).

58. D. H. Kaelble and E. H. Cirlin, *J. Polymer Sci. A-2* **9**, 363 (1971).

59. D. R. Fitchmun and S. Newman, *J. Polymer Sci. A-2* **8**, 1545 (1970).

60. T. K. Kwei, H. Schonhorn, and H. L. Frisch, *J. Applied Phys.* **38**, 2512 (1967).

61. D. Jenckel, E. Teege, and W. Hinrichs, *Kolloid Z.* **129**, 19 (1952).

62. R. J. Barriault and L. F. Gronholts, *J. Polymer Sci.* **18**, 3933 (1955).

63. R. K. Eby, *J. Appl. Phy.* **35**, 2720 (1964).

64. H. Schonhorn, *J. Polymer Sci. B* **2**, *465 (1964); H.* Schonhorn, *Macromolecules* **1**, 145 (1968).

65. H. M. Zupko, Unpublished data.

66. R. E. Cuthrell, *J. Appl. Polymer Sci.* **11**, 1495 (1967).

67. A. C. Zettlemoyer, A. Chand, and E. Gamble, *J. Amer. Chem. Soc.* **72**, 2752 (1950).

68. W. Thompson, *Physica* **26**, 890 (1960).

69. J. W. Hightower and R. H. Emmett, *J. Polymer Sci. A* **2**, 1647 (1964).

70. R. J. Salloum and R. E. Eckert, *J. Polymer Sci. A-2* **10**, 909 (1972).

71. D. P. Graham, *J. Phys. Chem.* **68**, 2788(1964).

72. J. E. Lohr and J. J. Scholz, *J. Colloid Sci.* **20**, 846 (1956).

73. R. F. Hoburg, G. S. Handler, and J. J. Scholz, *J. Colloid Interface Sci.* **27**, 642 (1968).

74. J. W. Whalen, W. H. Wade, and J. J. Porter, *J. Colloid Interface Sci.* **24**, 379 (1967).

75. J. J. Chessick, F. H. Healy, and A. C. Zettlemoyer, *J. Phys. Chem.* **60**, 1345 (1956).

76. D. P. Graham, *J. Phys. Chem.* **69**, 4387 (1965).

77. J. W. Whalen, *J. Colloid Interface Sci.* **28**, 443 (1968).

78. D. P. Graham, *J. Phys. Chem.* **66**, 1815 (1962).

79. S. H. Maron, M. E. Elder, and I. N. Ulevitch, *J. Colloid Sci.* **9**, 89 (1954).

80. H. Gerrens, in *Polymer handbook* (J. Brandrup and E. H. Immergut, eds.), Interscience, New York (1966), p. II-399.

81. T. R. Paxton, *J. Colloid Interface Sci.* **31**, 19 (1969).
82. R. C. Mast and L. Benjamin, *J. Colloid Interface Sci.* **31**, 31 (1969).
83. J. Paleos, *J. Colloid Interface Sci.* **31**, 7 (1969).
84. W. F. Taylor, *J. Catalysis* **16**, 20 (1970); *J. Phys. Chem.* **74**, 2250 (1970).
85. A. H. Ellison and W. A. Zisman, *J. Phys. Chem.* **58**, 506 (1954).
86. G. L. Gaines, Jr., Unpublished results.
87. T. Seimiya, H. Yamamoto, and T. Sasaki, *J. Polymer Sci. A-2* **8**, 595 (1970).
88. G. L. Gaines, Jr. *Polymer Eng. Sci.* **12**, 1 (1972).
89. H. Edwards, *J. Appl. Polymer Sci.* **12**, 2213 (1968).
90. J. R. J. Harford and E. F. T. White, *Trans. J. Plastics Inst.* **1969** (February), 53.
91. R. H. Dettre and R. E. Johnson, Jr., *J. Colloid Interface Sci.* **21**, 367 (1966).
92. R. J. Roe, V. L. Bacchetta, and P. M. G. Wong, *J. Phys. Chem.* **71**, 4190 (1967).
93. R. J. Roe, *J. Phys. Chem.* **72**, 2013 (1968).
94. T. Sakai, *Polymer* **6**, 659 (1965).
95. T. Hata, *Hyomen* (Surface, Japan) **6**, 281 (1968).
96. B. Vonnegut, *Rev. Sci. Instr.* **13**, 6 (1942).
97. H. M. Princen, I. Y. Z. Zia, and S. G. Mason, *J. Colloid Interface Sci.* **23**, 99 (1967).
98. H. T. Patterson, K. H. Hu, and T. H. Grindstaff, *ACS Polymer Preprints*, **11**, 1299 (1970).
99. D. C. Chappelear, *ACP Polymer Preprints* **5**, 363 (1964).
100. J. T. Davies and J. Llopis, *Proc. Roy. Soc. A* **227**, 537 (1955).
101. J. M. Andreas, E. A. Hauser, and W. B. Tucker, *J. Phys. Chem.* **42**, 1001 (1938).
102. D. N. Staicopolus, *J. Colloid Sci.* **17**, 439 (1962); **18**, 793 (1963).
103. D. N. Staicopolus, *J. Colloid Sci.* **23**, 453 (1967).
104. H. M. Princen, in *Surface and Colloid Science* (E. Matijevic, ed.), Vol. 2 Wiley–Interscience, New York (1969), p. 72.
105. D. G. Le Grand and G. L. Gaines, Jr., *J. Colloid Interface Sci.* **31**, 162, 430 (1969).
106. G. W. Bender and G. L. Gaines, Jr., *Macromol.* **3**, 128 (1970).
107. D. G. Le Grand and G. L. Gaines, Jr., *J. Colloid Interface Sci.* **42**, 181 (1973).
108. G. W. Bender, D. G. Le Grand, and G. L. Gaines, Jr., *Macromol.* **2**, 681 (1969).
109. G. L. Gaines, Jr. and D. G. Le Grand *J. Colloid Interface Sci.* (in press).
110. A. J. G. Allan, *J. Colloid Sci.* **14**, 206 (1959).
111a. N. L. Jarvis, R. B. Fox, and W. A. Zisman, in *Contact Angle, Wettability and Adhesion*, Advan. Chem. Series, No. 43 (1964), p. 317.
111b. R. C. Bowers, N. L. Jarvis, and W. A. Zisman, *Ind. Eng. Chem. Prod. Res. Develop.* **4**, 86 (1965).
112a. D. G. Le Grand and G. L. Gaines, Jr., *ACS Polymer Preprints* **11**, 442 (1970).
112b. M. J. Owen and T. C. Kendrick, *Macromol.* **3**, 458 (1970).
113a. D. L. Bailey, I. H. Petersen, and W. G. Reid, in *Proc. Fourth Intl. Congr. Surf. Activity* Gordon and Breach, New York (1967), Vol. 1, p. 173.
113b. B. Kanner, W. G. Reid, and I. H. Petersen, *Ind. Eng. Chem. Prod. Res. Develop.* **6**, 88 (1967).

113c. T. C. Kendrick, B. M. Kingston, N. C. Lloyd, and M. J. Owen, *J. Colloid Interface Sci.* **24**, 135, 141 (1967).

114. G. L. Gaines, Jr. and G. W. Bender, *Macromol.* **5**, 82 (1972).

115. R. J. Roe, *J. Phys. Chem.* **69**, 2809 (1965).

116. S. Wu, *J. Colloid Interface Sci.* **31**, 153 (1969).

117. H. Schonhorn, *J. Chem. Phys.* **43**, 2041 (1965).

118. J. H. Hildebrand and R. L. Scott, *The Solubility of Non-Electrolytes*, Reinhold, New York (1950), pp. 402, 431.

119. H. Schonhorn, *J. Chem. Engr. Data* **12**, 524 (1967).

120. D. J. Roe, *Proc. Nat. Acad. Sci.* **56**, 819 (1966).

121. D. Patterson and A. K. Rastogi, *J. Phys. Chem.* **74**, 1067 (1970).

122. S. Matsuoka and B. Maxwell, *J. Polymer. Sci.* **32**, 131 (1958).

123. K. S. Siow and D. Patterson, *Macromol.* **4**, 26 (1971).

124. C. W. Stewart and C. A. von Frankenberg, *J. Poly. Sci. A-2* **6**, 1686 (1968).

125. D. G. Le Grand and G. L. Gaines, Jr., *ACS Polymer Preprints* **11**, 1306 (1970).

126. R. Aveyard, *Trans. Faraday Soc.* **63**, 2778, (1967).

127. A. K. Rastogi and L. E. St. Pierre, *J. Colloid Interface Sci.* **31**, 168 (1969).

128. L. A. Girifalco and R. J. Good, *J. Phys. Chem.* **61**, 904 (1957); **64**, 561 (1960).

129. F. M. Fowkes, *J. Phys. Chem.* **66**, 32 (1962); *Ind. Eng. Chem.* **56**(12), 40 (1964); in *Contact Angle, Wettability and Adhesion*, Adv. Chem. Series, No. 43 (1964), p. 99.

130a. S. Wu, *Polymer Preprints* **11**, 1291 (1970)

130b. R. E. Johnson, Jr. and R. H. Dettre, in *Surface and Colloid Science* (E. Matijevic, ed.), Vol. 2, 285 Wiley–Interscience, New York (1969), Vol. 2, p. 285.

130c. M. Rosoff, in *Physical Methods in Macromolecular Chemistry* (B. Carroll, ed.) Marcel Dekker, New York (1969), Vol. 1, p. 69.

130d. R. J. Good and E. Elbing, *Ind. Eng. Chem.* **62**,(3), 54 (1970).

131. E. Helfand and Y. Tagami, *Polymer Letters* **9**, 741 (1971); *J. Chem. Phys.* **56**, 3592 (1972).

132. J. J. Bikerman, *Physical Surfaces*, Academic Press, New York (1970).

133. N. K. Adam and G. Jessop, *J. Chem. Soc.* **1925**, 1863.

134. N. K. Adam and R. S. Morrell, *J. Soc. Chem. Ind. (London)* **53**, 255T (1934).

135. C. H. Bosanquet and H. Hartley, *Phil. Mag.* **42**, (6), 456 (1921).

136. F. M. Fowkes and W. D. Harkins, *J. Am. Chem. Soc.* **62**, 3377 (1940).

137. R. Ablett, *Phil. Mag.* **46**, 244 (1923).

138. J. Leja and G. W. Poling, Preprint, International Mineral Processing Congress, London, April 1960.

139. A. W. Adamson, *Physical Chemistry of Surfaces*, 2nd ed, Wiley, New York (1967), p. 356.

140. I. Langmuir and V. J. Schaefer, *J. Am. Chem. Soc.* **59**, 2400 (1937).

141. T. Fort, Jr. and H. T. Patterson, *J. Colloid Sci.* **18**, 217 (1963).

142. W. C. Jones and M. C. Porter, *J. Colloid Interface Sci.* **24**, 1, (1967).

143. N. K. Adam and G. E. P. Elliot, *J. Chem. Soc.* **1926**, 2206.

144. A. W. Neumann, in *Proc. 4th International Congr. Surface Activity* (Brussels) (1964).

145. R. E. Johnson, Jr. and R. H. Dettre, *J. Colloid Sci.* **20**, 173 (1965).

146. H. Schonhorn, *Nature* **210**, 896 (1966).

147. C. A. Sutula, R. Hautala, R. A. Dalla Betta, and L. A. Michel, Abstracts, 153rd Meeting, Am. Chem. Soc., April 1967.

148. R. A. Dalla Betta, C. A. Sutula, R. Hautala, and L. A. Michel, Abstracts, 153rd Meeting Am. Chem. Soc., April 1967.

149. J. F. Padday, *J. Colloid Interface Sci.* **28**, 557 (1968).

150. F. D. Petke and B. R. Ray, *J. Colloid Interface Sci.* **31**, 216 (1969).

151. H. Schonhorn, *J. Adhesion* **1**, 38 (1969).

152. J. R. Dann, *J. Colloid Interface Sci.* **32**, 302, 321 (1970).

153. W. J. Murphy, M. W. Roberts, and J. R. H. Ross, *J. Chem. Soc. Faraday Trans. I*, **68**, 1190 (1972).

154. F. M. Fowkes, in *Treatise on Adhesion and Adhesives* (R. L. Patrick ed.), Marcel Dekker, New York (1967), p. 325.

155. R. J. Good, in *Treatise on Adhesion and Adhesives* (R. L. Patrick, ed.), Marcel Dekker, New York (1967), p. 9.

156. A. H. Ellison and W. A. Zisman, *J. Phys. Chem.* **58**, 260 (1954).

157. D. W. Van Krevelen, *Properties of Polymers*, Elsevier, Amsterdam, (1972), pp. 100–101.

158. F. Schulman and W. A. Zisman, *J. Colloid Sci.* **7**, 465 (1952).

159. D. Hoernschemeyer, *J. Phys. Chem.* **70**, 2628 (1966).

160. A. G. Pittman and B. A. Ludwig, *J. Polymer Sci. A-1* **7**, 3053 (1969).

161. D. H. Kaelble, *Physical Chemistry of Adhesion*, Wiley–Interscience, New York (1971).

162. S. Wu, *J. Phys. Chem.* **74**, 632 (1970).

163. J. E. Melrose, *J. Colloid Interface Sci.* **28**, 403 (1968).

164. D. K. Owens and R. C. Wendt, *J. Appl. Polymer Sci.* **13**, 1741 (1969).

165. S. Wu, *J. Polymer Sci. C* **34**, 19 (1971).

166. J. Cortes, *J. Colloid Interface Sci.* **45**, 209 (1973).

167. J. N. Israelachvili, *J. Chem. Soc., Faraday Trans. II* **69**, 1729 (1973).

168. J. F. Padday and N. D. Uffindell, *J. Phys. Chem.* **72**, 1407 (1968).

169. K. Kanamaru, *Kolloid.-Z.* **192**, 51 (1963); E. D. Shchukin, Y. V. Goryunov, G. I. Den'shchikovaz, N. V. Pertsov, and B. D. Summ, *Kolloid. Zh.* **25**, 108 (1963).

170. H. Schonhorn, H. L. Frisch, and T. K. Kwei, *J. Appl. Phys.* **37**, 4967 (1966).

171. R. H. Dettre and R. E. Johnson, Jr., *J. Adhesion* **2**, 61 (1970).

172. H. Van Oene, Y. F. Chang, and S. Newman, *J. Adhesion*, *1*, 54 (1969).

173. H. Van Oene, *J. Adhesion* **3**, 1 (1972).

174. S. Strella, *J. Appl. Phys.* **41**, 242 (1970).

175. V. V. Arslanov, T. I. Ivanova, and V. A. Ogarev, *Doklady Akad. Nauk SSSR* **198**, (5), 1113 (1971).

176. S. Newman, *J. Colloid Interface Sci.* **26**, 219 (1968).

177. B. W. Cherry and C. M. Holmes, *J. Colloid Interface Sci.* **29**, 174 (1969).

178. A. Kishimoto, H. Ueno, and H. Hasegawa, in *7th Symp. Adhesion and Adhesives*, Adhesion Society of Japan, (June 1969), p. 71.

179. T. K. Kwei, H. Schonhorn and H. L. Frisch, *J. Colloid and Interface Sci.* **28**, 543 (1968).

180. W. W. Y. Lau and C. M. Burns, *J. Colloid Interface Sci.* **45**, 295 (1973).

181. T. D. Blake and J. M. Haynes, *J. Colloid Interface Sci.* **30**, 421 (1969).
182. W. D. Bascom, R. L. Cottington, and C. R. Singleterry, in *Contact Angle, Wettability and adhesion*, Adv. Chem. Series, No. 43 (1964), p. 355.
183. F. P. Bowden and D. Tabor, *Friction and Lubrication of Solids*, Part II, Oxford University Press (1964).
184. M. W. Pascoe and D. Tabor, *Research Suppl.* **15**, 8 (1955); D. Tabor, *Wear* **1**, 9 (1958); C. Rubenstein, *Wear* **2**, 296 (1959); N. Adams, *J. Appl. Polymer Sci.* **7**, 2075 (1963); D. I. James, R. H. Norman, and A. R. Payne, S.C.I. Monograph No. 5, (1959), p. 233.
185. S. C. Cohen and D. Tabor, *Proc. Roy Soc. A* **291**, 186 (1966).
186. K. R. Makinson and D. Tabor, *Proc. Roy. Soc. A* **281**, 49 (1964); D. G. Flom and N. T. Porile, *J. Appl. Phys.* **26**, 1088 (1955).
187. D. Atack and D. Tabor, *Proc. Roy. Soc. A* **246**, 539 (1958).
188. H. G. Howell and A. S. Lodge, *Proc. Phys. Soc. B* **67**, 89 (1954).
189. C. M. Pooley and D. Tabor, *Nature* **273**, 88 (1972).
190. R. C. Bowers, W. C. Clinton, and W. A. Zisman, *Ind. Eng. Chem.* **46**, 2416 (1954).
191. T. Fort, Jr., *J. Phys. Chem.* **66**, 1136 (1962).
192. D. K. Owens, *J. Appl. Polymer Sci.* **8**, 1465 (1964); **14**, 185 (1970).
193. B. J. Briscoe, V. Mustafaev, and D. Tabor, *Wear* **19**, 399 (1972).
194. R. C. Bowers, N. L. Jarvis, and W. A. Zisman, *I & EC Prod. Res. Devel.* **4**, 86 (1965).
195. D. J. Alner (ed.), *Aspects of Adhesion*, University of London Press (1965), Vol. 1.
196. J. J. Bikerman, *The Science of Adhesive Joints*, Academic Press, New York (1961).
197. M. J. Bodna, (ed.), *Symposium on Adhesives for Structural Applications*, Wiley–Interscience, New York (1962).
198. F. P. Bowden and D. Tabor, *The Friction and Lubrication of Solids*, Clarendon Press: Oxford University Press, Parts I (1st ed. 1950), Part II (1st ed. 1964).
199. D. D. Eley (ed.), *Adhesion*, Oxford University Press (1961).
200. G. Epstein, *Adhesive Bonding of Metals*, Reinhold, New York (1954).
201. W. H. Guttman, *Concise Guide to Structural Adhesives*, Reinhold, New York (1961).
202. R. Houwink and G. Salomon, (eds.), *Adhesion and Adhesives*, Elsevier, New York (1965), Vols. 1, 2.
203. R. L. Patrick (ed.), *Treatise on Adhesion and Adhesives*, Marcel Dekker, New York Vols. 1 (1966), 2 (1969), 3 (1973).
204. H. A. Perry, *Adhesive Bonding of Reinforced Plastics*, McGraw-Hill, New York (1959).
205. F. W. Reinhart and I. G. Callomon, *Survey of Adhesion and Adhesives*, Wright Air Development Center, Ohio WADC Tech. Rept. 58–450, 1959.
206. I. Skeist (ed.), *Handbook of Adhesives*, Reinhold, New York (1962).
207. P. Weiss (ed.), *Adhesion and Cohesion*, Elsevier, Amsterdam (1962).
208a. *Adhesion* ASTM. Spec. Tech. Publ. No. 360, Philadelphia, Pennsylvania (1964).
208b. *Symposium, Haftsysteme and Haftfestigkeit*, April 1964, Dechema-Monograph Vol. 51 (1965).

209. *Adhesion and Adhesives, Fundamentals and Practice*, Soc. Chem. Ind., London (1954).
210. *Contact Angle, Wettability and Adhesion* (Kendall Award Symposium), Adv. in Chem. Ser. No. 43, Am. Chem. Soc., Washington (1964).
211. D. J. Angier, in *Chemical Reactions of Polymers* (E. M. Fettes, ed.) (High Polymers, Vol. 19), Interscience, New York (1964), p. 1009.
212. H. Schonhorn and R. H. Hansen, *J. Appl. Polymer Sci.* **12**, 1231 (1968).
213. D. A. Olsen and A. J. Osteraas, *J. Polymer Sci. A-1* **7**, 1913, 1921, 1927 (1969).
214. A. J. Osteraas and D. A. Olsen, *J. Appl. Polymer Sci.* **13**, 1537 (1969).
215. D. A. Olsen and A. J. Osteraas, *J. Appl. Polymer Sci.* **13**, 1523 (1969).
216. J. R. Hollahan, B. B. Stafford, R. D. Falb, and S. T. Payne, *J. Appl. Polymer Sci.* **13**, 807 (1969).
217. R. I. Leininger, R. D. Falb, and G. A. Grode, *Ann. N.Y. Acad. Sci.* **146**, 11 (1968).
218. J. R. Caldwell and W. J. Jackson, Jr., *J. Polymer Sci. C* **24**, 15 (1968).
219. A. W. Neumann and W. Tanner, *J. Colloid Interface Sci.* **34**, 1 (1970).

Index